Practical Methods for Biocatalysis and Biotransformations 3

Practical Methods for Biocatalysis and Biotransformations 3

Edited by

JOHN WHITTALL

Manchester Interdisciplinary Biocentre (MIB),
The University of Manchester, UK

PETER W. SUTTON

GlaxoSmithKline Research and Development Limited, UK

WOLFGANG KROUTIL

Department of Chemistry, Organic and Bioorganic Chemistry,
University of Graz, Austria

WILEY

Library of Congress Cataloging-in-Publication Data

Practical methods for biocatalysis and biotransformations 3 / edited by John Whittall, Manchester Interdisciplinary Biocentre (MIB), The University of Manchester, UK, Peter W. Sutton, GlaxoSmithKline Research and Development Limited, UK, Wolfgang Kroutil, Department of Chemistry, Organic and Bioorganic Chemistry, University of Graz, Austria
  pages cm
 Includes bibliographical references and index.
 ISBN 978-1-118-60525-7 (cloth)
 1. Enzymes–Biotechnology. 2. Biocatalysis. 3. Biotransformation (Metabolism) 4. Organic compounds–Synthesis. I. Whittall, John, editor. II. Sutton, Peter (Peter W.), editor. III. Kroutil, Wolfgang, 1972- editor.
  TP248.65.E59P73 2016
  660.6'34–dc23

                                                                           2015024267

A catalogue record for this book is available from the British Library.

ISBN: 9781118605257

Set in 10/12pt, TimesLTStd-Roman by Thomson Digital, Noida, India.

Printed in Singapore by C.O.S. Printers Pte Ltd

1   2016

Contents

List of Contributors

Syed T. Ahmed School of Chemistry, Manchester Institute of Biotechnology, The University of Manchester, UK

Ian Archer Ingenza Ltd, Roslin BioCentre, UK

Frances H. Arnold Division of Chemistry and Chemical Engineering, California Institute of Technology, USA

Robert Ashe AM Technology, UK

Lara Babich Van't Hoff Institute for Molecular Sciences, University of Amsterdam, The Netherlands

Jan-E. Bäckvall Department of Organic Chemistry, Arrhenius Laboratory, Stockholm University, Sweden

Maria Bawn Prozomix Limited, UK

Beatrice Bechi School of Chemistry, Manchester Institute of Biotechnology, The University of Manchester, UK

Gary Black Northumbria University, Department of Applied Science, UK

Fabrizio Bonina Institute of Pharmaceutical Sciences, Albert Ludwigs University of Freiburg, Germany

Uwe T. Bornscheuer Institute of Biochemistry, Department of Biotechnology and Enzyme Catalysis, University of Greifswald, Germany

Elisabetta Brenna Department of Chemistry, Material and Chemical Engineering "G. Natta," Polytechnic University of Milan, Italy

Aleksandra Bury Van't Hoff Institute for Molecular Sciences, University of Amsterdam, The Netherlands

Andrada But Biobased Commodity Chemistry, Wageningen University, The Netherlands

Simon Charnock Prozomix Limited, UK

Bi-Shuang Chen Department of Biotechnology, Delft University of Technology, The Netherlands

Yong-Jun Chen Department of Chemical and Biological Engineering, Zhejiang University, China

Pere Clapés Biotransformation and Bioactive Molecules Group, Institute of Advanced Chemistry of Catalonia (IQAC-CSIC), Spain

Thomas Classen Institute of Bioorganic Chemistry, Heinrich Heine University Düsseldorf, Germany

Marine Debacker Clermont University, Blaise Pascal University, ICCF, Clermont-Ferrand, France; CNRS, UMR 6296, France

Tom Desmet Centre for Industrial Biotechnology and Biocatalysis, Faculty of Bioscience Engineering, Ghent University, Belgium

Karel De Winter Centre for Industrial Biotechnology and Biocatalysis, Faculty of Bioscience Engineering, Ghent University, Belgium

Griet Dewitte Centre for Industrial Biotechnology and Biocatalysis, Faculty of Bioscience Engineering, Ghent University, Belgium

Alba Díaz-Rodríguez Department of Organic and Inorganic Chemistry, Asturias Institute of Biotechnology, University of Oviedo, Spain

Carola Dresen Institute of Pharmaceutical Sciences, Albert Ludwigs University of Freiburg, Germany

Richard Duncan Prozomix Limited, UK

Marc Dürrenberger Department of Inorganic Chemistry, University of Basel, Switzerland

Tadashi Ema Division of Chemistry and Biotechnology, Graduate School of Natural Science and Technology, Okayama University, Japan

Ulrike Engel Institute of Process Engineering in Life Sciences, Section II: Technical Biology, Karlsruhe Institute of Technology, Germany

Roman S. Esipov Shemyakin and Ovchinnikov Institute of Bioorganic Chemistry, Russian Academy of Sciences, Russia

Kurt Faber Department of Chemistry, Organic and Bioorganic Chemistry, University of Graz, Austria

Christopher C. Farwell Division of Chemistry and Chemical Engineering, California Institute of Technology, USA

James Finnigan Prozomix Limited, UK

Christine Fuchs Department of Chemistry, Organic and Bioorganic Chemistry, University of Graz, Austria

Michael Fuchs Department of Chemistry, Organic and Bioorganic Chemistry, University of Graz, Austria

Anna Fryszkowska Merck Research Laboratories, USA

Eduardo García-Junceda Department of Bioorganic Chemistry, Institute of General Organic Chemistry, Spain

Gilda Gasparini AM Technology, UK

Francesco G. Gatti Department of Chemistry, Material and Chemical Engineering "G. Natta," Polytechnic University of Milan, Italy

Edzard M. Geertsema Department of Pharmaceutical Biology, Groningen Research Institute of Pharmacy, University of Groningen, The Netherlands

Laura Getrey DECHEMA Research Institute, Germany

Diego Ghislieri School of Chemistry, Manchester Institute of Biotechnology, The University of Manchester, UK

Silvia M. Glueck Austrian Centre of Industrial Biotechnology, Department of Chemistry, Organic and Bioorganic Chemistry, University of Graz, Austria

Michael Golden AstraZeneca, Chemical Development, UK

Animesh Goswami Chemical Development, Bristol-Myers Squibb, USA

Vicente Gotor Department of Organic and Inorganic Chemistry, Asturias Institute of Biotechnology, University of Oviedo, Spain

Vicente Gotor-Fernández Department of Organic and Inorganic Chemistry, Asturias Institute of Biotechnology, University of Oviedo, Spain

Johannes Gross Austrian Centre of Industrial Biotechnology, Department of Chemistry, Organic and Bioorganic Chemistry, University of Graz, Austria

Christine Guérard-Hélaine Clermont University, Blaise Pascal University, ICCF, Clermont-Ferrand, France; CNRS, UMR 6296, France

Zhiwei Guo Chemical Development, Bristol-Myers Squibb, USA

Karl P. J. Gustafson Department of Organic Chemistry, Arrhenius Laboratory, Stockholm University, Sweden

Helen C. Hailes Department of Chemistry, Christopher Ingold Laboratories, University College London, UK

Ulf Hanefeld Department of Biotechnology, Delft University of Technology, The Netherlands

Steven P. Hanlon F. Hoffmann-La Roche Ltd., Switzerland

Aloysius F. Hartog Van't Hoff Institute for Molecular Sciences, University of Amsterdam, The Netherlands

Bernhard Hauer Institute of Technical Biochemistry, University of Stuttgart, Germany

Rachel S. Heath School of Chemistry, Manchester Institute of Biotechnology, The University of Manchester, UK

Virgil Hélaine Clermont University, Blaise Pascal University, ICCF, Clermont-Ferrand, France; CNRS, UMR 6296, France

Susanne Herter Institute of Biochemistry, Department of Biotechnology and Enzyme Catalysis, University of Greifswald, Germany

Matthew R. Hickey Chemical Development, Bristol-Myers Squibb, USA

Michael Hofer Fraunhofer Institute for Interfacial Engineering and Biotechnology, Institute branch Straubing, BioCat – Bio-, Chemo- and Electrocatalysis, Germany

Frank Hollmann Department of Biotechnology, Delft University of Technology, The Netherlands

Dirk Holtmann DECHEMA Research Institute, Germany

Karen Holt-Tiffin Dr Reddy's Laboratories Ltd, Chirotech Technology Centre, UK

Roger M. Howard Pfizer Ltd, Chemical Research & Development, UK

Gjalt Huisman Codexis Inc, USA

Shahed Hussain School of Chemistry, Manchester Institute of Biotechnology, The University of Manchester, UK

Syed Masood Husain Institute of Pharmaceutical Sciences, Albert-Ludwigs-Universität Freiburg, Germany

Todd K. Hyster Department of Inorganic Chemistry, University of Basel, Switzerland; Department of Chemistry, Colorado State University, USA

Ed Jones C-Tech Innovation Ltd, UK

Predrag Jovanovic Department of Organic Chemistry, Faculty of Pharmacy, University of Belgrade, Serbia

Shusuke Kamata Division of Chemistry and Biotechnology, Graduate School of Natural Science and Technology, Okayama University, Japan

Elena Kasparyan Institute of Pharmaceutical Sciences, Albert Ludwigs University of Freiburg, Germany

Hans Kierkels DSM Innovative Synthesis BV, The Netherlands

Matthias Kittelmann NovartisPharma AG, Switzerland

Livia Knörr Department of Inorganic Chemistry, University of Basel, Switzerland

Valentin Köhler Department of Inorganic Chemistry, University of Basel, Switzerland

Pieter de Koning Dr Reddy's Laboratories Ltd, Chirotech Technology Centre, UK

Robert Kourist Junior Research Group for Microbial Biotechnology, Ruhr-University Bochum, Germany

Thomas Krieg DECHEMA Research Institute, Germany

Wolfgang Kroutil Department of Chemistry, Organic and Bioorganic Chemistry, University of Graz, Austria

Jim Lalonde Codexis Inc, USA

Eleanor D. Lamming Department of Chemistry, Christopher Ingold Laboratories, University College London, UK

Alexander Lang General Biochemistry, Dresden University of Technology, Germany

Iván Lavandera Department of Organic and Inorganic Chemistry, Asturias Institute of Biotechnology, University of Oviedo, Spain

Friedemann Leipold School of Chemistry, Manchester Institute of Biotechnology, The University of Manchester, UK

Marielle Lemaire Clermont University, Blaise Pascal University, ICCF, Clermont-Ferrand, France; CNRS, UMR 6296, France

Jerôme Le Nôtre Biobased Commodity Chemistry, Wageningen University, The Netherlands

Shu-Ming Li Institute of Pharmaceutical Biology and Biotechnology, Philipp University of Marburg, Germany

Zhi Li Department of Chemical and Biomolecular Engineering, National University of Singapore, Singapore

Jack Liang Codexis Inc, USA

Benjamin Lichman Department of Biochemical Engineering, University College London, UK

Mike Liebhold Institute of Pharmaceutical Biology and Biotechnology, Philipp University of Marburg, Germany

Ji Liu Department of Chemical and Biomolecular Engineering, National University of Singapore, Singapore

Sarah L. Lovelock School of Chemistry, Manchester Institute of Biotechnology, The University of Manchester, UK

Ruth Lloyd Prozomix Limited, UK

Sumire Honda Malca Institute of Technical Biochemistry, University of Stuttgart, Germany

Francisco Marquillas Interquim SA, R&D Department, Spain

Oliver May DSM Innovative Synthesis BV, The Netherlands

Rebecca E. Meadows AstraZeneca, Chemical Development, UK

Elise Meulenbroeks DSM Innovative Synthesis BV, The Netherlands

Xiao Meng Department of Chemical and Biological Engineering, Zhejiang University, China

Yufeng Miao Department of Pharmaceutical Biology, Groningen Research Institute of Pharmacy, University of Groningen, The Netherlands

Marko D. Mihovilovic Institute of Applied Synthetic Chemistry, Vienna University of Technology, Austria

Igor A. Mikhailopulo Institute of Bioorganic Chemistry, National Academy of Sciences, Belarus

Daniel Mink DSM Innovative Synthesis BV, The Netherlands

Gordana Minovska Institute of Molecular Genetics and Genetic Engineering, University of Belgrade, Serbia

Anatoly I. Miroshnikov Shemyakin and Ovchinnikov Institute of Bioorganic Chemistry, Russian Academy of Sciences, Russia

Daniela Monti Institute of Molecular Recognition Chemistry (CNR), Italy

Thomas S. Moody Almac, Department of Biocatalysis and Isotope Chemistry, UK

Keith R. Mulholland AstraZeneca, Chemical Development, UK

Michael Müller Institute of Pharmaceutical Sciences, Albert Ludwigs University of Freiburg, Germany

Jan Muschiol Institute of Biochemistry, Department of Biotechnology and Enzyme Catalysis, University of Greifswald, Germany

Francesco G. Mutti School of Chemistry, Manchester Institute of Biotechnology, The University of Manchester, UK

James H. Naismith Centre for Biomolecular Science, University of St Andrews, UK

Yasuko Nakano Division of Chemistry and Biotechnology, Graduate School of Natural Science and Technology, Okayama University, Japan

Tanja Narancic Institute of Molecular Genetics and Genetic Engineering, University of Belgrade, Serbia

Bettina M. Nestl Institute of Technical Biochemistry, University of Stuttgart, Germany

Tristan Nicke General Biochemistry, Dresden University of Technology, Germany

Jasmina Nikodinovic-Runic Institute of Molecular Genetics and Genetic Engineering, University of Belgrade, Serbia

Mathias Nordblad DTU Chemical Engineering, Department of Chemical and Biochemical Engineering, Technical University of Denmark, Denmark

Nikolin Oberleitner Institute of Applied Synthetic Chemistry, Vienna University of Technology, Austria

Elaine O'Reilly School of Chemistry, Manchester Institute of Biotechnology, The University of Manchester, UK; School of Chemistry, University of Nottingham, UK

Fabio Parmeggiani Department of Chemistry, Material and Chemical Engineering "G. Natta," Polytechnic University of Milan, Italy

Eugenio P. Patallo General Biochemistry, Dresden University of Technology, Germany

Bharat P. Patel Chemical Development, Bristol-Myers Squibb, USA

Teresa Pellicer Interquim SA, R&D Department, Spain

Xavier Pérez Javierre Universitat Ramon Llull, Institut Químic de Sarrià, Laboratory of Biochemistry, Spain

Antoni Planas Universitat Ramon Llull, Institut Químic de Sarrià, Laboratory of Biochemistry, Spain

Christin Peters Institute of Biochemistry, Department of Biotechnology and Enzyme Catalysis, University of Greifswald, Germany

Mathias Pickl Department of Chemistry, Organic and Bioorganic Chemistry, University of Graz, Austria

Jörg Pietruszka Institute of Bioorganic Chemistry, Heinrich Heine University Düsseldorf, Jülich, Germany; IBG-1: Biotechnology Research Center Jülich, Germany

Gerrit J. Poelarends Department of Pharmaceutical Biology, Groningen Research Institute of Pharmacy, University of Groningen, The Netherlands

Stefan Polnick General Biochemistry, Dresden University of Technology, Germany

Marta Pontini School of Chemistry, Manchester Institute of Biotechnology, The University of Manchester, UK

Nicolas Poupard Clermont University, Blaise Pascal University, ICCF, Clermont-Ferrand, France; CNRS, UMR 6296, France

Sarah M. Pratter Institute of Biotechnology and Biochemical Engineering and Institute of Biochemistry, Graz University of Technology, Austria

Yu-Yin Qi Prozomix Limited, UK; Northumbria University, Department of Applied Science, UK

Xinhua Qian Chemical Development, Bristol-Myers Squibb, USA

Jelena Radivojevic Institute of Molecular Genetics and Genetic Engineering, University of Belgrade, Serbia

Hemalata Ramesh DTU Chemical Engineering, Department of Chemical and Biochemical Engineering, Technical University of Denmark, Denmark

Tamara Reiter Austrian Centre of Industrial Biotechnology, Department of Chemistry, Organic and Bioorganic Chemistry, University of Graz, Austria

Hans Renata Division of Chemistry and Chemical Engineering, California Institute of Technology, USA

Verena Resch Department of Chemistry, Organic and Bioorganic Chemistry, University of Graz, Austria

Andrew S. Rowan Almac, UK

Tomislav Rovis Department of Chemistry, Colorado State University, USA

Jens Rudat Institute of Process Engineering in Life Sciences, Section II: Technical Biology, Karlsruhe Institute of Technology, Germany

Florian Rudroff Institute of Applied Synthetic Chemistry, Vienna University of Technology, Austria

Alessandro Sacchetti Department of Chemistry, Material and Chemical Engineering "G. Natta," Polytechnic University of Milan, Italy

Takashi Sakai Division of Chemistry and Biotechnology, Graduate School of Natural Science and Technology, Okayama University, Japan

Israel Sánchez-Moreno Clermont University, Blaise Pascal University, ICCF, Clermont-Ferrand, France; CNRS, UMR 6296, France

Johan P. M. Sanders Biobased Commodity Chemistry, Wageningen University, The Netherlands

Johann H. Sattler Department of Chemistry, Organic and Bioorganic Chemistry, University of Graz, Austria

Michael A. Schätzle Institute of Pharmaceutical Sciences, Albert-Ludwigs-Universität Freiburg, Germany

Daniel Scheps Institute of Technical Biochemistry, University of Stuttgart, Germany

Markus Schober Department of Chemistry, Organic and Bioorganic Chemistry, University of Graz, Austria

Melanie Schölzel Institute of Bioorganic Chemistry, Heinrich Heine University Düsseldorf, Germany

Jens Schrader DECHEMA Research Institute, Germany

Joerg H. Schrittwieser Department of Chemistry, Organic and Bioorganic Chemistry, University of Graz, Austria

Martin Schürmann DSM Innovative Synthesis BV, The Netherlands

Elinor L. Scott Biobased Commodity Chemistry, Wageningen University, The Netherlands

Frank Seela Laboratory of Bioorganic Chemistry and Chemical Biology, Center for Nanotechnology, Germany

Volker Sieber Fraunhofer Institute for Interfacial Engineering and Biotechnology, Institute branch Straubing, BioCat – Bio-, Chemo- and Electrocatalysis, Germany; Technical University Munich, Germany

Robert C. Simon Department of Chemistry, Organic and Bioorganic Chemistry, University of Graz, Austria

Christopher Squire AstraZeneca, Chemical Development, UK

Vladimir A. Stepchenko Institute of Bioorganic Chemistry, National Academy of Sciences, Belarus

Harrie Straatman DSM Innovative Synthesis BV, The Netherlands

Grit D. Straganz Institute of Biotechnology and Biochemical Engineering and Institute of Biochemistry, Graz University of Technology, Austria

Harald Strittmatter Fraunhofer Institute for Interfacial Engineering and Biotechnology, Institute branch Straubing, BioCat – Bio-, Chemo- and Electrocatalysis, Germany

Christoph Syldatk Institute of Process Engineering in Life Sciences, Section II: Technical Biology, Karlsruhe Institute of Technology, Germany

Anna Szekrenyi Biotransformation and Bioactive Molecules Group, Institute of Advanced Chemistry of Catalonia (IQAC-CSIC), Spain

Lixia Tang School of Life Science and Technology, University of Electronic Science and Technology of China, China

Steve J. C. Taylor Celbius Ltd, CUBIC, Cranfield University, UK

Michael Toesch Department of Chemistry, Organic and Bioorganic Chemistry, University of Graz, Austria

Hai Giang Tran Centre for Industrial Biotechnology and Biocatalysis, Faculty of Bioscience Engineering, Ghent University, Belgium

Nicholas J. Turner School of Chemistry, Manchester Institute of Biotechnology, The University of Manchester, UK

Toby J. Underwood Royal Society of Chemistry, UK

Michael A. van der Horst Van't Hoff Institute for Molecular Sciences, University of Amsterdam, The Netherlands

Johan F. T. van Lieshout Van't Hoff Institute for Molecular Sciences, University of Amsterdam, The Netherlands

Karl-Heinz van Pée General Biochemistry, Dresden University of Technology, Germany

Oscar Verho Department of Organic Chemistry, Arrhenius Laboratory, Stockholm University, Sweden

Lydia S. Walter Institute of Pharmaceutical Sciences, Albert Ludwigs University of Freiburg, Germany

Simon Waltzer Institute of Pharmaceutical Sciences, Albert Ludwigs University of Freiburg, Germany

Liang Wang Department of Chemical and Biological Engineering, Zhejiang University, China

John M. Ward Department of Biochemical Engineering, University College London, UK

Thomas R. Ward Department of Inorganic Chemistry, University of Basel, Switzerland

Nicholas J. Weise School of Chemistry, Manchester Institute of Biotechnology, The University of Manchester, UK

Andrew S. Wells AstraZeneca, Chemical Development, UK

Ron Wever Van't Hoff Institute for Molecular Sciences, University of Amsterdam, The Netherlands

John Whittall Manchester Interdisciplinary Biocentre (MIB), The University of Manchester, UK

Peter William General Biochemistry, Dresden University of Technology, Germany

Yvonne M. Wilson Department of Inorganic Chemistry, University of Basel, Switzerland

Margit Winkler acib GmbH, Austria

Roland Wohlgemuth Sigma-Aldrich, Research Specialties, Switzerland

Michael Kwok Y. Wong Chemical Development, Bristol-Myers Squibb, USA

John M. Woodley DTU Chemical Engineering, Department of Chemical and Biochemical Engineering, Technical University of Denmark, Denmark

Jian-Ping Wu Department of Chemical and Biological Engineering, Zhejiang University, China

Christiane Wuensch Austrian Centre of Industrial Biotechnology, Department of Chemistry, Organic and Bioorganic Chemistry, University of Graz, Austria

Gang Xu Department of Chemical and Biological Engineering, Zhejiang University, China

Li-Rong Yang Department of Chemical and Biological Engineering, Zhejiang University, China

Daiki Yoshida Division of Chemistry and Biotechnology, Graduate School of Natural Science and Technology, Okayama University, Japan

Ferdinand Zepeck Sandoz GmbH, Austria

Xuechen Zhu School of Life Science and Technology, University of Electronic Science and Technology of China, China

Abbreviations

A5P	D-Arabinose-5-phosphate
ABP	Acyl-Co-A binding protein domain
ABTS	2,2′-Azino-bis(3-ethylbenzthiazoline-6-sulfonic acid
ACN	Acetonitrile
ACP	Acyl-carrier protein domain
AcOEt	Ethyl acetate
Acyl-Co-A	Acetyl coenzyme A
AD	4-Androstene-3,17-dione
ADD	1,4-Androstadiene-3,17-dione
ADH	Alcohol dehydrogenase
ADH-hT	ADH from *Bacillus stearothermophilus*
ADP	Adenosine diphosphate
δ-ALA	δ-Aminolevulinic acid
AP	Area percentage
API	Active pharmaceutical ingredient
ArR-ωTA	ω-Transaminase from *Aspergillus terreus*
AST	Arylsulfotransferase
ATHase	Artificial transfer hydrogenases
ATP	Adenosine triphosphate
*Av*PAL	*Anabaena variabilis* phenylalanine ammonia lyase
a_w	Water activity
BBE	Berberine bridge enzyme
BDC	Benzoic acid decarboxylase
BHT	2,6-Di-tert-butyl-4-methylphenol
BIA	Benzylisoquinoline alkaloid
BM3	CYP102A1 from *Bacillus megaterium*
*Bm*GDH	*Bacillus megaterium* glucose 1-dehydrogenase
Boc	*t*-Butoxycarbonyl
BSA	Bovine serum albumin
BSTR	Batch-stirred tank reactor
BVMO	Baeyer–Villiger monooxygenase
b.y.	Baker's yeast
CALA	*Candida antarctica* lipase A
CALB	*Candida antarctica* lipase B
CDW	Cell dry weight
CEBR	Continuous expanded-bed reactors

CFBR	Continuous fluidized-bed reactor
CH$_3$-IMH	D,L-5-(3-Indolylmethyl)-3-N-methylhydantoin
CHMO	Baeyer–Villiger monooxygenase from *Acinetobacter sp.*
CLEA	Cross-linked enzyme aggregates
CLEC	Cross-linked enzyme crystals
CPE	Chloro-1-phenylethanol
CPFR	Continuous plug-flow reactor
CPO	Chloroperoxidase from *C. fumago*
CPR	Cytochrome P450 reductase
CRED	Carbonyl reductase
CSTR	Continuous stirred-tank reactor
CV-ωTA	ω-Transaminase from *Chromobacterium violaceum*
CYP	Cytochrome P450 monooxygenase
DAAO	D-amino acid oxidase
DAB	1,4-Diaminobutane
dC	2′-Deoxycytidine
DCC	Dicyclohexylcarbodiimide
DCM	Dichloromethane
DCP	Dicholoro-2-propanol
DCU	Dicyclohexylurea
ddAHU7P	1,2-Dideoxy-D-arabino-hept-3-ulose 7-phosphate
DEAE	Diethylaminoethanol (group in ion-exchange resin)
dF6P	1-Deoxy-D-fructose-6-phosphate
dG	2′-Deoxyguanosine
DHA	Dihydroxyacetone
DHAK	Dihydroxyacetone kinase
DHAP	Dihydroxyacetone phosphate
DHK	Dihydroxyacetone kinase
dH$_2$O	Distilled water
DIPE	Diisopropylether
DIPEA	Di-isopropyl ethylamine
DKR	Dynamic kinetic resolution
DMAP	4-Dimethylaminopyridine
DMAPP	Dimethylallyl pyrophosphate
DMF	Dimethylformamide
DMSO	Dimethyl sulfoxides
DNA	Deoxyribonucleic acid
DSP	Downstream processing
DTT	Dithiothreitol
EDA	Ethyl diazoacetate
E	Enantiomeric ratio
EBA	Expanded-bed adsorption
EDTA	Ethylenediaminetetraacetic acid
ee	Enantiomeric excess
EEHP	Ethyl 2-ethoxy-3-(4-hydroxyphenyl)propanoate
ELSD	Evaporative light-scattering detector

ER	Ene-reductases
ESI-MS	Electrospray ionization – mass spectrometry
Et$_2$O	Diethyl ether
EtOAc	Ethyl acetate
F6P	D-Fructose-6-phosphate
FAD	Flavin adenine dinucleotide
FADH$_2$	Flavin adenine dinucleotide, reduced form
FBR	Fluidized-bed reactor
FDH	Formate dehydrogenase
FID	Flame ionization detection
FMN	Flavin mononucleotide
FMO	Flavin monooxygenase enzyme
FPLC	Fast protein liquid chromatography
FSA	D-Fructose-6-phosphate aldolase
FTIR	Fourier-transform infrared spectroscopy
αG1P	α-D-Glucose 1-phosphate
G3P	D-Glyceraldehyde-3-phosphate
GA	Glycolaldehyde
GABA	γ-Aminobutyric acid
GC	Gas chromatography
GC-FID	Gas chromatography – flame ionization detection
GC-MS	Gas chromatography – mass spectrometry
GDE	Gas diffusion electrode
GDH	Glucose dehydrogenase
Gly-Gly	Glycyl-glycine buffer
GM	Glucose-milk medium
GPC	Gel permeation chromatography
GPDH	α-Glycerophosphate dehydrogenase
HA	Hydroxyacetone
Hal	Halogenase
HapD	Thiamine diphosphate-dependent enzyme from *Hahella chejuensis*
HB	Hydroxybutanone
Hct	Hectochlorin biosynthesis enzyme
HEPES	4-(2-Hydroxyethyl)piperazine-1-ethanesulfonic acid
Hhe	Halohydrin dehalogenase
HLADH	Horse liver alcohol dehydrogenase
HNB	Hefe–Nährbouillon
4-HPAA	4-Hydroxyphenylacetaldehyde
HPLC	High-performance liquid chromatography
HPLC-DAD	High-performance liquid chromatography with diode-array detection
HPLC-RI	High-performance liquid chromatography with refractive index
HRP	Horse radish peroxidase
IL	Ionic liquid
INT	Iodonitrotetrazolium salt
IPA	Isopropyl alcohol
IPAc	Isopropyl acetate

IPTG	Isopropyl β-D-1-thiogalactopyranoside
IRED	Imine reductase
IS	Internal standard
ISPR	*In situ* product removal
KDO	3-Deoxy-D-*manno*-oct-2-ulosonic acid
α-KG	α-Ketoglutarate
KPi	Potassium phosphate buffer
KR	Kinetic resolution
KRED	Ketoreductase
LB	Luria–Bertani medium
LDH	Lactose dehydrogenase
LK-ADH	ADH from *Lactobacillus kefir*
LSADH	*Leifsonia* sp. alcohol dehydrogenase
M9-N	M9 minimal salts microbial growth medium
MALDI	Matrix-assisted laser desorption/ionization
MAO-N	Monoamine oxidase from *Aspergillus niger*
MBR	Membrane bioreactor
MCF	Siliceous mesocellular foam
MeCN	Acetonitrile
MeOH	Methanol
MES	2-(N-Morpholino)ethanesulfonic acid
MOPS	4-Morpholinepropanesulfonic acid
MS	Mass spectrometry
MPLC	Medium-pressure liquid chromatography
MPTA	α-Methyl-α-trifluoromethylphenylacetic acid
MTBE	Methyl *t*-butyl ether
MWCO	Molecular weight cut-off
m/z	Mass-to-charge ratio
NAAAR	*N*-Acetyl amino acid racemase
NAD+	β-Nicotinamide adenine dinucleotide
NADH	β-Nicotinamide adenine dinucleotide, reduced form
NADPH	β-Nicotinamide adenine dinucleotide 2′-phosphate, reduced form
NADP+	β-Nicotinamide adenine dinucleotide 2′-phosphate
NCS	(*S*)-Norcoclaurine synthase
Ni-NTA	Nickel-nitrilotriacetic acid
nm	Nanometer
*N*CβPhe	*N*-Carbamoyl-β-phenylalanine
NMR	Nuclear magnetic resonance spectroscopy
OD	Optical density
ω-OHFA	ω-Hydroxy fatty acid
OxM	Oxygenase medium
4-OT	Oxalocrotonate tautomerase
OYE	Old yellow enzyme
P450cam	Cytochrome P450 camphor-hydroxylating
PAD	Phenolic acid decarboxylase
PAGE	Polyacrylamide gel electrophoresis

PAL	Phenylalanine ammonia lyase
PAPS	Adenosine-3′-phospho-5′-phosphosulfate
PAS	Sulfatase from *Pseudomonas aeruginosa*
PBR	Packed-bed reactor
PCR	Polymerase chain reaction
PE	Petrol ether
PEP	Phosphoenolpyruvic acid
PFAM	Protein family
PheDU	Phenyldihydrouracil
PhoN-Sf	Phosphatase from *Shigella flexneri*
PigD	Thiamine diphosphate-dependent enzyme from *Serratia marcescens*
PK	Pyruvate kinase
PLP	Pyridoxal 5′-phosphate
PMMA	Poly(methyl methacrylate)
PMSF	Phenylmethylsulfonyl fluoride
PNP	Purine nucleoside phosphorylase
PPi	Pyrophosphate anion $P_2O_7^{4-}$
PPM	Phosphopentomutase
PPTase	4′-Phosphopantetheinyl transferase
PTFE	Polytetrafluoroethylene
PYR	Pyruvate
prnA	Tryptophan 7-halogenase
R&D	Research and development
rac	Racemic
Rf	Retention factor
*Rg*PAL	*Rhodotorula glutinis* phenylalanine ammonia lyase
RK	Ribokinase
*Rm*Qred	Quinuclidinone reductase of *Rhodotorula mucilaginosa*
ROH	Generic alcohol
RPE	Ribulose-5-phosphate epimerase
RPI	Ribose-5-phosphate isomerase
PVDF	Polyvinylidene difluoride
pyrH	Tryptophan 5-halogenase
*Rm*Qred	Quinuclidinone reductase of *Rhodotorula mucilaginosa*
rpm	Revolutions per minute
rt	Room temperature
Rt	Retention time
Sav	Streptavidin
SDR	Short-chain alcohol dehydrogenase
SDS	Sodium dodecyl sulfate
SeAAS	Thiamine diphosphate dependent enzyme from *Saccharopolyspora erythraea*
SFC	Supercritical fluid chromatography
Sfp	Phosphopanthetheinyl moiety transfer protein from *Bacillus subtilis*
SFPR	Substrate feeding product removal
SIM	Single-ion monitoring

TA	Transaminase
TA-CV	ω-Transaminase from *Chromobacterium violaceum*
TB	Terrific broth
TEAA	Triethylamine acetate
TEMPO	2,2,6,6-Tetramethylpiperidine-1-oxyl
TFA	Trifluoroacetic acid
THIQ	Tetrahydroisoquinoline
THNR	Tetrahydroxynaphthalene reductase
TLC	Thin-layer chromatography
TOF	Time-of-flight
TPI	Triosephosphate isomerase
Tris	Tris(hydroxymethyl)aminomethane
TycF	Thioesterase from *Bacillus brevis*
U	Units
UF	Ultrafiltration
UHPLC-UV	Ultra high-performance liquid chromatography, ultraviolet
UV	Ultraviolet
VCD	Vibrational circular dichroism
VCPO	Vanadium chloroperoxidase
v/v	Volume/volume
vvm	Gas volume flow per unit of liquid volume per minute (vessel volume per minute)
wcw	Wet cell weight
wt	Wild type
w/v	Weight/volume
w/w	Weight/weight
2xYT	2xYT Medium for microbial growth
YPD	Yeast extract peptone dextrose medium

1

Considerations for the Application of Process Technologies in Laboratory- and Pilot-Scale Biocatalysis for Chemical Synthesis

Hemalata Ramesh,[1] Mathias Nordblad,[1] John Whittall,[2] and John M. Woodley[1]

[1]*Department of Chemical and Biochemical Engineering,*
Technical University of Denmark, Denmark
[2]*Manchester Interdisciplinary Biocentre (MIB), The University of Manchester, UK*

1.1 Introduction

The development and implementation of an efficient new biocatalytic process relies upon successful communication between the scientists establishing the chemical reaction (organic chemists, process chemists, analysts, etc.), those developing the biocatalyst (microbiologists, biochemists and molecular biologists, analysts, etc.), and those scaling up the process (process, biochemical, and chemical engineers). The working relationship between the first two groups has strengthened enormously in recent years, but nevertheless successful scale-up also requires process engineering involvement from an early stage. In the pharmaceutical industry, it is easy to argue that the rate of attrition of new target molecules is such that any consideration for scale-up should be delayed for as long as possible. However, the reality is that to address the process aspects too late is equally problematic. The problem is exacerbated by many of the chemical reactions of greatest commercial interest in transforming non-natural substrates. In some cases, the selectivity of an enzyme is not compromised, but its activity is nearly always found to be lower than on a

Practical Methods for Biocatalysis and Biotransformations 3, First Edition.
Edited by John Whittall, Peter W. Sutton, and Wolfgang Kroutil.

comparable natural substrate. Additionally, the conditions under which these enzymes are expected to operate in industry are also very often far from those found in nature, further affecting their activity and stability. Therefore, this necessitates improvements not only to the biocatalyst, but also to the reactor and process, such that a suitable system can be designed and implemented for scale-up. This demands effective communication and dialog between the various scientists at an early stage of process development.

This chapter is written with the intent of giving chemists, biologists, and engineers a basic idea of the concepts involved in implementing a biocatalytic reaction for development, scale-up, and, ultimately, production of a target product. Frequently, information relevant to these fields is scattered, and therefore a deliberate attempt has been made here to bring it together into a single compilation to help in disseminating the available knowledge to those working in all aspects of biological chemical conversions.

It is hoped that this will give a better understanding to scientists working in unifying these fields for efficient process development. For example, a biologist should be able to use this chapter to help understand the importance of setting commercial targets to measure the success of the biocatalysts they have developed. Likewise, process chemists can appreciate the key differences between the application of chemical catalysis and biological catalysis. Additionally, this chapter aims to guide chemists and biologists in designing experiments to obtain relevant data that might help in a smooth transition from a laboratory proof-of-concept to a scalable chemical synthesis with product isolation. Engineers will also abstract the differences between biochemical and conventional chemical transformations.

The aim of this chapter is to provide readers with an understanding of the important tools and technologies available for use in biocatalysis. Specifically, the technologies that can be implemented at laboratory and pilot scale will be addressed. Quantitative information will be provided when possible for application of these technologies, which will hopefully guide the reader to make educated decisions on how to efficiently operate their processes. The purpose, therefore, is not to answer all questions, but to give a quick overview of the different characteristics and considerations for the said technologies.

Finally, it is of vital importance to acknowledge that this text is based on the contributions and experiences of many scientists and engineers (both in academia and in industry) from different spheres involved in the establishment of fundamental and applied research of the discussed technologies.

1.2 Process Intensification and Proposed Scale-Up Concept

The arguments for the application of biocatalysis as a catalytic tool in organic synthesis and production are numerous, but are perhaps most usually focused on the exquisite selectivity that biocatalysts offer [1]. Clearly, the rationale for implementation depends upon the industrial sector and the value of the product to be produced.

One of the central challenges in the development and implementation of new enzymatic processes in industry is translating an established laboratory-scale reaction into a commercial process. The first step in that journey should be to establish suitable conditions for the reaction, in particular the required selectivity and product purity. This is mainly the work of organic and process chemists. The enhancement of enzyme properties is also a major preoccupation at this point, carried out by molecular biologists. In order to make a process that can meet the demands of industry, scale-up also needs to be considered. As we have suggested in several recent publications, it is best to address this in two steps: first, by improving the process via enzyme modification and process intensification, and second by considering scale-up by volume increase [2]. This two-part philosophy builds confidence at an early stage that the process is indeed scalable and helps to test the limits of the process technology, both at laboratory and at pilot scale.

Biological conversions using enzyme(s) or enzymes in cells (in a non-fermentative state) are also known as "biocatalytic reactions," and are typically carried out in aqueous media. Biocatalysts have evolved to work on particularly low concentrations of natural substrates, so as to make them highly efficient in nature. However, in industrial applications, biocatalysts are often subject to non-natural environments, such as reaction solutions with high concentrations of substrate or product. Both the activity (reaction rate) and stability (maintenance of reaction rate over time) of the biocatalyst are affected by this. However, both parameters are also critical for the performance of the biocatalytic process. The rationale behind operating the reaction at high concentrations of substrate and product is the need to meet the minimal process metrics (in order to fulfill the required commercial targets). Reaction (and process) yield, as well as space–time yield, are determined by the commercial targets of a particular product. In other words, a certain supply rate of product will be required for economic feasibility (either in an existing facility or in a dedicated facility). These commercial targets will, in turn, also set the process metrics. For a biological conversion with isolated (immobilized) enzymes or resting whole cells, the process metrics can be defined by the product concentration and the biocatalyst yield (linked to the allowable cost for the biocatalyst). For biological conversions with growing cells (fermentations), space–time yield replaces the biocatalyst yield metric (since the time required for cell growth limits the process). The various commercial targets and process metrics are defined as follows:

Reaction yield (in engineering terms – same as "conversion" in laboratory terms) is a measure of the mass of product formed per mass of substrate consumed (usually expressed in units such as $g.g^{-1}$). The reaction yield may also be expressed on a molar basis. Process yield (rather than reaction yield or conversion) can be used to take into account losses in the downstream recovery of the product following the reaction. It is important to consider process yield in order to ensure sufficient product is made to achieve the necessary commercial metrics. Together with the difference in value between the reaction substrate and product, the reaction (and process) yield will determine the value added to the substrate as a result of the reaction. It is therefore the paramount commercial metric.

Reaction yield = mass of product formed/mass of substrate consumed

Biocatalyst yield is the mass of product formed per mass of biocatalyst provided (expressed in units such as $g.g^{-1}$). In many cases, biocatalysts are recovered and then recycled following a reaction. Hence, the biocatalyst yield should reflect the cumulative mass of product formed over all the batches in which the biocatalyst is used. Together with the absolute cost of the biocatalyst (determined by the supplier or the fermentation conditions), the biocatalyst yield will determine the cost contribution of the biocatalyst to the final operating cost. In many cases, this is essential to ensuring the cost of the manufacturing process is sufficiently low for the product value to meet market expectations.

$$Biocatalyst\ yield = mass\ of\ products\ formed/mass\ of\ biocatalyst\ provided$$

Product concentration is the mass of product formed per reaction volume (usually expressed in units such as $g.L^{-1}$). The product concentration defines the scale of the downstream product recovery process and will therefore determine both the operating cost and the capital cost for dedicated plants.

$$Product\ concentration = mass\ of\ product\ formed/volume\ of\ reactor$$

Space–time yield is the mass of product formed per reaction volume per time (usually expressed in units such as $g.L^{-1}.h^{-1}$) and is a measure of the capacity of a process. For a given production rate, the space–time yield thus defines the scale of the reactor to achieve a given commercial target. For a process based on growing cells (fermentation), the space–time yield is largely governed by the time required for the growth of the cells, which thus determines the cost contribution of the fermentation to the overall conversion process.

$$Space - time\ yield = mass\ of\ product\ formed/volume\ of\ reactor/time$$

There are as yet no accepted guidelines for the minimum process metrics, and in any case they are, of course, in large part dependent on the potential production volume (market size), required purity, and value of the product (e.g., bulk chemical, speciality chemical, or small-molecule pharmaceutical). Nevertheless, some example values are given in Table 1.1 as a first guide. The required biocatalyst yield is closely linked to the allowable cost for the biocatalyst. The guidelines in Table 1.1 are for immobilized enzymes and the required yields are consequently high (see Section 1.3.1).

In order to achieve such process metrics, it is clear that in almost all cases a strategy of biocatalyst improvement (via targeted protein engineering, biocatalyst modification, and improved biocatalyst production) should be complemented by the implementation of process intensification options into the reaction (and process) of interest. The purpose of the following sections is to introduce the reader to the available process-intensification options. With that in mind, the sections have been formulated by first describing the

Table 1.1 *Example guidelines for minimum process metrics (biocatalyst yield and product concentration) for immobilized enzyme-catalyzed reactions in different industrial sectors.*

Product market	Product value ($.kg^{-1})	Biocatalyst yield (kg.kg^{-1})	Product concentration (kg.m^{-3})
High value (e.g., pharmaceuticals)	500	50	50
Medium value (e.g., flavors)	100	500	150
Low value (e.g., bulk chemicals)	1	5000	300

technologies available and then outlining some of the considerations required for the implementation of each. The purpose is not to give answers to the question of how a process should be operated, but rather to provide a guide for selection of the most suitable options in which to invest time, effort, and money in further research in a given case. Necessarily, the chapter does not aim to be comprehensive.

1.3 Enabling Technologies

Enabling technologies are those technologies that allow the process to be implemented. While biocatalyst immobilization (Section 1.3.1) is not essential to implementation, consideration of the options is nearly always required. On the other hand, all reactions will need to be operated in an available reactor, or, more rarely, a dedicated reactor (Section 1.3.2). Understanding the implications of using a given option is of great importance to achieving the required process metrics for given commercial targets.

1.3.1 Biocatalyst Immobilization

Immobilization is the process of attaching soluble enzyme, or alternatively whole cells, on to or into larger particles of inert support materials, with the primary aim of facilitating an easy separation so as to remove (and frequently recover and recycle) the biocatalyst from the product stream. A secondary objective is to improve the operational stability of the biocatalyst. Biocatalyst immobilization is a process of significant importance and plays a central role in the operational performance of a biocatalyst. Immobilization can be applied to both enzymes and whole cells. Very many (in fact, several hundred) techniques, which can broadly be classified into carrier immobilization, carrier-free immobilization, and entrapment, are available for this purpose [3,4]. Carrier-immobilization and carrier-free-immobilization methods will be outlined here.

Attaching the biocatalyst to a support material brings the advantage of separation by simple filtration (e.g., using microfiltration, or even sieves – so-called "Johnson" screens – for larger particles). Such separation facilitates the first step in the downstream process, enabling complete removal of the biocatalyst (which is especially required in pharmaceutical manufacture). Provided the biocatalyst has sufficient stability (this is often improved on immobilization), it can potentially be recovered for subsequent recycle, enabling an increase in biocatalyst yield. The insoluble format of the immobilized enzyme makes the use of continuous-packed-bed and fluidized-bed reactors possible.

Another aspect of immobilization is that it allows for specific control of the biocatalyst's microenvironment. For example, the hydrophilic/hydrophobic balance of the immobilized biocatalyst can be changed by altering the support material. Finally, multi-enzyme and chemoenzymatic cascades can be made possible using such technology to co-immobilize them on to a single support as a multifunctional catalyst, when both (bio)catalysts have similar stability (provided that the activities are balanced). When this is not the case, immobilization can allow the separation of (bio)catalysts inside a reactor (by compartmentalization).

The stated advantages of separating the biocatalyst from the bulk reaction media are applicable to the majority of biocatalytic reactions, most of which are carried out in aqueous media. For those biocatalytic reactions that are performed in organic or biphasic media, the formulation of an immobilized biocatalyst is essential, since the biocatalysts cannot generally be dissolved in such systems.

1.3.1.1 General Considerations for Implementation

The first and most important consideration is that the process of immobilization itself may lead to a loss of enzyme activity. The conditions for immobilization are often harsh, and potentially up to 50% of the activity can be lost in the preparation step [5]. Furthermore, the carrier must be added, and this, combined with operational costs for immobilization, adds considerably to the final cost of the biocatalyst [6]. Hence, biocatalyst recovery and recycle, while clearly advantageous, may in many cases also be a necessity, in order to recover the losses caused by the immobilization procedure, as well as the cost of the support material in the formulation. In some rare cases, recycle of the carrier (support) material may also be possible, even if the enzyme is not stable enough to be recycled.

A further complication with immobilized biocatalysts is the potential for substrate mass transfer limitations. During operation, the substrate(s) needs access to the enzyme, which is rarely on the surface of the support alone, but usually in pores throughout the carrier particle. In general, a large surface area, typically greater than $100\,m^2.g^{-1}$, is required for immobilization [7,8]. Consequently, very small particles (with an average diameter less than 50 μm) would be required were surface immobilization used alone, in order to provide sufficient activity per volume, and this would make filtration problematic, defeating the primary purpose of the immobilization. For this reason, most particles are larger (with an average diameter of 150–500 μm), implying that the rate of substrate transport into the support (and the rate of product out of the support) can potentially be rate-limiting. In fast reactions, with a low-porosity structure (sometimes used to give the required strength to the support), diffusional limitations can therefore be expected [9]. The measured reaction rate in such cases may be lower than the equivalent rate measured with soluble enzyme, depending upon the K_m of the enzyme. Mass transfer limitations can also be detrimental to the selectivity of the biocatalyst.

1.3.1.2 Carrier-Bound Supported Enzymes

1.3.1.2.1 Adsorption. Adsorption refers to the binding of the enzyme (or, potentially, whole-cell biocatalysts) on to a porous support (or carrier) via physical interactions such as hydrogen bonds, van der Waals forces, and hydrophobic interactions [10]. The maximum adsorption of a protein on a hydrophobic carrier usually occurs around its isoelectric pH (pI value). A schematic representation of the adsorption process is shown in Figure 1.1.

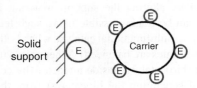

Figure 1.1 *Schematic representation of enzyme immobilization by adsorption (E, enzyme).*

In such cases, the activity loss is minimal provided conformational changes are avoided. The method is also simple and versatile [11]. The amount of protein that can be loaded is somewhere between 2 and 50 mg of protein per gram of support. The support is usually between 100 and 200 µm in diameter. A classical industrial example of immobilization by adsorption is the immobilization of *Candida antarctica* B-lipase on a polymeric carrier (Novozym 435).

Considerations for Implementation.   With high protein loading, some steric hindrance may lead to lower measured reaction rates compared to the equivalent measurements using soluble enzyme. Perhaps more serious is that desorption (sometimes referred to as "leaching") of protein is possible: this means not only that the enzymatic activity will be reduced, but also that some protein can pass downstream. For pharmaceutical applications, this is a serious limitation, and usually necessitates an extra ultrafiltration step immediately downstream of the reactor, prior to the other product recovery operations. However, potential desorption is very dependent upon operating media. For example, in organic media, adsorption works particularly well, since the protein does not leach from the surface.

1.3.1.2.2 Covalent Binding.

A second group of immobilization methods can be classified as those based on covalent binding of the enzyme to the support material. The carrier is bound to the enzyme by means of functional groups (i.e., the amino acid residues) of the protein. Care should be taken that use of these residues does not interfere with the active site or the substrate binding sites of the enzyme. For glycosylated enzymes, there is also the option of coupling using their carbohydrate moiety. The most common modes of covalent binding are diazo-coupling, peptide bond formation, and alkylation or arylation. Other, less common methods include Schiff's base formation and amidation. A schematic representation of enzyme immobilization by covalent binding is shown in Figure 1.2.

Clearly, covalent immobilization establishes a permanent bond between the enzyme and the support, meaning that the enzyme is usually stabilized by maintaining its tertiary structure intact and the potential for desorption of the protein is eliminated. There are many support materials that can be used, provided the material can be activated so as to form a covalent bond with the enzyme. Eupergit C is an excellent example of enzyme immobilization by covalent bonding [12]. An application of this type has been demonstrated in the production of N-acetylneuraminic acid using N-acetylneuraminic acid aldolase

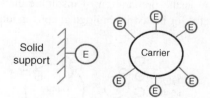

Figure 1.2 *Schematic representation of enzyme immobilization by covalent binding (E, enzyme).*

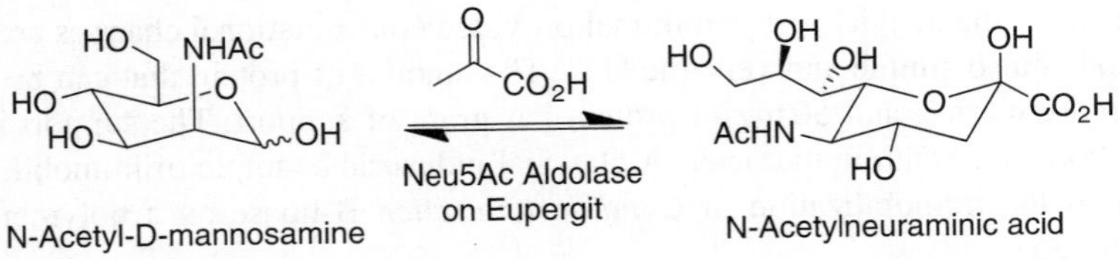

Scheme 1.1 *Synthesis of N-acetylneuraminic acid using supported enzyme.*

immobilized in Eupergit C. The immobilized enzyme was reused for at least nine cycles without significant loss in activity (Scheme 1.1) [13].

Considerations for Implementation. Potentially, the enzyme can change conformation during covalent bond formation, leading to a significant loss in activity and/or selectivity. It is important to consider that the carrier needs to be chemically activated prior to formation of the covalent bond. In some cases, the coupling agent can deactivate the enzyme of interest, lowering the applicability.

1.3.1.2.3 Ionic Binding. Ionic binding is a method of immobilization that exploits the ionic interaction between the carrier and the support to facilitate binding. Both cationic and anionic exchangers can be used as the support. A schematic diagram of such enzyme immobilization is depicted in Figure 1.3.

Clearly, one of the major advantages of using ionic interactions to attach the enzyme to the support is that there are already many cheap and readily available ion-exchange resins on the market that can potentially be used. While the binding forces are stronger than those of physical adsorption, they are not as strong as covalent binding.

Considerations for Implementation. The binding stability (balance of adsorption and desorption) is affected by the pH and the ionic strength of the reaction medium, which must be well understood and controlled. Reactions in which the ionic strength of the medium changes (e.g., reactions where an acid or base is added to maintain constant pH changes on account of the reaction) need to be examined with particular care. Likewise, for whole-cell immobilization, cell age, pH, ionic strength, and surface charges can all affect the performance of the immobilization procedure.

1.3.1.3 Carrier-Free Immobilization

Carrier-free immobilization is the preparation of insoluble enzymes (without the use of carriers), which can be useful for industrial biological conversions. Several methods exist,

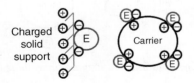

Figure 1.3 *Schematic representation of enzyme immobilization by ionic binding (E, enzyme).*

including cross-linked enzyme aggregates (CLEAs), cross-linked enzyme crystals (CLECs), flocculation, and aggregation [14].

1.3.1.3.1 Cross-Linked Enzyme Aggregates (CLEAs™). An alternative and potentially easier approach to conventional immobilization of a carrier support material is to aggregate the proteins. Proteins usually aggregate in the presence of salts, non-ionic polymers, or water-miscible organic solvents. Subsequently, the aggregates can be cross-linked to render them permanently insoluble. Such cross-linked insoluble aggregates are termed "CLEAs" [15,16].

Protein aggregation is a simple immobilization technique amenable to rapid optimization. There are few process steps for catalyst formulation, and it can potentially be used to combine purification and immobilization in a single unit operation [17]. It is a relatively low-cost method, since it avoids the use of expensive carrier supports and leads to a short development time. Additionally, high catalyst activity can be achieved, since the enzyme activity is concentrated as an insoluble aggregate with very high protein content. Likewise, catalyst stability can be very high. Furthermore, the technique can stabilize quaternary structures of multi-meric enzymes [18].

Recently, it has been shown that simple multi-enzyme immobilization is also possible. This results in combinations of enzymes that can be immobilized together and have potential for use in cascade reactions (combi-CLEAs) [19]. Inevitably, typical biocatalyst aggregate sizes are quite small – usually with diameters between 5 and 50 μm – but larger sizes are also available [20]. Nevertheless, biocatalyst recycle is relatively easy using either filtration or the magnetic attraction of iron trapped in the aggregates [21].

Considerations for Implementation. Enzymes that have a low number of surface-reactive amino groups lead to the formation of mechanically unstable CLEAs due to poor cross-linking, and this can also lead to leaching of the enzyme during operation. Additionally, a new immobilization protocol has to be developed for the aggregation and cross-linking of each enzyme. In some cases, this may require some purification of the crude protein prior to CLEA formation. The glutaraldehyde linker in high concentrations tends to cause a loss of enzyme activity, due either to chemical modification of the functional groups or to denaturation induced by derivatization [15]. As with other immobilizations, the optimal size of the CLEA particles is a trade-off between their being small enough to minimize substrate mass transfer limitations and their being large enough to facilitate process handling. Although crucial for enzyme activity, control of CLEA size is difficult. Also, the CLEA particles may not always be able to maintain their mechanical stability under agitated conditions.

1.3.2 Reactor Options

As with conventional chemical synthesis, several reactor options are at the disposal of a process engineer wishing to implement and develop a biological conversion for larger-scale application. Both whole cells and enzymes (free and immobilized) have been used in several reactor configurations. The characteristics of these reactor designs will be discussed in this section. In an industrial setting (especially in the pharmaceutical industry), the reactor is often already defined, and the reactions need to be fitted to the equipment. In such cases, it is also useful to understand the characteristics of each reactor type, in order to understand compromises and trade-offs that have to be considered. Finally, from the perspective of the process chemist and biologist, the reaction and biocatalyst characteristics have a profound influence on the choice of equipment to be used, or, alternatively, on the degree of compromise required to fit into existing equipment, in any given case.

1.3.2.1 Ideal Reactors

Classical chemical reaction engineering classifies reactors into groups dependent upon hydrodynamic properties and mode of operation. Three "ideal" reactor configurations are defined based on the concept of "ideal" hydrodynamics (well-mixed or plug-flow) and mode of operation (batch or continuous). At large scale, such "ideality" does not exist, but nevertheless the classification provides a useful basis for reactor characterization. In a reactor with well-mixed hydrodynamics, the reaction mixture is agitated so that the concentration, temperature, and pH are identical throughout the vessel [22–24]. For a continuous system, this also implies the leaving concentrations are the same as those in the tank, which has important implications for achievable yield and kinetics. For ideal plug-flow hydrodynamics, there is no mixing in the direction of flow through the reactor, and substrate and product concentration are a function of distance traveled (also expressed as "residence time"). Hence, three types of ideal reactor can be distinguished: batch stirred-tank reactor (BSTR), continuous stirred-tank reactor (CSTR), and continuous plug-flow reactor (CPFR). Each comes with a number of advantages and disadvantages related to, for example, volumetric efficiency, space–time yield, and achievable conversion, which will be discussed in the following subsections.

1.3.2.2 Modes of Operation

Most chemical processes can be characterized as either batch or continuous processes. The two clearly differ in that the former outputs product in discreet volumes – batches – and the latter delivers product continuously. However, the two modes of operation are also different in that the continuous process normally operates at steady state (reaction composition at a given point in the process does not change with time), whereas a batch process is dynamic. Both modes of operation come with several advantages and disadvantages. The initial investigation and reaction development for biocatalytic processes is almost exclusively conducted in batch or fed-batch mode.

There is often an interest in modifying processes to make them suitable for continuous operation as they are implemented for full-scale production. Among the clear advantages of continuous operation are the simplified process control for a system that operates at steady state and the potential for improved productivity, since the downtime for filling and emptying vessels (which can represent a significant part of the overall processing time in large-scale processes) is removed. However, a continuous process also comes with a number of challenges. Truly continuous operation requires dedicated equipment and either a high operational catalyst stability or a method for continuously resupplying the process with fresh catalytic activity. Starting up and balancing multiple process steps can also be a complicated task. Finally, continuous operation inherently makes it more difficult to separate product into distinct groups, or batches. As a consequence, errors or contamination in the process become more difficult to trace than in a batch process, and such problems can therefore potentially be more costly in a continuous system.

1.3.2.3 Well-Mixed Reactor Hydrodynamics

Regardless of the mode of operation (batch, fed-batch, or continuous) in well-mixed reactors, the composition of the tank contents is homogenous [25]. This has an important implication: the concentration of any given species is the same at all points in the reactor, meaning there is no concentration gradient across any part of the reactor.

1.3.2.3.1 Stirred Tanks.

By far the most common type of well-mixed reactor is the stirred tank. Such reactors are very flexible but also have some important features to be borne in mind if they are to be used for biocatalytic reactions. For example, a maximum loading of immobilized biocatalyst of approximately 10% by volume can be tolerated in a conventional stirred-tank reactor. Higher biocatalyst loadings lead to a higher rate of particle attrition (due to a higher frequency of collision), which, aside from affecting both the activity and the stability of the biocatalyst, can also create a significant problem for downstream biocatalyst filtration for removal, or for recovery and subsequent reuse [26,27]. The aspect ratio of the reactor (ratio of tank height to tank diameter) also affects bulk mixing. High aspect ratios result in poorer overall mixing, so a ratio of unity is normally applied [24]. However, in cases where mass transfer of poorly water-soluble gases (such as oxygen) into the reaction is required, the aspect ratio is increased to as high as 3 in order to increase the residence time of the gas phase in the reactor, and consequently the uptake of gaseous solutes. Typically, reactors are baffled in order to prevent vortex formation and improve the stirring efficiency of the power input. Four baffles are normally used, with a dimension of around 1/10 the diameter of the vessel [24]. When solids are present in the medium (or an immobilized biocatalyst/CLEA is used), the agitator speed must be sufficient to allow good suspension of the particles. The minimum agitation speed required to keep all the immobilized particle in suspension is termed just off-speed limit [28]. Reactors may use more than one impeller (dependent on the scale), with a typical spacing of between 2/3 T and T, and where $D/T \approx 0.4$ (D and T are the impeller and tank diameters, as illustrated in Figure 1.5). Power input for stirring should be around $1–1.5 \text{ W.L}^{-1}$. Indeed, for reactors up to a volume of around 500 L, a power input of $1–1.5$ W. L^{-1} gives a well-mixed reaction volume [24]. Nevertheless, as scale increases, mixing time (inversely proportional to the agitator speed) will also increase for a given power input. Stirred-tank reactors can be operated in batch, fed-batch, or continuous mode of operation. The advantages and disadvantages of these modes are discussed later in this section.

Considerations for Implementation

- **Mixing type:** Axial mixing is typically used for reactors handling solids, while radial mixing is extensively used for two-liquid phase systems [29]. A schematic representation of the types of mixing can be found in Figure 1.4.
- **Impeller position:** When handling multiphasic systems, the impeller must be placed in the phase that is to be under continuous operation conditions.
- **Motor type:** In order to give some flexibility, the motor used will typically be a variable speed motor.
- **Aspect ratio:** Mixing time increases with the height-to-width ratio of the tank, hence it is always beneficial to keep this ratio close to unity. However, in some systems (where gas is required, for example) it may be necessary to use higher ratios [23]. In such cases, it is common practice to place multiple impellers on the shaft.

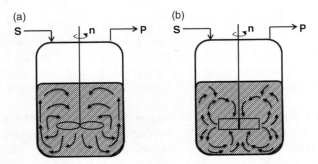

Figure 1.4 *Stirred reactors with (a) axial (down) flow and (b) radial flow (S, substrate; P, product; n, number of rotations per unit time).*

- **Mechanical stress on biocatalyst particles:** Mechanical stress is usually high on immobilized enzymes, due to agitation, especially at higher biocatalyst loadings [30–32]. This stress can be limited by reducing the mixing speed/power input. However, this will result in poorer overall mixing.
- **Heat transfer:** Heat transfer in the reactor is usually achieved through a jacket fitted around the reactor. Although maintenance of the correct temperature is important, control is rarely an issue in biocatalytic reactions, with the exception of enzyme-catalyzed polymerization (which can be exothermic).

1.3.2.3.2 Batch Stirred-Tank Reactors (BSTRs). BSTRs are stirred-tank reactors that are operated in batch mode. All the reactants (substrates) are charged into the reactor at the start of the operation and allowed to react. Upon completion, the product is recovered (Figure 1.5). The conversion achieved is a function of the batch operation time, and in principle complete conversion is possible.

This type of reactor can be used for kinetically slow reactions. Many enzymatic reactions typically fall into this category. The reactor may also be operated in fed-batch mode, meaning that some (or all) of the substrates are fed to the reactor in a controlled manner, at

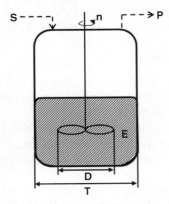

Figure 1.5 *Schematic diagram of a batch stirred-tank reactor (BSTR) (E, immobilized enzyme; S, substrate; P, product; D, diameter of the stirrer; T, diameter of the tank reactor; n, number of rotations per unit time).*

constant or variable feed rate [33]. This is particularly useful for substrates that are inhibitory or toxic to the biocatalyst. Likewise, the mixing means that this configuration offers ease of handling for multiphasic (gas–liquid, solid–liquid, or liquid–liquid) reaction mixtures. The reactor type can accommodate all types of biocatalyst and is relatively easy to scale up. Finally, the BSTR is relatively straightforward to adapt at industrial scale, as it is commonly available and many industrial processing sites have suitable vessels already available for use with minimal capital investment.

Considerations for Implementation

- **Batch shift:** Sometimes it is desirable for the batch reaction to fit within a regular 6- or 8-hour working shift. This dictates the target space–time yield for the process and must be considered.
- **Catalyst deactivation:** It is often desirable to reuse a biocatalyst in multiple batches, especially for immobilized formulations (due to the added cost of the immobilization procedure). However, the performance of the catalyst will typically vary from batch to batch, due to a gradual loss of activity over time. This activity loss can be compensated for in several ways. One method is to add fresh catalyst to each batch; this approach is limited by the amount of fresh biocatalyst that can be added to the reactor (without exceeding loading limitations). Alternatively, the reactor can be operated for longer time periods, to account for lower activity as a function of time. In order to manage biocatalyst deactivation, it may also be possible to increase the operating temperature in the reactor, but this is limited by the thermal stability (or otherwise) of the biocatalyst and reaction components [34]. Loss of a batch of immobilized biocatalyst can have serious economic implications, since it prevents reuse, significantly increasing the cost contribution of the biocatalyst.

1.3.2.3.3 Continuous Stirred-Tank Reactors (CSTRs).

CSTRs use the same type of reaction vessel as BSTRs but operate with a continuous reactant feed and product removal stream (Figure 1.6). For an ideal reactor, this implies operation at the exit substrate concentration. A direct consequence of this is that reaching complete conversion is not

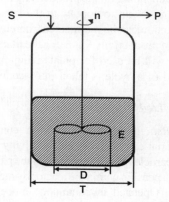

Figure 1.6 *Schematic diagram of a continuous stirred-tank reactor (CSTR) (E, immobilized enzyme; S, substrate; P, product; D, diameter of the stirrer; T, diameter of the tank reactor; n, number of rotations per unit time).*

possible (since some substrate must be present for the reaction to take place). It also indicates that there is a trade-off between achievable conversion and reaction rate (dependent upon the value of the K_m of the enzyme) (see, for example, [35]).

As explained earlier, compared to the equivalent BSTR, the CSTR is usually larger (dependent upon the K_M value of the enzyme of interest), since the kinetics are not as favorable. Nevertheless, the saving on downtime (filling, emptying, and cleaning) can be significant. This is emphasized in batch reactors with a relatively short operating time (e.g., those limited by product inhibition). In cases where the kinetics are poor in a CSTR (i.e., the exit concentration of substrate is at a value that is just a fraction of the enzyme K_M), the reaction may be improved by simulating plug-flow operation, achieved by operating several CSTRs in series (although, for economies of scale, it is unlikely more than three will be used). Interestingly, retrofit of existing batch reactors into CSTRs should be fairly straightforward, provided pumps are available. Such systems can easily cope with multiphasic media, although care should be taken about feed and draw streams in relation to the phase ratio [32]. The CSTR is also highly efficient in terms of man power and control, since the system is in steady state and typically exhibits a first-order response to any perturbations.

Considerations for Implementation. The main disadvantages of a CSTR are that it generally operates at a lower average reaction rate than a BSTR and that it by definition cannot achieve equilibrium conversion. Another aspect that should be considered early in process development is the method of catalyst recycling and reuse, if this is required for the economic feasibility of the process. When a non-immobilized biocatalyst is employed, it must be separated from the product stream and recycled to the reactor in a way that does not inactivate the biocatalyst. Even so, both immobilized and non-immobilized biocatalysts will gradually lose activity over time. In order to manage biocatalyst deactivation, one of the following two methods should be adopted [36]:

1. **Add fresh biocatalyst to the system:** The design of the CSTR must then take into account that fresh biocatalyst will be added. Clearly, there will be a volumetric limit to how much fresh biocatalyst can be added to the system.
2. **Lowering the flow rate:** Depending on the productivity required from the plant, the flow rate can be reduced to some extent to cope with the loss of biocatalytic activity. This means that identical conversion can be maintained, thereby simplifying the downstream process. Ideally, an average productivity value for the plant over the whole of the usable lifetime of the biocatalyst will be used for plant sizing calculations, such that reduced throughput toward the end of a cycle is taken into account.

1.3.2.3.4 Alternative Well-Mixed Reactors

Continuous Fluidized-Bed Reactors (CFBRs). In a continuous fluidized-bed reactor (CFBR), mixing of solid material can be achieved by driving a gas (or, in principle, liquid) phase through solid or porous particles at such a speed that the particles become "fluidized." The particles are therefore held in suspension within the reactor by means of the fluid passing through the system (Figure 1.7) [37]. Operated in continuous mode, such reactors provide good opportunity for multiphasic systems.

Such reactors have the advantage of high biocatalyst loading, without the pressure-drop problems of packed-bed reactors. However, the scientific literature reports only a few cases

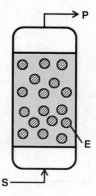

Figure 1.7 *Schematic diagram of a fluidized-bed reactor (FBR) (E, immobilized enzyme; S, substrate; P, product).*

of biocatalytic fluidized beds [38–40]. Since the immobilized enzymes are suspended in liquid, the system should be treated with care to prevent the enzyme being washed away. Conventional fluidized beds use solids suspended in gases, where the density difference between phases is much greater. Hence, while small particles of biocatalyst can be used (which would give a high-pressure drop in a packed-bed reactor), there is always the risk of wash-out where biocatalyst is lost from the reactor. Small particles have the benefit of having no diffusional limitations but are hard to retain in the reactor. Extremely small immobilized enzyme particles (around 10 µm) may be used with a fluidization aid. On the other hand, large biocatalyst particles, which would not be suited to use in a CSTR, due to attrition, may be suitable in a CFBR. Due to the "fluid-like" behavior, the handling of solids is easy, so these reactors are also well suited to reactions with solid substrates or products [34,41]. Likewise, viscous substrates and products can be used.

Considerations for Implementation. A CFBR is a well-mixed reactor, implying kinetics similar to those of a CSTR. In principle, if rated correctly, the biocatalyst will be retained in the reactor without the need for a sieve. Nevertheless, care should be taken, since small immobilized biocatalyst particles can easily be removed in the effluent stream, necessitating further ultrafiltration downstream. While some immobilized biocatalyst particle attrition can be expected, far higher loadings of immobilized biocatalyst can be used than in the equivalent stirred-tank reactor. Reactors are usually fitted with a perforated plate at the bottom to facilitate distribution of fluids. In general, operating such reactors requires extensive knowledge and experience, which also makes scale-up potentially problematic.

Continuous Packed-Bed Reactors (CPBRs). Continuous packed-bed reactors (CPBRs) are tubular reactors filled with (bio)catalyst particles, which are retained by means of a filter [42]. The feed is often pumped through the bottom of the reactor to enable any intrinsic gas bubbles to escape from the column (Figure 1.8). The material flows through the column at the same velocity, and parallel to the axis of the column, without back-mixing (hence, plug flow). Since the biocatalyst is fixed inside the column, the residence time is a function of the position of the material in the column: the longer the column, the higher the conversion. However, the pressure drop across the column also increases in proportion with column length, limiting the

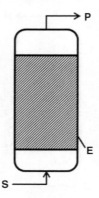

Figure 1.8 *Schematic diagram of a continuous packed-bed reactor (CPBR) (E, immobilized enzyme; S, substrate; P, product).*

maximum length of the column [20]. Hence, for scale-up, the width of the column can be increased in order to increase throughput without having to use a large number of columns. Care must be taken to increase the width in such a way that plug flow is maintained. The implication of having plug flow is that there is no concentration gradient across the width of the column. One example of the use of such a reactor is presented by Marrazzo and co-workers [43], and there are numerous others in the scientific literature.

The space–time yield is high, as there is a significant amount of biocatalyst in the reactor [33]. Space–time yields up to six times greater than those found in a stirred-tank reactor can be achieved. Likewise, the kinetic profile is favorable, and therefore shorter residence times are required than in the equivalent CSTR. The kinetic profile is identical to that in a BSTR, replacing the reaction time dimension with reactor length. The high concentration of biocatalyst means that reactors are smaller for a given conversion, and hence investment is lower (in terms of both capital and operating costs) than in the CSTR option. Additionally, no stirring is required, although pumps are necessary to pump reactants through the reactor. Immobilized biocatalyst particles will be exposed to less mechanical stress, since there is no stirring and the particles are stationary in the column. On the other hand, the particles must instead be able to withstand a (high) pressure drop, meaning they must be rigid and maintain their structure under pressure.

One successful commercial implementation of a biocatalytic PBR system is the transesterification process developed by Novozymes. The process uses a lipase immobilized on silica particles to exchange fatty acids between the triglycerides of different oils and fats, providing a superior product to that made in the alternative chemical processes.

Considerations for Implementation. Mass transfer limitations are prominent due to the absence of mixing, and it therefore appears attractive to use small particles. On the other hand, small particles result in a high pressure drop. A compromise must therefore be reached. Given the high space–time yield, this is usually achievable. Without a good distribution system at the inlet of the reactor, there is a risk of so-called "channeling" or "tunneling" of material through the column, which will lead to deviations from plug flow and a reduction in the kinetic performance of the reactor. As stated previously, dependent upon the size and porosity of the particles, a high pressure drop across the column is possible. The ideal particle size for the immobilized particles used in packed-bed reactors (to ensure low back pressure) ranges from 200 to 400 μm [44]. Because the CPBR is a closed system, it is difficult to add or remove material as the reaction medium travels through the column. This is typically handled by splitting the process into multiple CPBRs in series. Coupling with *in situ* product removal (ISPR) is possible via an external loop, since the biocatalyst is retained in the reactor. However, control of pH (and temperature) is more difficult, as there will be a gradient in the column for reactions undergoing a pH change. Nevertheless, such a system can be operated with an external loop fitted to a small stirred tank, where acid or base can be added to neutralize changes in the column. Clearly, the pH change over the column must be sufficiently small to be effectively managed by the biocatalyst. Byproduct accumulation in the column is a common problem, and can be particularly significant when the byproduct causes inhibition of the biocatalyst. CSTRs are not good for multiphasic systems, since distribution leads to varying phase ratios, channeling, and even blockage of the column. When filling the reactor with biocatalyst, allowances should be made for biocatalyst swelling. The calculated 60% volumetric occupation by immobilized biocatalyst assumes no swelling. Swelling can also cause an increase in pressure drop over the column.

Continuous Expanded-Bed Reactors (CEBRs).   Continuous expanded-bed reactors (CEBRs) are tubular reactors that are operated as fluidized beds, except at lower velocities. In order to maintain plug flow. the immobilized biocatalyst particles are usually of a variety of sizes and/or densities, so that they each find their correct suspension position in the reactor. Hence, the biocatalysts, based on their sizes and densities, align themselves in the reactor in such a way that there is a gradient of particle size and density along the length of the column: the larger particles are at the bottom of the column, the smaller particles are at the top (Figure 1.9). The flow of the fluid in the column then follows plug flow.

Interestingly, since there is voidage in the column, solid particles can be used in the feed; these would block a CPBR. Hence, a prefiltration step is not required when using a CEBR,

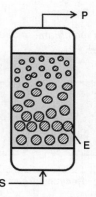

Figure 1.9   *Schematic diagram of a continuous expanded-bed reactor (CEBR) (E, immobilized enzyme; S, substrate; P, product).*

unlike a CPBR. In principle, particles of similar sizes and densities can be used, held in place via a magnetic field if they are constructed around iron particles, although there are only rather limited reports of such systems in the scientific literature. Integration with ISPR and other product recovery operations is also possible using adsorbent or absorbent resins in the bed, together with immobilized particles. Particle sizes are between 50 and 400 μm and densities between 1.1 and 1.3 g.mL^{-1} [45].

Considerations for Implementation.   CEBRs are sensitive to operating conditions. While a CEBR has significant benefits over a CFBR, such as reduced particle attrition, it is still essential to use care and carry out small-scale tests to ensure plug-flow distribution in the reactor.

1.3.2.3.5 Membrane Bioreactors (MBRs).   Membrane bioreactors (MBRs) have found applications in a myriad of fields, including the petrochemical, water-treatment, food, and pharmaceutical industries. Such reactors have also been applied in bioprocesses [46–48]. The use of these reactors exploits the fact that reaction and separation processes can be combined. Hence, the motivation for their development is the cost saving derived from the reduced number of processing stages. They can be used in two main applications: (i) where the membrane acts as a support upon which the enzyme is immobilized (Figure 1.10) and (ii) where the membrane is used for separation of the product integrated with the reaction (Figure 1.11). Several configurations of MBR are available for use, although their details are not discussed here.

   There are many advantages of such a reactor type, dependent upon configuration, but primarily it is a simple system to operate. Normal operation is in continuous mode. High

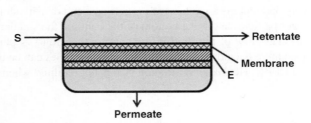

Figure 1.10   *Schematic diagram of a membrane bed reactor containing immobilized catalyst (E, immobilized enzyme; S, substrate).*

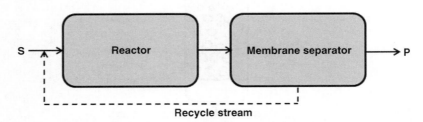

Figure 1.11   *Schematic diagram of a membrane bioreactor (MBR), used for separation integrated with reaction (E, immobilized enzyme; P, product).*

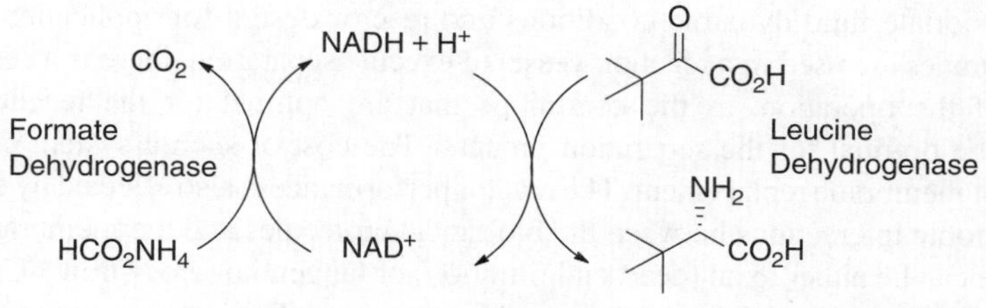

Scheme 1.2 *Synthesis of tert-leucine with membrane separation.*

yields (and potentially high selectivity) can be expected, since the basic concept allows reaction and separation steps to be integrated. Likewise, the system can be integrated with an ISPR [49]. Potentially, biocatalyst yield can be enhanced by maintaining high reactor productivity and increasing the usable biocatalyst lifetime by eliminating direct contact between the biocatalyst and inhibitory compounds. The biocatalyst can be separated to allow reuse and recycle, improving biocatalyst yield [50,51]. In a similar manner, expensive co-factors can also be retained for regeneration and recycle. Enzymes grafted on to the membrane surfaces can sometimes be more stable and resistant toward organic solvents than their soluble counterparts. Operation is usually via ultrafiltration membranes, with an average pore size around 0.1 to several micrometers. In such a system, the enzymes can be retained. In order to scale up, it is desirable to keep the membrane surface area-to-reaction volume ratio constant.

Degussa's dehydrogenase technology for the continuous production of (*S*)-*tert*-leucine is a classic example of a membrane reactor in which the membrane is used to retain the biocatalyst in the reactor (Scheme 1.2) [52].

Another example of a continuous process for the synthesis of (*R*)-3-(4-fluorophenyl)-2-hydroxy propionic acid was demonstrated by Pfizer: a multi-kilogram-scale synthesis with a space–time yield of 560 g.L.day^{-1} (Scheme 1.3) [53].

Considerations for Implementation.   Membrane fouling and concentration polarization mean that flux (area-based flow rate) is subject to reduction as a function of time (lowering of flux is seen in [48]). Care should be taken to design (and rate) the system with this in mind. The molecular size of all components in the system (substrate(s), product(s), enzyme(s), and co-factor(s)) compared to the pore size of the membrane should be taken into account, since the performance of the reactor is in large part determined by the mass transfer across the membrane (or retention by the membrane) of these components. Such a reactor will therefore

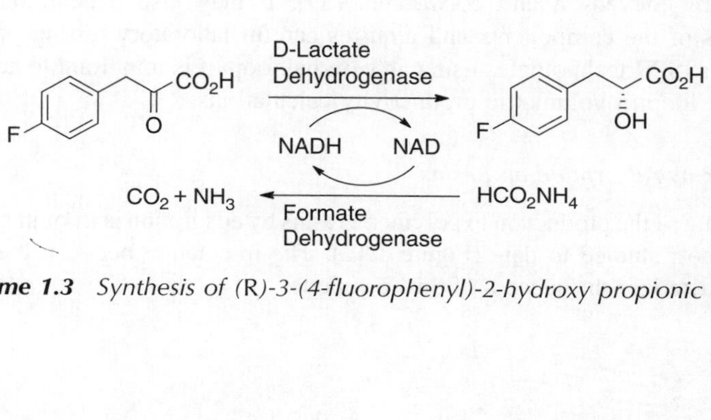

Scheme 1.3 *Synthesis of (R)-3-(4-fluorophenyl)-2-hydroxy propionic acid.*

require appropriate fluid-dynamic conditions and reactor design for application. However, when membranes are used in a reaction vessel to execute separation, there is a certain loss in flexibility of the operation, as the conditions that are optimal for the reaction will not necessarily be optimal for the separation process. The cost of such a system can be high, due to cost of membrane replacement. The reactor performance is also affected by electrostatic and hydrophobic interactions between the biological molecules and the membrane. Flow of the substrate can be either axial (dead-end filtration) or tangential (cross-flow). Cross-flow is more suitable for large-scale applications. Membrane fouling is a common problem in operating such reactors, and consequently membrane cleaning and sterilizing techniques between consecutive operations are critical to the success of these reactors.

1.4 Enhancing Technologies

1.4.1 *In Situ* Product Removal (ISPR)

In order to meet the required productivities for successful application of an industrial biocatalytic process, reactions need to be operated at high substrate and therefore high product concentration. However, this situation is often unsuitable for the biocatalyst, and a lower product concentration needs to be maintained in its vicinity in order to overcome toxic and inhibitory effects. In principle, an increase in productivity and yield can also be achieved by shifting the equilibrium of thermodynamically unfavored reactions, although this is far from straightforward [54,55]. Another, relatively common situation is that the reaction product(s) are unstable under the operating conditions, necessitating immediate removal to avoid yield loss. Many ISPR techniques are available for the removal of products from the site of the reaction. This section will briefly discuss the different possibilities (for more extensive reviews, see [56–58]).

1.4.1.1 Considerations for Implementation

The application of ISPR is not widespread, and the choice of method will depend on the combination of product and substrate/impurity properties that gives the largest driving force for separation. Hence, application is complicated when substrates and products have very similar properties, which is often the case in biocatalysis. For thermodynamically unfavorable reactions, it will be essential to remove products more effectively than substrates in order to shift the equilibrium, and hence selectivity is of utmost importance. One successful example of the shift of thermodynamic equilibrium using selective removal is described by Stevenson and co-workers [59]. It may also depend on the relative concentrations of the components and requires careful laboratory testing prior to implementation. For ISPR technologies using an external loop, it is important to account for the volume in the loop in volumetric productivity calculations.

1.4.1.2 ISPR by Adsorption on Resins

ISPR by binding of the product on to polymeric resins by adsorption is a common technique – perhaps the most studied to date (Figure 1.12). The interaction between the product and the resin characterizes the success of this technique.

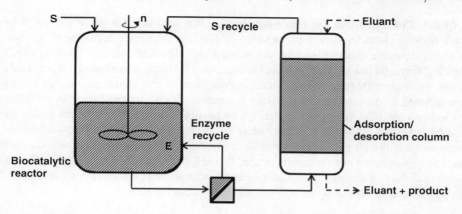

Figure 1.12 *Schematic diagram of a scheme using ISPR with adsorbent resins in an external loop (E, immobilized enzyme; S, substrate; n, number of rotations per unit time).*

Such a system is in principle highly selective, with high mechanical (when operated as an external loop) and chemical stability, rapid adsorption kinetics, sterilizability (provided that the resin can withstand sterilization conditions), and easy regeneration and recycle. Adsorption usually follows the Langmuir adsorption isotherm. Both fluidized and expanded-bed reactors can be used. However, capacity is limited (typically ~40 mg.g^{-1} of resin), which is a significant limitation in the production of small molecules.

Eli Lilly has adopted the use of resins (XAD-7) for substrate supply and product removal strategies for the production of (3,4-methylenedioxyphenyl)-2-propanol using whole cells expressing alcohol dehydrogenase. The process was run at 300 L scale, where it achieved 96% yield, >99.9% ee, and a space–time yield of 75 g.L.day^{-1} (Scheme 1.4) [60].

A further example of the use of resins for substrate supply was adopted by Sigma-Aldrich for the production of lactone using whole-cell Baeyer-Villiger monooxygenase, Over 200 g of the combined lactone was produced (Scheme 1.5) [61].

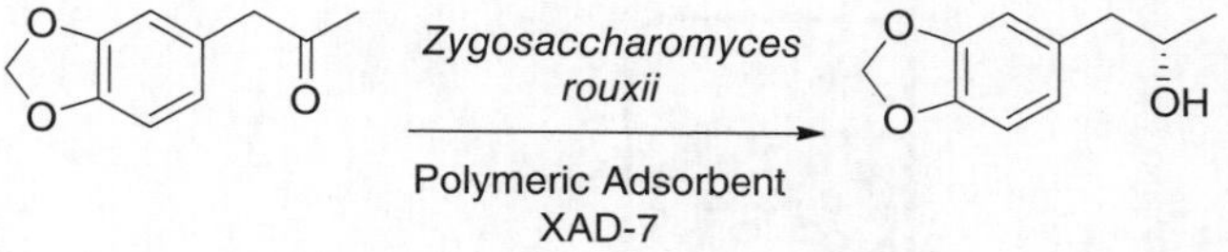

Scheme 1.4 *Synthesis of (S)-3,4-methylene-dioxypheyl isopropanol.*

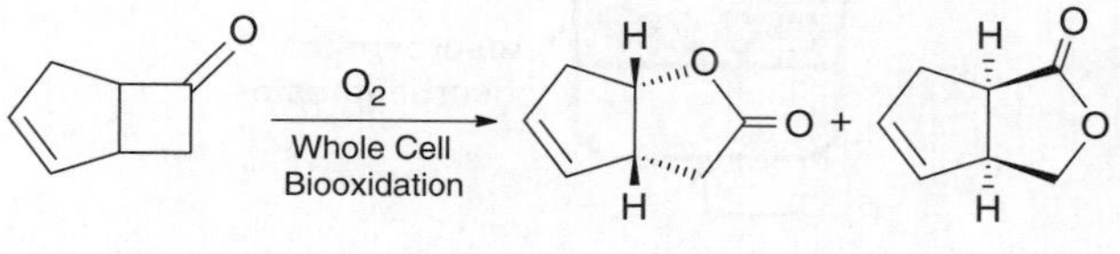

Scheme 1.5 *Biooxidation of bicyclo[3.2.0]hept-2-en-6-one.*

1.4.1.2.1 Considerations for Implementation.   It is desirable for the adsorbent to have the following characteristics: high capacity for the target molecule, a favorable adsorption isotherm, low non-specific binding, mechanical and chemical stability, biocompatibility, sterilizability, and low cost (or, at least, the capability for being regenerated and recycled). It is not uncommon that the substrate and product have similar characteristics, so thorough research into the adsorption characteristics is necessary for the successful implementation of the technology [62]. Adsorption is affected by the pH of the media, so the pH effects must also be evaluated, although it is not necessary that the pH optimum for adsorption is the same as the optimum pH for the biological conversion. The amount of adsorbent material that can be employed in a reaction is limited by the type of reactor. Molecular imprinting of polymeric resin or directed evolution of enzyme to operate at the conditions favorable for adsorption can also be adopted to increase the process efficiency.

1.4.1.3   ISPR Using Expanded-Bed Adsorption (EBA)

Expanded-bed adsorption (EBA) uses an expanded-bed reactor and combines solid–liquid separation with adsorptive purification. The reactor design is similar to that of a CEBR, except that the immobilized enzymes are replaced with adsorbent resins, on to which the product will bind (Figure 1.13) [63,64].

   This technique combines separation and purification, decreasing the number of downstream processing steps and thereby increasing the potential yield of the process.

1.4.1.3.1 Considerations for Implementation.   Reduction in adsorbent binding capacity due to binding of other impurities (cell debris, etc.) can be a problem, but, in principle, "dirty" feed streams can be used. The inlet may include perforated plates of metal mesh to produce back pressure in the system, which might clog with solids. For high flow rates or a higher viscosity of the feed solution, an increase in adsorbent size or density is desirable. It will probably be necessary to evaluate the effects of impurities on the adsorption of the resins prior to their use. The choice of appropriate adsorbent may be difficult. Laboratory experimental measurements are essential prior to scale-up for the pilot plant. Elution of the EBA is usually

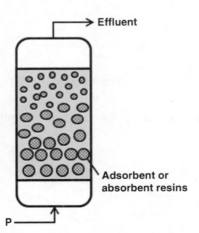

Figure 1.13 *Schematic diagram of expanded-bed adsorption (EBA) for ISPR (P, product).*

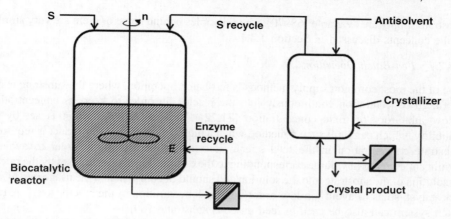

Figure 1.14 *Schematic diagram of an ISPR scheme using crystallization (E, immobilized enzyme; S, substrate; n, number of rotations per unit time).*

done in packed-bed mode to reduce the amount of eluent required. Equipment and adsorbents that can handle a large amount of feedstock have to be developed.

1.4.1.4 ISPR by Crystallization

Crystallization of the product can be coupled with the reaction in order to precipitate the product when the concentration reaches above the saturation limit (Figure 1.14).

In cases where super-saturation is reached, product crystallization occurs, decreasing the product concentration and thereby increasing the productivity and yield (this is especially useful for unstable products or in cases of product inhibition) [65]. Crystallization can provide a valuable tool in industrial biocatalysis by simplifying product recovery. A successful example of crystallization as an ISPR technology has been established by DSM in a multi-thousand-ton production of aspartame by thermoase/thermolysin [66,67].

1.4.1.4.1 Considerations for Implementation.   When organic solvents are used for precipitation prior to crystallization, the number of processing steps increases, which decreases the yield of the product. This also leads to high waste generation and hence higher production costs. When product inhibition occurs below the solubility limit, *in situ* crystallization may not be viable and the use of an external loop should be considered. The rate of crystallization must be able to keep up with the production rate in order for the process to be efficient. Accumulation of byproducts can interfere with crystallization. In cases where a co-solute is used, the effect of the solute on the product purity must be evaluated.

1.4.2 Substrate Feeding Strategies

Biocatalysts are often required to work at high substrate concentrations, which seldom correspond to their natural working environment. This can lead to reduced activity or performance. Additionally, they may be inhibited at high substrate concentrations. One solution to this problem is to supply the substrates such that their concentration is kept below the inhibitory or toxic concentration. It is therefore of vital importance to select and adopt substrate feeding strategies for different bioprocesses, as appropriate. Many strategies for substrate supply are available [68–70]. The principles behind some of them are very similar to the concepts discussed in Section 1.4.1.

1.4.2.1 Fed-Batch Operation

One of the most common supply methods is fed-batch operation, where the substrate is fed at a high concentration continuously into the reactor, in order to keep its concentration below inhibitory or toxic concentration [71,72]. The feed concentration is set by its solubility, which can limit such a method when a high concentration of product is required. Substrate concentration in the feed stream is typically kept high, in order to avoid a significant increase in volume corresponding to the dosing. The feed rate is set by the rate of dissolution of the substrate into the solution and should of course match the usage rate by the biocatalyst, so as to avoid too low or too high a concentration in the reactor itself. A fed-batch system can also be used to feed gaseous substrates [60].

Fed-batch processes are widely implemented in the industry. When the solubility of the substrate is low, solid feeding can be adopted. Lonza has adopted feeding of solids for the conversion of nicotinic acid to 6-hydroxynicotinate, catalyzed by whole cells expressing nicotinic acid hydroxylase. By adopting this feeding strategy, Lonza achieved a product concentration of 75 g.L^{-1} and a reaction yield of more than 90% [73].

1.4.2.1.1 Considerations for Implementation. The dosing profile needs to be adjusted and altered according to the intended use of the fed-batch operation. For instance, if the solubility profile changes with the composition of the reaction mixture, this should be designed into the dosing profile.

1.4.3 Non-Conventional Media

The natural environment for the great majority of biocatalysts is an aqueous solution. Unfortunately, many of the substrates and products that are of interest for industrial biocatalysis have low solubility in water, or can be inhibitory or indeed toxic to the biocatalyst at the concentrations required for a commercially viable bioprocess. One method of addressing these problems is to use a non-aqueous solvent, either in a single liquid reaction phase with neat solvent or water and a co-solvent, or in a two-liquid phase system [74,75].

1.4.3.1 Single Non-Conventional Liquid Phase Systems

Replacing water with another reaction solvent offers a number of potential advantages in a biocatalytic reaction. Using a water-soluble co-solvent can greatly improve substrate solubility, but it also stresses the stability of the catalyst, as exemplified by the Sitagliptin process developed by Codexis and Merck (Scheme 1.6) [76].

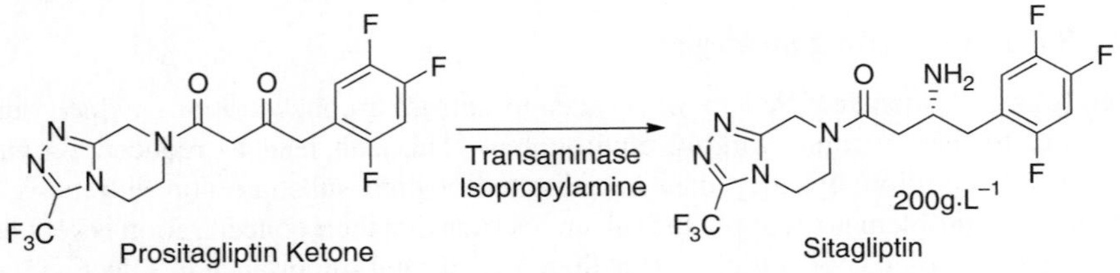

Prositagliptin Ketone → Transaminase Isopropylamine → Sitagliptin 200g·L^{-1}

Scheme 1.6 *Synthesis of sitagliptin-6 g.L^{-1} enzyme in 50% DMSO with 92% yield.*

Another option is to replace water completely as a solvent. In many cases, this can drastically change the thermodynamic reaction equilibrium. This has allowed the use of hydrolytic enzymes in synthetic chemistry through catalysis of the reverse reaction [77].

Biocatalytic processes in non-conventional media typically involve the use of highly polar (water-miscible) or non-polar (only sparingly soluble in water) organic solvents. Another option is supercritical CO_2, which circumvents some of the hazards of solvent use.

1.4.3.1.1 Considerations for Implementation.
Selection of organic solvents can prove a major challenge. Proteins are typically very difficult to dissolve in non-aqueous solvent without loss of their structural functionality. Because of this, successful applications tend to use water and highly polar co-solvents (such as DMSO or DMF) or, alternatively, neat hydrophobic solvents that interact little with the protein (log P typically higher than 2). It is important to note that, since free enzyme formulations (or cells, for that matter) will not dissolve in neat solvents, they need to be formulated as immobilized catalysts to be efficient in such systems [78,79].

Many of the organic solvents that are useful for biocatalytic conversions are flammable, and thus introduce a complication in the process. Additionally, they can be hazardous to both health and the environment, and care must therefore be taken to limit both waste and emissions. These challenges are avoided in a process based on supercritical CO_2, but the construction of such a process is more complex (and costly) due to the high pressures involved.

1.4.3.2 Aqueous–Organic Two-Liquid Phase Systems

Aqueous–organic two-liquid phase systems may be used for *in situ* substrate supply where the water-solubility of the substrate is so low as to preclude fed-batch operation. In principle, this approach can simultaneously be used for product recovery via extractive ISPR [61].

Substrate supply is driven by mass transfer from the organic to the aqueous phase. In principle, neat (poorly water-soluble) liquid substrates can be used as the second liquid phase. However, in many cases (e.g., when the substrate toxicity limit is lower than the solubility limit or where the substrate is solid), it is useful to use an organic solvent in which to dissolve the substrate (and from which to extract the product).

1.4.3.2.1 Considerations for Implementation.
The considerations for single non-conventional liquid phase systems also apply here. Additionally, when using two-liquid phase systems, emulsions may form, which can prove hard to break downstream, making product recovery and/or biocatalyst recovery and recycle difficult. Dissolved levels of solvent can affect the biocatalyst (i.e., when the solvent has low log P values), but interfacial effects have also been found to be important.

1.4.3.3 Aqueous–Ionic Liquid Two-Liquid Phase Systems

Biocatalysis often takes place in aqueous media, but some of the organic components (usually the substrate and products) exhibit low solubility levels in water. The solubility of such reactants needs to be improved, often by using water-immiscible organic solvents (see Section 1.4.3.1). Ionic liquids (ILs) can also be used in place of the organic solvent, again forming a two-liquid phase system. ILs are essentially organic salts that are liquid at (or near to) room temperature. The use of ILs has attracted significant interest in recent years [80,81].

This system is considered a potential "green" alternative to organic solvents, dependent on the type of IL. Likewise, many ILs have been found to be biocompatible. The synthesis of ILs is such that tailoring an IL toward the need of the process by choosing an appropriate cation and anion becomes a real possibility.

1.4.3.3.1 Considerations for Implementation.   It is important to consider that the availability and cost of ILs today mean that recycle is essential. Furthermore, the anion of an IL may cause conformational change of the enzyme and lead to a loss in activity. Likewise, the interaction between water and IL can cause complications in the intended reaction system and make separation of IL from the product difficult.

1.4.4   Oxygen Supply Strategies

Biocatalytic oxidation has been gaining importance in synthetic chemistry owing to its high selectivity compared to its chemical counterparts. Supply of molecular oxygen is a key part of biocatalytic oxidation reactions, and although oxygen is a substrate for such biocatalysts (e.g., oxidases, monooxygenases, dioxygenases), it is a special substrate and is hence dealt with separately here. The major advantage of supplying gas to the liquid is the high mass transfer to the liquid via the interface between the gas and liquid phase. On the other hand, oxygen compromises the stability of many biocatalysts, either through interfacial effects or through chemical modification (e.g., oxidation of amino acid residues). It should be noted that oxygen can be introduced into a reactor either as pure gas or in the form of air to drive the reaction. This choice is more of a strategic decision for each reaction, and therefore will not be separately discussed. Several reactor configurations have been used for oxygen supply, and some of them will be discussed in this section. It should be noted that the oxygen supply strategies and the available literature catering to biocatalysis and bioprocesses are scarce. However, it is rather easier to get information on fermentation, and therefore, in some of these cases, the reactors were used for fermentation or growth of cells and are discussed here to give an idea of the potential alternatives available for bioprocess engineers.

1.4.4.1   Surface Aeration

Surface aeration is a method by which oxygen/air is transferred into the bulk liquid phase through the gas–liquid interface at the top of the reactor. Therefore, there is no bulk gas transfer through the bulk liquid. This kind of transfer is always present in reactors unless special modifications have been made to the reactor design to avoid it. It is commonly used in laboratories when biocatalysis requiring oxygen is carried out using shake-flasks or vials. Figure 1.15 represents surface aeration in a batch reactor with agitation.

The efficiency of this method of aeration depends on the mass transfer limitations of the gas and the surface area: mass transfer occurs through the surface of the reactor. It is more useful for small-scale reactors – the reactor size limitation will depend on the oxygen requirement for the system, but it is clearly very limited.

1.4.4.1.1 Considerations for Implementation.   Such a method is obviously unsuitable for anything other than the laboratory. It is nevertheless widely used, and experimentalists should be careful about oxygen limitation leading to low observed reaction rates.

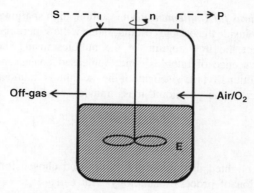

Figure 1.15 *Surface aeration with agitation in a batch reactor (S, substrate; P, product; n, number of rotations per unit time).*

1.4.4.2 Sparged Aeration

Sparged aeration is a method by which oxygen/air is introduced at the bottom of the reactor either through distribution plates or using a sparger. A sparger is more commonly used in stirred-tank reactors, while distribution plates are commonly found in bubble-column reactors.

Sparged aeration is a highly convenient method of oxygen supply for reactors above 1 L working volume. In a stirred system, the stirrer will break the air bubbles so that they are small, and thereby create a large interfacial area for mass transfer. The driving force will be determined in large part by the concentration of oxygen in the liquid in equilibrium with the gas. Although in such systems the liquid is well mixed, it can often be the case that the gas phase flows through the reactor in plug-flow mode. This has a significant effect on the oxygen transfer that is possible.

1.4.4.2.1 Considerations for Implementation.
Correlations for oxygen mass transfer potential versus gas flow rate and mixing are rather poor in general, but as a guide, reactors should operate with a gas input of around 0.5–1.0 vvm (volume/volume/minute) and a stirrer power input of about 1.0–1.5 W.L^{-1}. Maximum oxygen transfer rates of around 100 mmol.L^{-1}.h^{-1} can be expected when air is supplied [82].

1.4.4.3 Bubble-Column Reactors

Bubble-column reactors are simple reactors in which the gas is sparged through the bottom. The motion of the bubble from the source (usually distribution plates) through the reactor causes the mixing, avoiding any need for stirring (see also [83,84]).

Mass transfer in such a system depends on the superficial gas velocity and the initial bubble size, both of which are affected by the diameter of the orifice (of the sparger). Industrial bubble-column bioreactors have capacities up to 400 m^3.

The flow direction of the liquid has little effect on the gas hold-up, although a downflow bubble column is preferred when longer gas phase residence times are desired. Bubble-column reactors offer reasonable biocatalyst stability, since there is no stirrer. Due to the lack of moving parts, maintenance costs are low.

1.4.4.3.1 Considerations for Implementation. Often, high sparging rates are required to achieve turbulent flow inside the bubble column. Indeed, the superficial gas velocity and the reactor diameter affect the flow regime of the bubbles inside the reactor. A further consideration is coalescence of bubbles, since increased coalescence will decrease the aeration efficient. Additionally, the viscosity of the medium is frequently important, since it has a significant effect on the gas–liquid mass transfer.

1.5 Conclusion

The technologies highlighted in this chapter have been chosen for practical reasons to illustrate the integration of process technology with early-stage considerations for bio-catalytic process development. For more detailed descriptions of particular technologies, the reader is advised to consult the references herein.

References

1. Pollard, D.J. and Woodley, J.M. (2007) *Trends in Biotechnology*, **25**, 66–73.
2. Lima-Ramos, J., Neto, W., and Woodley, J.M. (2013) *Topics in Catalysis*, **57**, 301–320.
3. Prenosil, J.E., Kut, Ö.M., Dunn, I.J., and Heinzle, E. (2009) Biocatalysis, 2. Immobilized biocatalysts, in *Ullmann's Encyclopedia of Industrial Chemistry*, Wiley-VCH Verlag GmbH and Co. KGaA., p. 477.
4. Garcia-Galan, C., Berenguer-Murcia, Á., Fernandez-Lafuente, R., and Rodrigues, R.C. (2011) *Advanced Synthesis and Catalysis*, **353**, 2885–2904.
5. Nara, T.Y., Togashi, H., Sekikawa, C. *et al.* (2010) *Journal of Molecular Catalysis B: Enzymatic*, **64**, 107–112.
6. Tufvesson, P., Lima-Ramos, J., Nordblad, M., and Woodley, J.M. (2011) *Organic Process Research and Development*, **15**, 266–274.
7. Cantone, S., Ferrario, V., Corici, L. *et al.* (2013) *Chemical Society Reviews*, **42**, 6262–6276.
8. Bayne, L., Ulijn, R.V., and Halling, P.J. (2013) *Chemical Society Reviews*, **42**, 9000–9010.
9. Rotticci, D., Norin, T., and Hult, K. (2000) *Organic Letters*, **2**, 1373–1376.
10. Cao, L. (ed.) (2005) *Carrier-Bound Immobilized Enzymes – Principles, Applications and Design*, John Wiley & Sons, Weinheim.
11. Jesionowski, T., Zdarta, J., and Krajewska, B. (2014) *Adsorption*, **20**, 801–821.
12. Boller, T., Meier, C., and Menzler, S. (2002) *Organic Process Research & Development*, **6**, 509–519.
13. Mahmoudian, M., Noble, D., Drake, C.S. *et al.* (1997) *Enzyme and Microbial Technology*, **20**, 393–400.
14. Cao, L., Langen, L., and Sheldon, R.A. (2003) *Current Opinion in Biotechnology*, **14** (4), 387–394.
15. Sheldon, R. (2007) *Biochemical Society Transactions*, **35** (6), 1583.
16. Sheldon, R.A. (2011) *Applied Microbiology and Biotechnology*, **92**, 467–477.
17. Sheldon, R.A. and van Pelt, S. (2013) *Chemical Society Reviews*, **42**, 6223–6235.
18. Fernandez-Lafuente, R. (2009) *Enzyme and Microbial Technology*, **45**, 405–418.
19. Chmura, A., Rustler, S., Paravidino, M. *et al.* (2013) *Tetrahedron: Asymmetry*, **24**, 1225–1232.
20. Sheldon, R.A. (2011) *Organic Process Research and Development*, **15** (1), 213–223.
21. Sheldon, R.A., Sorgedrager, M.J., and Kondor, B. (2012) International Patent Application WO2012/023847A2.

22. Fogler, H.S. (1999) *Elements of Chemical Reaction Engineering*, 3rd edn, Prentice-Hall, Inc., New Jersey.
23. Levenspiel, O. (1999) *Industrial & Engineering Chemistry Research*, **38**, 4140–4143.
24. Schuler, M.L. and Kargi, F. (2001) *Bioprocess Engineering: Basic Concepts*, 2nd edn, Prentice-Hall, Inc., New Jersey.
25. Nienow, W.A. (2012) Stirred tank reactors, in *Ullmann's Encyclopedia of Industrial Chemistry*, Wiley-VCH Verlag GmbH & Co., Weinheim, pp. 416–432.
26. Woodley, J.M. and Lilly, M.D. (1994) Biotransformation reactor selection and operation, in *Applied Biocatalysis* (eds J.M.S. Cabral, D. Best, L. Boross, and J. Tramper), Harwood Academic, Switzerland, pp. 371–393.
27. Lee, T.S., Turner, M.K., and Lye, G.J. (2002) *Biotechnology Progress*, **18**, 43–50.
28. Zhu, Y. and Wu, J. (2002) *Canadian Journal of Chemical Engineering*, **80**, 1–6.
29. Paul, E.L., Atiemo-obeng, V.A., and Kresta, S.M. (eds) (2004) *Handbook of Industrial Mixing: Science and Practice*, John Wiley & Sons, Inc., New Jersey.
30. Hilterhaus, L., Thum, O., and Liese, A. (2008) *Organic Process Research and Development*, **12**, 618–625.
31. Wiemann, L.O., Nieguth, R., Eckstein, M. *et al.* (2009) *ChemCatChem*, **1**, 455–462.
32. Regan, D.L., Dunnill, P., and Lilly, M.D. (1974) *Biotechnology and Bioengineering*, **XV**, 333–343.
33. Woodley, J.M. (2012) Reaction and process engineering, in *Enzyme Catalysis in Organic Synthesis: A Comprehensive Handbook* (eds K. Drauz, H. Gröger, and O. May), 3rd edn, Wiley-VCH Verlag & Co, KGaA, Weinheim, Germany, pp. 217–248.
34. Nielsen, P.M., Brask, J., and Fjerbaek, L. (2008) *European Journal of Lipid Science and Technology*, **110**, 692–700.
35. Carleysmith, S.W. and Lilly, M.D. (1979) *Biotechnology and Bioengineering*, **21**, 1057–1073.
36. Woodley, J.M. and Lilly, M.D. (1994) Biotransformation reactor selection and operation, in *Applied Biocatalysis* (eds J.M.S. Cabral, D. Best, L. Boross, and J. Tramper), Harwood Academic, Switzerland, pp. 371–393.
37. Werther, J. (2007) Fluidized-bed reactors, in *Ullmann's Encyclopedia of Industrial Chemistry*, Wiley-VCH Verlag GmbH & Co., Weinheim, pp. 320–366.
38. Bodalo, A., Gomez, J., Gomez, E. *et al.* (1995) *Enzyme and Microbial Technology*, **17**, 915–922.
39. Lieberman, R. and Ollis, D. (1975) *Biotechnology and Bioengineering*, **17**, 1401–1419.
40. Gómez, J.L., Bódalo, A., Gómez, E. *et al.* (2007) *Chemical Engineering Journal*, **127**, 47–57.
41. Cheryan, M., van Wyk, P.J., Olson, N.F., and Richardson, T. (1975) *Biotechnology and Bioengineering*, **XVII**, 585–598.
42. Eigenberger, G. and Ruppel, W. (2000) Fixed-bed reactors, in *Ullmann's Encyclopedia of Industrial Chemistry*, Wiley-VCH Verlag GmbH & Co., Weinheim, pp. 1–66.
43. Marrazzo, W.N., Merson, R., and McCoy, B. (1975) *Biotechnology and Bioengineering*, **17**, 1515–1528.
44. Liese, A. and Hilterhaus, L. (2013) *Chemical Society Reviews*, **42**, 6236–6249.
45. Freire, D.M.G. and Sant'Anna, G. (1990) *Applied Biochemistry and Biotechnology*, **26**, 23–34.
46. Giorno, L. and Drioli, E. (2000) *Trends in Biotechnology*, **18**, 339–349.
47. Wöltinger, J., Karau, A., Leuchtenberger, W., and Drauz, K. (2005) *Technology Transfer in Biotechnology*, **92**, 289–316.
48. Judd, S. (2008) *Trends in Biotechnology*, **26**, 109–116.
49. Rehn, G., Adlercreutz, P., and Grey, C. (2014) *Journal of Biotechnology*, **179**, 50–55.
50. Mendoza, L., Jonstrup, M., Hatti-Kaul, R., and Mattiasson, B. (2011) *Enzyme and Microbial Technology*, **49**, 478–484.
51. Luo, J., Nordvang, R.T., Morthensen, S.T. *et al.* (2014) *Bioresource Technology*, **166**, 9–16.

52. Wöltinger, J., Karau, A., Leuchtenberger, W., and Drauz, K. (2005) *Advances in Biochemical Engineering/Biotechnology*, **92**, 289–316.
53. Tao, J. and McGee, K. (2002) *Organic Process Research & Development*, **6**, 520–524.
54. Halim, M., Rios-Solis, L., Micheletti, M. *et al.* (2014) *Bioprocess and Biosystems Engineering*, **37**, 931–941.
55. Tufvesson, P., Bach, C., and Woodley, J.M. (2014) *Biotechnology and Bioengineering*, **111**, 309–319.
56. Freeman, A., Woodley, J.M., and Lilly, M.D. (1993) In situ product removal as a tool for bioprocessing. *Nature Biotechnology*, **11**, 1007–1012.
57. Lye, G.J. and Woodley, J.M. (1999) *Trends in Biotechnology*, **17**, 395–402.
58. Schügerl, K. and Hubbuch, J. (2005) *Current Opinion in Microbiology*, **8** (3), 294–300.
59. Calvin, S.J., Mangan, D., Miskelly, I. *et al.* (2012) *Organic Process Research and Development*, **16**, 82–86.
60. Zmijewski, M.J., Reinhard, M.R., Landen, B.E. *et al.* (1997) *Enzyme and Microbial Technology*, **20**, 494–499.
61. Doig, S.D., Avenell, P.J., Bird, P.A. *et al.* (2002) *Biotechnology Progress*, **18**, 1039–1046.
62. Mirata, M.A., Heerd, D., and Schrader, J. (2009) *Process Biochemistry*, **44**, 764–771.
63. Hjorth, R. (1997) *Trends in Biotechnology*, **15** (6), 230–235.
64. Hubbuch, J., Thömmes, J., and Kula, M.R. (2005) *Technology Transfer in Biotechnology*, **92**, 101–123.
65. Buque-Taboada, E.M., Straathof, A.J.J., Heijnen, J.J., and van dr Wielen, L.A.M. (2006) *Applied Microbiology and Biotechnology*, **71**, 1–12.
66. Hanzawa, S. (1999) Aspartame, in *Encyclopedia of Bioprocess Technology: Fermentation, Biocatalysis, and Bioseparation* (eds M.C. Flickinger and S.W. Drew), John Wiley & Sons Ltd., New York, pp. 201–210.
67. Oyama, K. (1992) The industrial production of aspartame, in *Chirality in Industry* (eds A.N. Collins, G.N. Sheldrake, and J. Crosby), John Wiley & Sons Ltd., New York, pp. 237–247.
68. Schmölzer, K., Mädje, K., Nidetzky, B., and Kratzer, R. (2012) *Bioresource Technology*, **108**, 216–223.
69. Kim, P.Y., Pollard, D.J., and Woodley, J.M. (2007) *Biotechnology Progress*, **23** (1), 74–82.
70. Straathof, A.J.J. (2003) *Biotechnology Progress*, **19** (3), 755–762.
71. Nordblad, M., Silva, V.T.L., Nielsen, P.M., and Woodley, J.M. (2014) *Biotechnology and Bioengineering*, **111**, 2446–2453.
72. Hagström, A.E.V., Törnvall, U., Nordblad, M. *et al.* (2011) *Biotechnology Progress*, **27**, 67–76.
73. Glöckler, R. and Roduit, J.P. (1996) *Chimia*, **50**, 413–415.
74. Valivety, R.H., Johnston, G.A., Suckling, C.J., and Halling, P.J. (1991) *Biotechnology and Bioengineering.*, **38** (10), 1137–1143.
75. Straathof, A.J. and Adlercreutz, P. (eds) (2003) *Applied Biocatalysis*, CRC Press.
76. Savile, C.K., Janey, J.M., Mundorff, E.C. *et al.* (2010) *Science*, **329**, 305–309.
77. Klibanov, A.M. (2001) *Nature*, **409**, 241–246.
78. Persson, M., Wehtje, E., and Adlercreutz, P. (2002) *Chembiochem*, **3**, 566–571.
79. Rees, D.G. and Halling, P.J. (2000) *Enzyme and Microbial Technology*, **27** (8), 549–559.
80. Sheldon, R.A., Lau, R.M., Sorgedrager, M.J. *et al.* (2002) *Green Chemistry*, **4** (2), 147–151.
81. van Rantwijk, F., Lau, M.R., and Sheldon, R.A. (2003) *Trends in Biotechnology*, **21** (3), 131–138.
82. Tindal, S., Carr, R., Archer, I.V., and Woodley, J.M. (2011) *Chemistry Today*, **29** (2), 60–61.
83. Shah, Y.T., Kelkar, B.G., Godbole, S.P., and Deckwer, W. (1982) *AIChE Journal*, **28**, 353–379.
84. Kulkarni, A.V. and Joshi, J.B. (2011) *Chemical Engineering Research and Design*, **89**, 1986–1995.

2

Cytochrome P450 (CYP) Progress in Biocatalysis for Synthetic Organic Chemistry

Thomas S. Moody[1] and Steve J.C. Taylor[2]
[1]Almac, Department of Biocatalysis and Isotope Chemistry, UK
[2]Celbius Ltd, CUBIC, Cranfield University, UK

2.1 Introduction

Synthetic chemists have long been seduced by the potential of the cytochrome P450 monooxygenase (CYP) enzyme family. These enzymes are ubiquitous in nature, and through the specialist chemistry that they perform they provide a range of essential functions for the organisms in which they are active. They are heme-containing enzymes that reduce molecular oxygen, using the co-factor NAD(P)H as a source of electrons. P450 enzymes are a widely studied class and their catalytic activity is well understood and well described elsewhere [1,2].

They perform a number of roles, ranging from degradation of carbon sources [3] to biosynthesis of complex natural products [4] to the detoxification of molecules that might be harmful to the cell [5]. They are found abundantly in both prokaryotes and eukaryotes.

In mammals, one of their primary functions is to provide a mechanism to rid the body of potentially harmful chemicals by transforming them into a state in which they can be more easily excreted. This is typically a hydroxylation reaction that makes the molecule more water-soluble and provides a handle for the addition of further solubilizing groups, such as a polar glucuronide or sulfate. In plants and fungi, CYPs are intimately involved in the synthesis of complex natural products, such as alkaloids and terpenes. The resulting molecules may have many functions, for example as growth hormones or antibiotics. In

Practical Methods for Biocatalysis and Biotransformations 3, First Edition.
Edited by John Whittall, Peter W. Sutton, and Wolfgang Kroutil.
© 2016 John Wiley & Sons, Ltd. Published 2016 by John Wiley & Sons, Ltd.

bacteria, CYPs find application in energy metabolism, where they are responsible for the initial hydroxylation of long-chain hydrocarbons, giving products that can enter further oxidation reactions, releasing energy for cell growth and metabolism [6,7].

The CYP superfamily can perform a broad range of oxidative chemistries, such as hydroxylation, epoxidation, dealkylation, and oxidation of nitrogen and sulfur atoms; this article will focus on hydroxylation [8–10]. The mechanism for these reactions is described elsewhere and is not part of this report [11,12]. The selective insertion of oxygen into a relatively inactive hydrogen–carbon bond is one of the main attractions of CYP chemistry, since it represents the potential to shorten otherwise lengthy synthetic sequences involving numerous protecting group manipulations. This is illustrated by the work of Bernhardt *et al.* in the hydroxylation of abietic acid [13]. Scheme 2.1 shows the impact of CYP hydroxyl-ation in the synthesis of 15-hydroperoxyabietic acid, which is of interest for study because it is a potent allergen. CYP biocatalysis transforms a ten-step synthesis into a two-stage process. This is extremely important in the drive for greener processes, and is ultimately a way to influence the E factor of any commercial process [14].

CYP biocatalysis is a relatively complex field and there is a need to simplify its application, while at the same time continuing the progress toward examples of robust industrial hydroxylation chemistry. The literature is dominated by examples of CYP biocatalysis having low productivity or forming complex mixtures of products, reflecting some of the problems that still exist in this area with regard to robust, selective bio-transformations [15]. However, it is worth noting that CYP hydroxylation has in fact been scaled up previously, with the hydroxylation of Compactin to Pravastatin providing an excellent example (discussed in more detail in Section 2.3).

The full industrial potential of this enzyme class is still some way from being realized, and represents a challenging system to get to work with under synthetic conditions [16]. For a start, many of these enzymes have large molecular weights and can be membrane-bound, which is problematic for expressing in microbial hosts with high protein solubility and good activity [17,18]. Many CYP enzymes are multi-component; for example, eukaryotic CYPs such as drug-metabolizing human liver CYPs and many plant and fungal CYPs implicated in secondary metabolism are two-component systems, consisting of a membrane-bound P450 domain that relies on another membrane protein – namely, a NAD(P)H-dependent diflavin [FAD/FMN] reductase (CPR) – for supply of the electrons necessary to drive catalysis. The majority of prokaryotic enzymes comprise three components, although some are self-contained and comprise a single polypeptide with all of the necessary oxidative and reductive functionality [13,19].

2.2 CYP Development

Much of the progress reported on CYPs has emerged from specialist academic groups that clone and engineer their own enzymes. Such work is broadly reported at two scales of application. First, many papers report CYP biocatalysis in which substrates are transformed at very low concentration and in low volume. Such work demonstrates the activities of new CYPs or of existing CYPs that have been engineered, or of new catalyzed reactions, where products are not actually isolated, but their presence is evidenced through analytical technology such as gas chromatography (GC), high-performance liquid chromatography

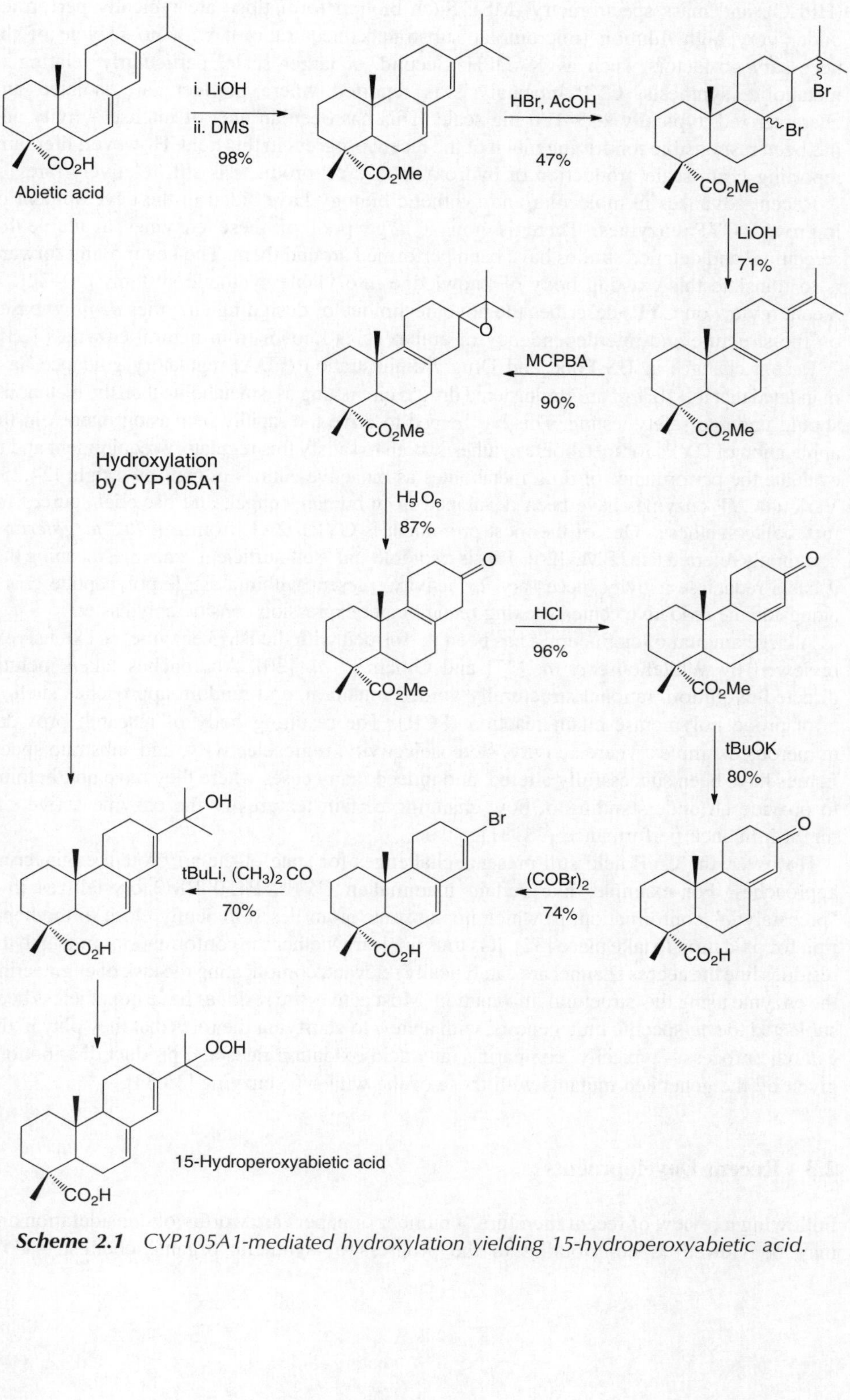

Scheme 2.1 *CYP105A1-mediated hydroxylation yielding 15-hydroperoxyabietic acid.*

(HPLC), and mass spectrometry (MS). Such biotransformations are typically performed under very high dilution (micromolar substrate concentration), with no recycle of the necessary co-factors, such as NADPH. Second, at larger scale, particularly relating to metabolite synthesis, CYP biocatalysis is reported where products are isolated and characterized, typically at 5–100 mg scale. This has been an area of intense activity and has been responsible for driving much of the recent progress in this field. However, literature reporting gram-scale production of hydroxylated CYP products is still relatively rare.

Recent advances in molecular and synthetic biology have aided in the development of improved CYP enzymes. There is now a large pool of these enzymes available for screening, and detailed studies have been performed around them. The key moving forward is to translate this existing body of knowledge into viable synthetic systems [20–22]. A recent review on CYPs describes the novel technique of designing enzymes *de novo* based on the structure and interdependency of amino acids known from natural enzymes [23].

Recent changes in US Food and Drug Administration (FDA) regulatory guidance have mandated that if >10% of an administered drug is circulating as a metabolite then the metabolite should undergo safety testing. This has helped to drive the rapidly increasing interest in the application of CYPs for metabolite synthesis, both to satisfy this regulatory requirement and to evaluate the performance of drug metabolites as bioactive entities in their own right [24,25]. Various CYP enzymes have been developed from human, fungal, and bacterial sources for metabolite synthesis. One of the most prominent is CYP102A1, from *Bacillus megaterium*, commonly referred to as BM3 [26]. This is a soluble and "self-sufficient" enzyme, meaning that it has a reductase activity (necessary for activity) present within a single polypeptide chain, alongside the oxidative center, making recombinant expression considerably easier.

A large amount of engineering has been performed with the BM3 enzyme, as extensively reviewed by Whitehouse *et al.* [27] and Gillam *et al.* [20]. Approaches taken include directed evolution, rational structurally guided mutation, and random approaches such as error-prone polymerase chain reaction (PCR). The resulting body of research provides numerous examples where activity, stereoselectivity, regioselectivity, and substrate specificities have been successfully altered, and indeed many cases where they have not, helping to provide an understanding of how changing certain features of the enzyme active site might influence performance [28–31].

However, the CYP field still presents challenges for state-of-the-art protein engineering approaches. For example, like certain mammalian CYPs, P450 BM3 crystallizes in a "precatalytic" conformation, in which no substrate atom lies sufficiently close to the heme iron for oxidation to take place [32]. It is thus unclear whether the conformations in which the residues line the access channel are catalytically relevant, complicating the task of engineering the enzyme using this structural information. Most active-site residues have nonetheless been subjected to site-specific mutagenesis, with a view to clarifying the roles that they play in the catalytic process – typically, comparing fatty acid oxidation rates and product distributions given by the generated mutants with those of the wild-type enzyme [33,34].

2.3   Recent Developments

Following a review of recent literature, a number of papers are worthy of consideration and may be viewed as contributing to the progress of synthetic organic chemistry CYP

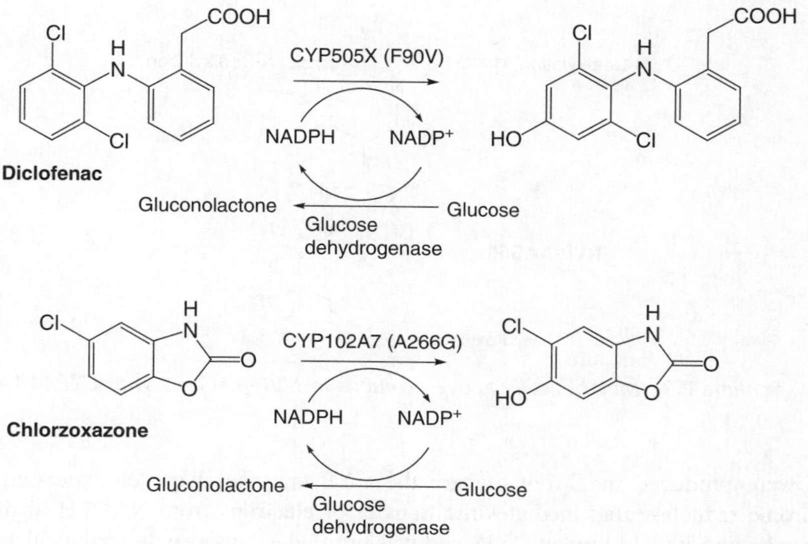

Scheme 2.2 *Examples of CYP oxidation for metabolite synthesis.*

biocatalysis. These either relate to pharmaceutical metabolite production or are more aligned with fine chemical synthesis.

The amino acid sequence of BM3 was utilized for the construction of a panel of CYP enzymes from both eukaryotic and prokaryotic sources [35]. These were expressed readily in *E. coli*, and, utilizing previously published data on certain mutations, a secondary panel of enzymes was created. Characterization of the panel led to its use in the synthesis of metabolites of Diclofenac and Chlorzoxazone (Scheme 2.2).

These reactions were carried out in a stirred, aerated biotransformation system using lyophilized cell-free extracts containing the CYP. Glucose dehydrogenase (GDH) was utilized to regenerate the co-factor NADPH, which is oxidized to NADP$^+$ during the hydroxylation reaction, by converting added glucose to gluconolactone. While the isolated product yields attained were modest (8%, 40 mg for Diclofenac and 12%, 8 mg for Chlozoxazone), they nevertheless provide an illustration of how preparatively useful amounts of metabolites can be readily synthesized from a commercially available panel of CYP enzymes.

A different approach to CYP expression was utilized by Bureik *et al.*, who expressed a number of human CYP enzymes in the fission yeast *Schizosaccharomyces pombe* [36]. This yeast, expressing the human CYP2C9, was used to biotransform Diclofenac in a shake-flask biotransformation into the hydroxylated metabolite, as shown in Scheme 2.2. The reactions used live cell-paste to convert 2 mM Diclofenac in a total of six 1 L-scale biotransformations to yield 2.8 g of 4′-hydroxydiclofenac (from 3.8 g of Diclofenac) after purification. The authors compared this method against four other literature methods and concluded that recombinant yeast showed some of the best characteristics, such as product yield.

In a similar vein, Luetz and co-workers at Novartis Institute for Biomedical Research described the production and application of a panel of recombinant human CYPs [37].

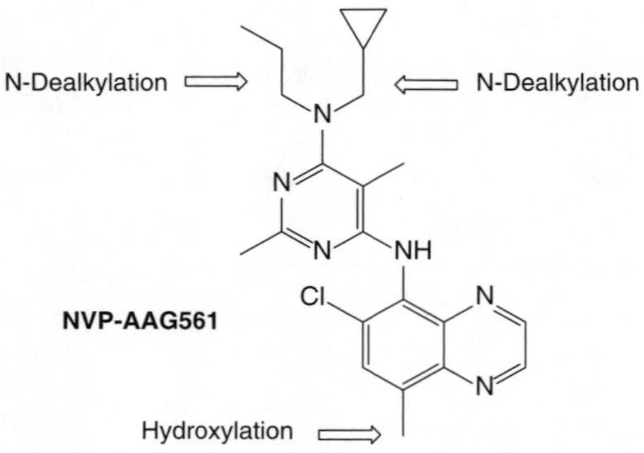

Scheme 2.3 *Metabolically active positions in NVP-AAG561 with CYP3A4.*

These were produced in *E. coli*, where the CYP enzymes were co-expressed with a cytochrome reductase that mediates the transfer of electrons from NADPH to the CYP. The panel comprised 14 human CYPs and was profiled extensively in analytical screening experiments against compounds of interest to Novartis. Interestingly, this showed that one particular CYP, namely CYP3A4, was able to form metabolites from 50 out of 60 compounds tested. This paper describes the typical methodology utilized for the bio-transformation of 250 mg of the drug candidate NVP-AAG561, giving rise to five metabolites, four of which were isolated. The molecule of interest is shown in Scheme 2.3. The methodology here utilized an oxygenated live recombinant cell-paste formulation of the CYP and exploited cellular turnover of the NADP co-factor. The CYP in this example performed both hydroxylation and *N*-dealkylation in various combinations to give five products. The major product comprised a single *N*-depropylation.

CYP enzymes have become available commercially for metabolite synthesis and other applications, and this, together with the rapid advances in the rational or semi-rational redesign of the enzymes, is starting to have an impact, bringing the technology to a wider audience for testing. One of the most prominently used enzymes in this class is the *Bacillus megaterium* BM3 enzyme (see Section 2.2). A useful review of the properties and applications of this enzyme is provided by Di Nardo and Gilardi, including recent attempts to increase substrate specificity by protein engineering, improve catalytic performance (K_M, k_{cat}, coupling efficiency) toward drugs, substitute the costly NADPH co-factor, and provide immobilization and scale-up of the process for industrial application [38]. Di Nardo and Gilardi conclude that great progress has been made, and that this CYP remains the focus of many R&D groups, but that much work is still required to increase its performance in terms of turnover and coupling efficiency.

While the application of CYPs for metabolite synthesis has featured heavily in the literature, substantial effort is directed toward their use in the production of pharmaceutical actives and fine chemicals. One of the best examples of CYP exploitation – if not the best – is in the bioconversion of Compactin to Pravastatin. Pravastatin, a cholesterol-lowering drug, is produced by a two-step process, in which the first stage is production of the precursor Compactin by industrial fermentation of *Penicillium citrinum*. This is hydroxylated to Pravastatin by a CYP that resides within *Streptomyces carbophilus* (Scheme 2.4) [39].

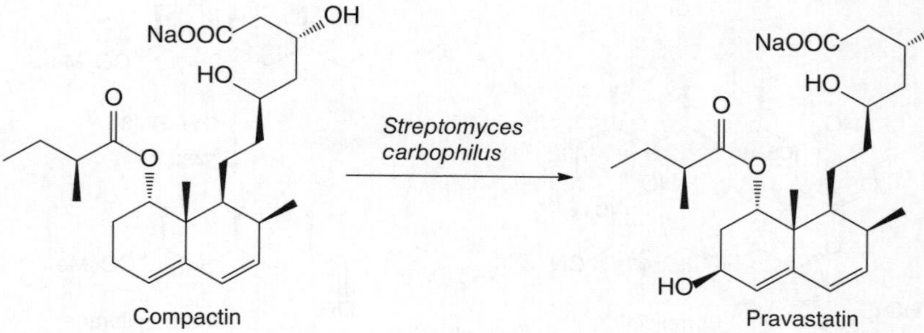

Scheme 2.4 *Biooxidation of Compactin to Pravastatin using CYP enzymes.*

More recent examples of CYP biocatalysis aimed at fine chemical and pharmaceutical synthesis are given later in this section. Returning to the CYP BM3 enzyme, Pietruszka reported the benzylic hydroxylation of various aromatic compounds, utilizing a double mutant of BM3 (F87A, L188C) for the benzylic hydroxylation [40] (Scheme 2.5). This biotransformation used a crude cell extract of the BM3 mutant, with recycle of NADPH by GDH. With a charge of 10 mM substrate, the product was isolated in 73% yield (86 mg) with a trace amount of the aldehyde formed from over-oxidation of the alcohol. The double mutant used in this reaction was 535-fold more active than the wild-type enzyme, and was discovered by targeted replacement of key residues around the active site using previously generated data on the effect of certain mutations within the active-site vicinity.

The polyketide (−)-borrelidin has been of substantial pharmaceutical interest, as it shows a range of biological activities. It is a complex natural product and its biosynthetic route, where the molecule is readily made by fermentation, has been challenging to emulate in synthetic chemistry. Laschat *et al.* report the synthesis of a key major C3–C11 fragment, and the route is notable for having included the application of CYP biocatalysis [41]. The structure of (−)-borrelidin and the key biocatalytic CYP step used in its synthesis are shown in Scheme 2.6. It is an encouraging sign that this type of biocatalysis is being considered by synthetic chemists engaged in natural product synthesis.

In this example, the hydroxylated product (comprising an 82 : 18 ratio of diastereomers) is used as a facile means for accessing the terminal alkene for subsequent ring construction. The hydroxylation was performed on 500 mg-scale using a purified wild-type BM3 expressed in *E. coli*, resulting in the isolation of product (184 mg) in 34% yield. Recycle of the NADH co-factor was achieved in this instance by the use of formate dehydrogenase (FDH) and sodium formate.

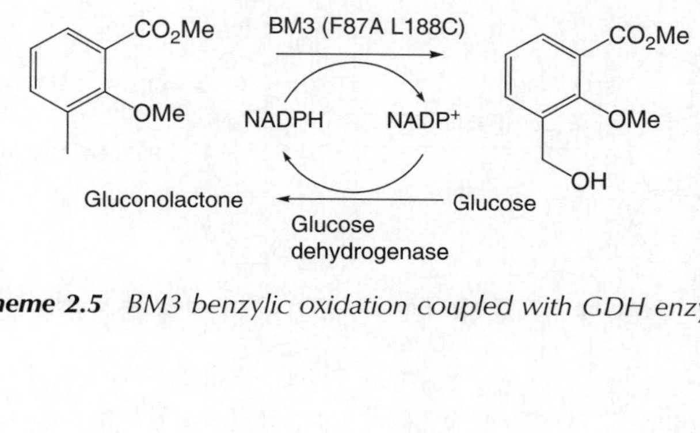

Scheme 2.5 *BM3 benzylic oxidation coupled with GDH enzyme.*

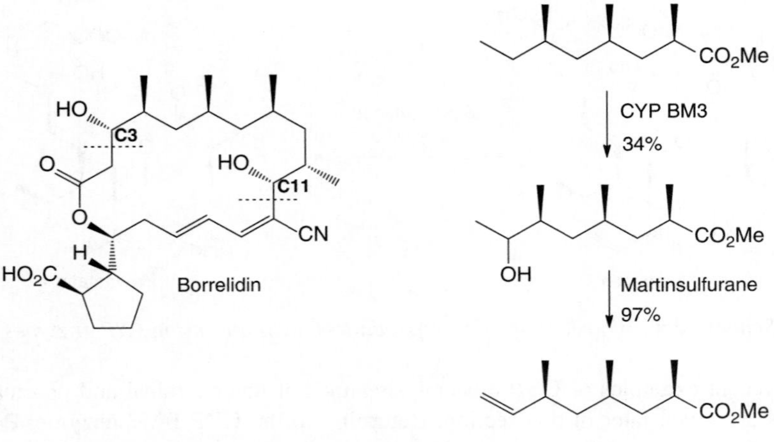

Scheme 2.6 *BM3 oxidation of aliphatic side chain en route to borrelidin.*

The co-expression of a multi-component CYP and GDH for cofactor recycle was developed by Li *et al.*, and applied to the biohydroxylation of alicyclic compounds [42]. This is a particularly good example of how CYP biocatalysis may progress, since it utilizes an easily accessible whole-cell biocatalyst with built-in co-factor recycle, demonstrates reasonable activity, and accumulates product to good concentration: nearly 20 mM in one example, $3.7 \, \text{g.L}^{-1}$ for (*S*)-*N*-benzyl-4-hydroxy-pyrrolidin-2-one (Scheme 2.7).

This system is based on the P450pyr monooxygenase from *Sphingomonas* sp. HXN-200, a class I P450 requiring ferredoxin and ferredoxin reductase for electron transfer, which has demonstrated hydroxylation at non-activated carbon atoms with unique substrate specificity, broad substrate range, and good to excellent regio- and stereoselectivity. This CYP had been previously mutated to incorporate three changes that greatly improve its stereoselectivity for these alicyclic substrates. The synthetic biology strategy here was to express the CYP and the ferredoxin reductase on one plasmid (pETDuet) and GDH and ferredoxin

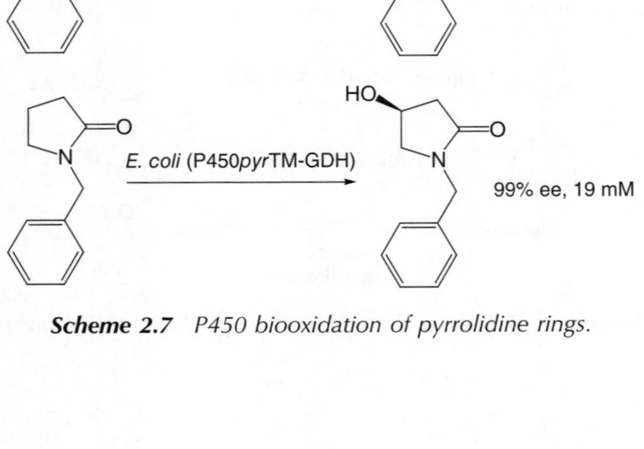

Scheme 2.7 *P450 biooxidation of pyrrolidine rings.*

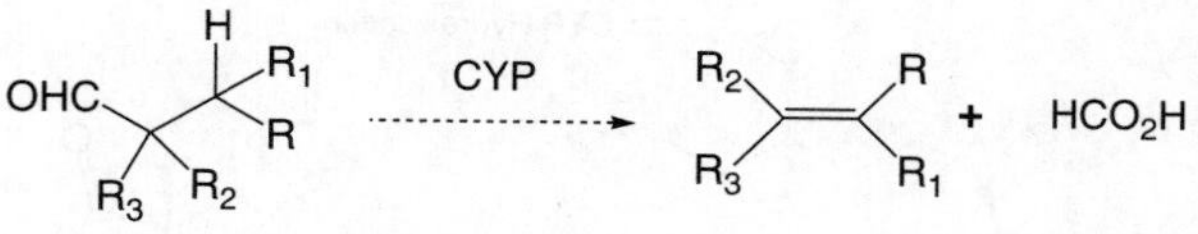

Scheme 2.8 *CYP-induced formation of alkenes and by-product formic acid.*

on another (pRSFDuet), since it was shown that *in vivo* co-expression of GDH gave a much more effective hydroxylation biocatalyst and that a high level of ferredoxin expression was beneficial.

This discussion has covered some recent applications of CYPs in the synthesis of metabolites and other molecules, using various CYPs and biocatalyst formulations. Rational catalyst evolution through bioinformatics and structural informed design is a strong theme, using previously gained structure-activity knowledge and the various synthetic biology databases that are now available. However, there is still good reason to continue to screen for new CYP sources and activities, such as fungi like wood-rotting basidiomycetes, each species of which can harbor hundreds of diverse CYP enzymes, which are largely uncharacterized and are therefore not identifiable from conventional bioinformatic interrogation. This subject is explored by Ichinose *et al.*, who elegantly describe the development of a 96-well plate functional screening system for testing of novel basidiomycete CYPs expressed in yeast [43].

In addition to continuing the development of new sources of CYPs and ways of expressing and utilizing them, new reaction types are of interest. While all of the examples cited so far result in stable hydroxylated products, this is not always the case. In some situations, the hydroxylated product is not stable and may spontaneously rearrange to give a new product. In other instances, rearrangement may occur within a reaction intermediate. A recent paper by Guengerich and Munro discusses some of these cases [44]. As an example, the formation of an alkene from an aldehyde is possible (or an aromatic ring, as in the biosynthesis of estrogen), where the aldehyde is cleaved from the molecule as formic acid (Scheme 2.8). This further extends the range of products achievable by CYP biocatalysis.

CYP biocatalysis is particularly interesting for application to certain classes of compounds, such as steroids. Steroids represent one of the major classes of drug, with around 300 steroid drugs in clinical use, and the hydroxylation of a steroid molecule can have profound influences on its biological properties, altering hydrophobicity, toxicity, and cellular location. Microbial steroid transformations are reviewed by Donova and Egorova, who note that current trends in the microbial CYP-catalyzed hydroxylation of steroids focus on achieving hydroxylation reactions of importance to the steroid industry and on creating novel hydroxylated intermediates for drug discovery [45]. Their review tabulates the recent hydroxylation literature, and it is evident that most of this has utilized wild-type whole-cell systems, although there are some examples utilizing engineered CYPs. It is worth noting that CYPs have been used indirectly at commercial scale for a number of years, in conjunction with other metabolic enzymes, for the bioconversion of phytosterols by microbial strains to provide the building blocks AD and ADD for steroid drugs (Scheme 2.9).

However, despite the use of CYPs as part of a larger metabolic sequence, to the author's knowledge, there have been no large-scale applications of isolated CYPs in steroid synthesis to date. It may be anticipated, however, that such reactions will be developed in due course.

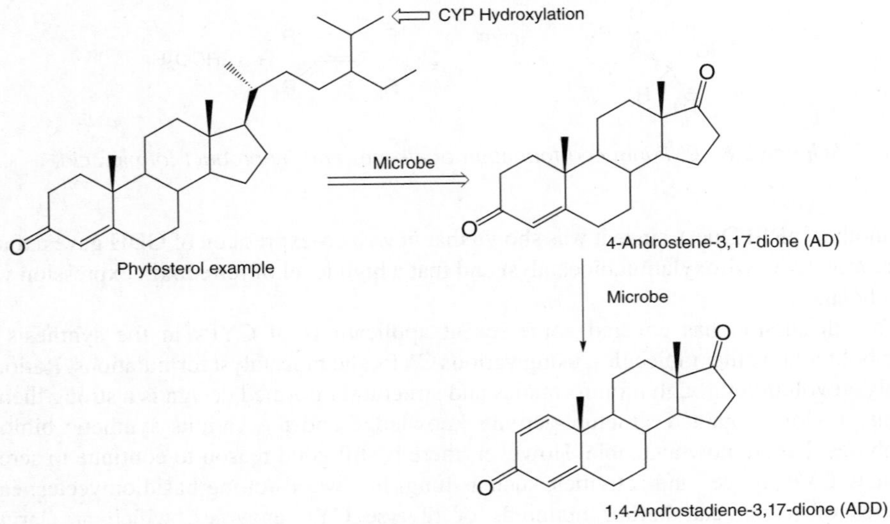

Scheme 2.9 *Steroidal oxidation using CYP enzymes.*

CYP chemistry is also of much interest in the synthesis of pharmaceuticals themselves. Perhaps one of the best examples of the application of current synthetic biology methodology to CYP enzymes is in Artemisinin biosynthesis, where a semi-synthetic route has been demonstrated (Scheme 2.10) [46].

Amorpha-4,11-diene, a precursor molecule accumulated by an engineered cell, is a challenging target for selective oxidation, and an epoxidation was required for a shorter semi-synthetic route. Wild-type P450 BM3 showed no activity toward amorpha-4,11-diene. Subsequently, F87 in the substrate binding pocket was identified as a key residue by docking the substrate into the active site of a predicted transition-state model. Upon the mutation of this residue to alanine, low activity against amorpha-4,11-diene was detected. To increase the size of the binding pocket and, hopefully, improve the yield of epoxide, saturation mutagenesis of key residues in the active site was then preformed. A mutant (F87A, R47L, Y51F, and A328L) was identified that was able to accumulate 250 mg.L^{-1} of the epoxide, which could then be transformed by high yielding chemistry to Artemisinin.

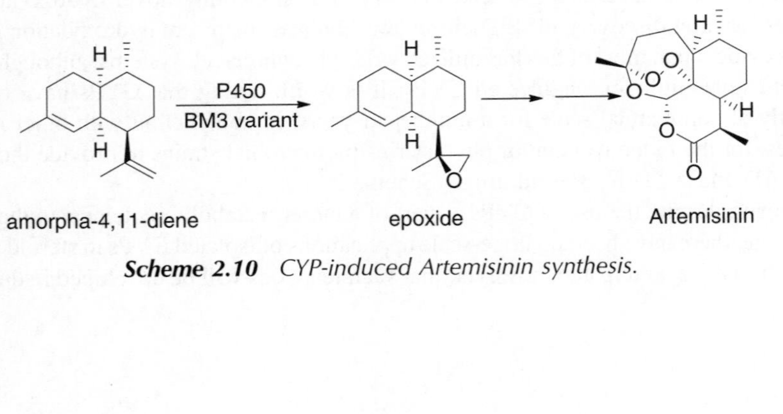

Scheme 2.10 *CYP-induced Artemisinin synthesis.*

This is an excellent example of how modeling and synthetic biology approaches can be used to redesign CYP enzymes, conceive novel metabolic pathways, and provide solutions to challenging chemical problems. As the pool of knowledge increases, particularly relating to structural information, it may be anticipated that this type of approach will rapidly become the method of choice for CYP-related chemistry. The increasing number of distinct CYP structures (>30) in the Protein Databank will continue to stimulate the rational redesign and mutagenesis of CYPs for improved or altered properties [47].

2.4 Conclusion

In summary, CYP biocatalysis continues to progress, and recent literature shows how synthetic biology is being exploited to facilitate this progress. Established CYPs are being evolved further for new and improved properties, and are being expressed and applied in new ways. New CYPs continue to emerge, realized by genomic-assisted screening technology. As high-value products, pharmaceuticals will continue to be the main driving force for CYP development, but, as examples of larger-scale syntheses with CYP emerge, we may see increasing interest in other industries, such as flavor and fragrances. This field of biocatalysis will continue to seduce synthetic chemists and can be expected to enter the mainstream of biocatalysis at some point in the future, to sit comfortably alongside other technologies such as hydrolases and ketoreductases.

References

1. Ortiz de Montellano, P.R. (2009) *Chemical Reviews*, **110**, 932–948.
2. Fasan, R. (2012) *ACS Catalysis*, **2**, 647–666.
3. Das, N. and Chandran, P. (2011) *Biotechnology Research International*. doi: 10.4061/2011/941810.
4. Chen, T.C., Sakaki, T., Yamamoto, K., and Kittaka, A. (2012) *Anticancer Research*, **32**, 291–298.
5. Li, A.F., Shen, G.L., Jiao, S.Y. *et al.* (2012) *Beijing Da Xue Xue Bao*, **44**, 431–436.
6. Cresner, B. and Petric, S. (2011) *Biochimica et Biophysica Acta (BBA) – Proteins and Proteomics*, **1814**, 29–35.
7. Lamb, D.C., Lei, L., Warrilow, A.G. *et al.* (2009) *Journal of Virology*, **83**, 8266–8269.
8. Zhang, J.D., Li, A.T., Yu, H.L. *et al.* (2011) *Journal of Industrial Microbiology and Biotechnology*, **38**, 633–641.
9. Lee, S.H., Kwon, Y.C., Kim, D.M., and Park, C.B. (2013) *Biotechnology and Bioengineering*, **110**, 383.
10. Roberts, K.M. and Jones, J.P. (2010) *Chemistry- A European Journal*, **16**, 8096–8107.
11. Ortiz de Montellano, P.R. (2009) *Chemical Reviews*, **110**, 932–948.
12. Fasan, R. (2012) *ACS Catalysis*, **2**, 647–666.
13. Janocha, S., Zapp, J., Hutter, M. *et al.* (2013) *Chembiochem*, **14**, 467–473.
14. Sheldon, R.A. (2008) *Chemical Communications*, 3352–3365.
15. Urlacher, V.B. and Girhard, M. (2012) *Trends in Biotechnology*, **30**, 26–36.
16. O'Reilly, E., Köhler, V., Flitsch, S.L., and Turner, N.J. (2011) *Chemical Communications*, **47**, 2490–2501.
17. Cojocaru, V., Balali-Mood, K., Sansom, M.S.P., and Wade, R.C. (2011) *PLoS Computational Biology*, **7**, e1002152. doi: 10.1371/journal.pcbi.1002152.

18. Sandee, D. and Miller, W.L. (2011) *Endocrinology*, **152**, 2904–2908.
19. Omura, T. (2010) *Journal of Biochemistry*, **147**, 297–306.
20. Gillam, E.M.J. (2008) *Chemical Research in Toxicology*, **21**, 220–231.
21. Romero, P.A. and Arnold, F.H. (2009) *Nature Reviews Molecular Cell Biology*, **10**, 866–876.
22. Armstrong, C.T., Watkins, D.W., and Anderson, J.L.R. (2013) *Dalton Transactions*, **42**, 3136–3150.
23. Kang, J.-Y., Ryu, S.H., Park, S.-H. *et al.* (2014) *Biocatalysis, Protein Engineering, and Nanobiotechnology*, **111**, 1313–1322.
24. Van, L.M., Sarda, S., Hargreaves, J.A., and Rostami-Hodjegan, A. (2009) *Journal of Pharmaceutical Sciences*, **98**, 763–771.
25. Locuson, C.W., Ethell, B.T., Voice, M.W. *et al.* (2009) *Drug Metabolism and Disposition*, **37**, 457–461.
26. Fulco, A.J. (1991) *Annual Review of Pharmacology*, **31**, 177–203.
27. Whitehouse, C.J.C., Bell, S.G., and Wong, L.-L. (2012) *Chemical Society Reviews*, **41**, 1218–1260.
28. Seifert, A., Vomund, S., Grohmann, K. *et al.* (2009) *ChemBioChem*, **10**, 853–861.
29. Tang, W.L., Li, Z., and Zhao, H. (2010) *Chemical Communications*, **46**, 5461–5463.
30. Agudo, R., Roiban, G.D., and Reetz, M.T. (2012) *Chembiochem*, **13**, 1465–1473.
31. Tsotsou, G.E., Sideri, A., Goyal, A. *et al.* (2012) *Chemistry - A European Journal*, **18**, 3582–3588.
32. Williams, P.A., Cosme, J., Vinkovic, D.M. *et al.* (2004) *Science*, **305**, 683–686.
33. Mosher, C.M., Hummel, M.A., Tracy, T.S., and Rettie, A.E. (2008) *Biochemistry*, **47**, 11725–11734.
34. Khan, K.K., He, Y.Q., Domanski, T.L., and Halpert, J.R. (2002) *Molecular Pharmacology*, **61**, 495–506.
35. Roland, W., Winkler, M., Schittmayer, M. *et al.* (2009) *Advanced Synthesis and Catalysis*, **351**, 2140–2146.
36. Drăgan, C.-A., Peters, F.T., Bour, P. *et al.* (2011) *Applied Biochemistry and Biotechnology*, **163**, 965–980.
37. Schroer, K., Kittelmann, M., and Luetz, S. (2010) *Biotechnology and Bioengineering*, **106**, 699–706.
38. Di Nardo, G. and Gilardi, G. (2012) *International Journal of Molecular Sciences*, **13**, 15901–15924.
39. Matsuoka, T., Miyakoshi, S., Tanzawa, K. *et al.* (1989) *European Journal of Biochemistry*, **184**, 707–713.
40. Neufeld, K., Marienhagen, J., Schwaneberg, U., and Pietruszka, J. (2013) *Green Chemistry*, **15**, 2408.
41. Theurer, M., El Baz, Y., Koschorreck, K. *et al.* (2011) *European Journal of Organic Chemistry*, **22**, 4241–4249.
42. Pham, S.Q., Gao, P., and Li, Z. (2013) *Biotechnology and Bioengineering*, **110**, 363–373.
43. Ichinose, H. (2013) *Biotechnology and Applied Biochemistry*, **60**, 71–81.
44. Guengerich, F.P. and Munro, A.W. (2013) *Journal of Biological Chemistry*, **288**, 17065–17073.
45. Marina, V., Egorova, D., and Egorova, O.V. (2012) *Applied Microbiology and Biotechnology*, **94**, 1423–1447.
46. Dietrich, J.A., Yoshikuni, Y., Fisher, K.J. *et al.* (2009) *ACS Chemical Biology*, **4** 261–267.
47. Grogan, G. (2011) *Current Opinion in Chemical Biology*, **15**, 241–248.

3

Use of Hydrolases and Related Enzymes for Synthesis

3.1 Continuous-Flow Reactor-Based Enzymatic Synthesis of Phosphorylated Compounds on a Large Scale

Lara Babich, Aloysius F. Hartog, Michael A. van der Horst, and Ron Wever

Van't Hoff Institute for Molecular Sciences, University of Amsterdam, The Netherlands

Phosphorylated compounds are central in the chemistry of natural processes [1]. Given the importance of these compounds, many phosphorylation methods, both chemical and biochemical, have been used [2–4]. We have developed a simple enzymatic method using recombinant acid phosphatase immobilized on beads and used in a small (0.5 mL) continuous-flow packed-bed reactor (Scheme 3.1) [5]. The phosphatase (PhoN-Sf) catalyzes transphosphorylation reactions in which the phosphate group from cheap pyrophosphate (PPi) is transferred to primary alcohol groups [3].

The feed of the reactor contains the activated phosphate donor pyrophosphate and a compound containing a primary alcohol group, and this solution is continuously pumped through the column packed with acid phosphatase immobilized on Immobeads. With this continuous method and a column containing only 0.5 mL of beads, grams of a variety of phosphorylated products can be obtained. The excess of phosphate produced is easily removed by precipitation with barium salts. The procedure is readily scaled for large-scale production of many phosphorylated compounds. In particular, carbohydrates are easily phosphorylated with nearly quantitative conversions. In the procedure that follows, only the method used to synthesize *N*-acetyl-D-glucosamine-6-phosphate is described. Reference [5] gives the procedures by which other phosphorylated compounds (e.g., glucose-6-phosphate) have been synthesized.

Practical Methods for Biocatalysis and Biotransformations 3, First Edition.
Edited by John Whittall, Peter W. Sutton, and Wolfgang Kroutil.
© 2016 John Wiley & Sons, Ltd. Published 2016 by John Wiley & Sons, Ltd.

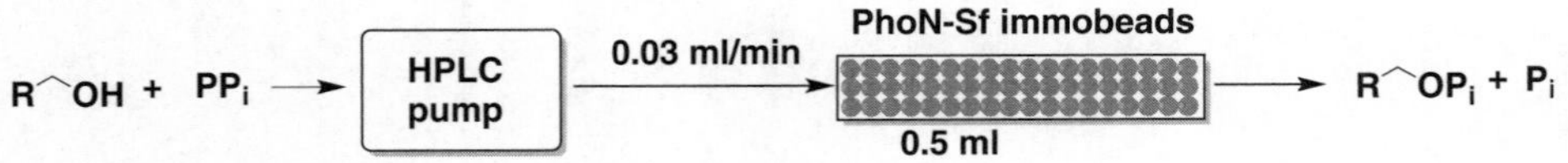

Scheme 3.1 *Scheme of the continuous-flow system with immobilized PhoN-Sf for the phosporylation of primary alcohols using PPi as a cheap phosphate donor.*

3.1.1 Materials and Equipment

- Recombinant acid phosphatase [6] (PhoN-Sf) immobilized on Immobeads-150
- A solution of disodium dihydrogen pyrophosphate (55.5 g.L^{-1}) and a solution of tetra sodium pyrophosphate (66.5 g.L^{-1}). By mixing the two salt solutions in a proper ratio, a 250 mM concentration at the desired pH (4–6) can be obtained.
- *N*-Acetyl-D-glucosamine (10 g)
- Barium acetate (50 g)
- Small empty high-performance liquid chromatography (HPLC) column
- HPLC pump
- Glass-filter funnel with glass equipment
- HPLC [5] with RI and UV detection
- 1 M NaOH Solution
- pH Meter

3.1.2 Immobilization of Acid Phosphatase on Immobeads

1. PhoN-Sf [6] (10 mg, 300 U) was added to Immobeads-150 (300 mg) in 1.25 M potassium phosphate, pH 8 (15 mL), and the mixture was shaken at 20 °C.
2. After 24 hours, the beads were washed twice with water (15 mL), diluted with 2 M glycine solution, pH 8.5 (2 volumes), and shaken at 20 °C for 24 hours to end-cap the unreacted epoxy groups of the beads.
3. The beads were then washed with water (2 × 4 mL), followed by 0.1 M potassium phosphate, pH 7.4 (2 × 4 mL) to remove glycine, and stored at 4 °C. The immobilized phosphatase is stable for months.

3.1.3 Production of *N*-acetyl-D-Glucosamine-6-Phosphate (Scheme 3.2)

1. A slurry of the acid phosphatase immobilized on Immobeads-150 was prepared and packed into a small HPLC column (3 cm in length, 0.46 cm in diameter, 0.5 mL volume).
2. 400 mL of a solution containing N-acetyl-D-glucosamine (8.8 g, 100 mM) in 250 mM PPi with a pH of 4.2 was prepared.

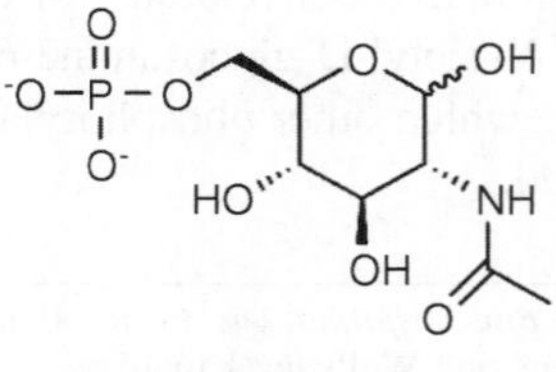

Scheme 3.2 *N-acetyl-D-glucosamine-6-phosphate.*

3. The substrate solution was then pumped through the column at room temperature at a rate of $1.8 \, mL.h^{-1}$.

4. After 9 days, the collected outlet solution (375 mL) containing 65 mM phosphorylated N-acetyl-D-glucosamine, PPi and Pi, was stirred at room temperature with 10% w/v barium acetate at pH 7 to remove free phosphate.

5. After 2 hours, the solution was filtered and the solid salt of barium phosphate was discarded.

6. The filtrate containing alcohol, non-phosphorylated alcohol, and acetate was adjusted to pH 9 with 1 M NaOH. Cold ethanol (4 volumes) was added, and the mixture was stirred and left standing at 4 °C for 1 day.

7. The solution was filtered and dried to afford the barium salt of N-acetyl-D-glucosamine-6-phosphate (10.3 g, 82%) in a purity of 85%. It is possible to improve the purity by dissolving in a small amount of water and precipitating with ethanol.

3.1.4 Conclusion

The broad substrate specificity of the acid phosphatase allows the phosphorylation of a variety of carbohydrates and various alcohols, using cheap pyrophosphate as the activated phosphate donor. In the previously described batch processes [3,4,6], soluble acid phosphatase was used to synthesize phosphorylated compounds. However, in a number of cases the hydrolytic action of the phosphatase leads to hydrolysis of the phosphorylated compounds once formed. Due to the physical separation in a flow system of the phosphorylated product from the immobilized enzyme, competing background hydrolysis will be prevented. Also, immobilization enables recycling of the catalyst and improves its stability, making the economics of the process much more attractive.

3.2 Deracemization of *sec*-Alcohols via Enantio-Convergent Hydrolysis of *rac*-Sulfate Esters

Michael Toesch, Markus Schober, and Kurt Faber
Department of Chemistry, Organic and Bioorganic Chemistry, University of Graz, Austria

Sulfatases make up a heterogenic class of enzymes, comprising three distinct subgroups acting through different mechanisms: aryl-, alkyl-, and α-ketoglutarate-dependent sulfatases [7,9]. From a synthetic viewpoint, aryl- and alkyl-sulfatases are most valuable, because they allow the stereoselective hydrolysis of *sec*-alkyl sulfate esters, enabling their kinetic resolution [7,8]. Most intriguing is the unique double-selectivity of this class of enzymes; that is, they are not only enantioselective – by preferring one substrate enantiomer over its counterpart – but also stereoselective with respect to their mechanism of action: sulfate ester hydrolysis via nucleophilic attack of [OH⁻] at sulfur liberates the *sec*-alcohol with unchanged configuration at the chiral carbon atom, while attack at C causes inversion. This latter feature has been successfully employed for the deracemization of *sec*-alcohols via enantio-convergent hydrolysis of their racemic sulfate esters (Scheme 3.3) [9,10]. Mechanistically, aryl-sulfatases are particularly well characterized, due to their involvement in Austin's disease (multiple sulfatase deficiency) in humans. Prominent examples include

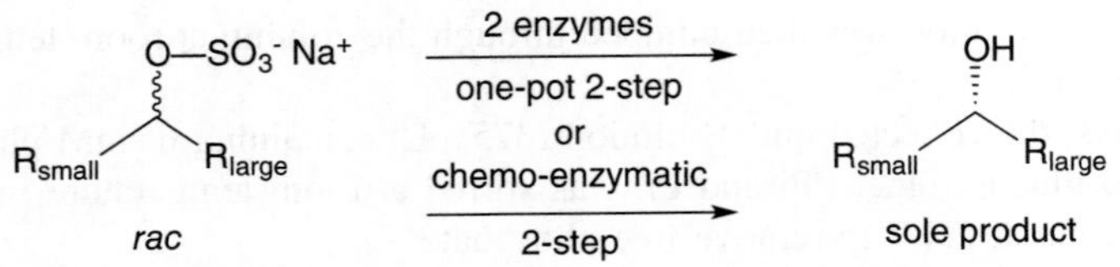

Scheme 3.3 *Enzymatic and chemoenzymatic deracemization of sec-alcohols via enantio-convergent hydrolysis of their corresponding monosulfate esters using retaining and inverting sulfatases.*

human aryl sulfatase A and the sulfatase from *Pseudomonas aeruginosa* (PAS) [11,12]. On alkyl sulfates, PAS acts via retention of configuration. Less is known about inverting alkyl sulfatases: the most thoroughly investigated enzyme from this group is Pisa1, from *Pseudomonas* sp. DSM6611, which acts through nucleophilic attack of a water molecule – activated by a binuclear Zn^{2+} cluster – on carbon [8,13].

The chemical hydrolysis of sulfate esters is only viable via acid catalysis through retention of configuration. It is initiated by protonation of the (negatively charged) sulfate ester, which provides HSO_4^- as a good leaving group and allows nucleophilic attack of H_2O at sulfur [14,15]. In contrast, inverting chemical hydrolysis via nucleophilic attack of $[OH^-]$ at C is a very slow process.

Overall, a racemic *sec*-sulfate ester may be deracemized to furnish the corresponding single stereoisomeric *sec*-alcohol as its sole product, by combining a retaining and an inverting sulfate ester hydrolysis step [10,13]. In an ideal case, two stereo-complementary sulfatases possessing opposite enantiopreference may be combined in a single-step process. If only one sulfatase is enantioselective, a two-step one-pot process is feasible, where the enantioselective sulfatase converts its preferred substrate enantiomer, while the non-reacted enantiomer is hydrolyzed by the non-selective enzyme in a second step (Procedure 1). If only an inverting sulfatase is available, the second (retaining) hydrolysis step is also feasible, via acid catalysis (Procedure 2).

3.2.1 Procedure 1: One-Pot Two-Step Deracemization Using Two Stereo-Complementary Sulfatases (Scheme 3.4)

3.2.1.1 Materials and Equipment

- Pisa1 (13 mg, 177 nmol, expressed and purified as previously described) [13]
- PAS (26 mg, 454 nmol, expressed and purified as previously described) [10]
- Substrate *rac*-**1a** (1 g, 4.4 mmol, synthesized as previously described) [16]
- Tris/HCl buffer (100 mM, pH 8.0, 200 mL)
- Na_2SO_4 anhydrous
- *t*-BuOMe (300 mL)
- Ethyl acetate (1 mL)
- Acetic anhydride (100 µL)
- DMAP (4-dimethylaminopyridine, 1 mg)
- Round-bottom flask (250 mL)
- Rotary shaker at 30 °C
- Eppendorf tubes

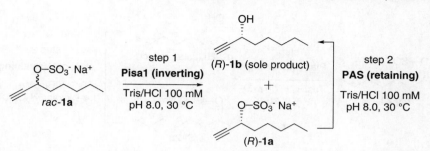

Scheme 3.4 *Stepwise deracemization of a propargylic sulfate ester (rac-**1a**) using Pisa1 for inversion (step 1) and PAS for retention (step 2) to yield (R)-**1b** as the sole product.*

3.2.1.2 Procedure

1. Substrate *rac*-**1a** (1 g, 4.4 mmol) was placed in a 250 mL round-bottom flask and dissolved in 100 mM Tris/HCl buffer, pH 8.0 (200 mL).
2. Purified Pisa1 (13 mg, 177 nmol) was added to the solution and the mixture was shaken for 24 hours at 120 rpm and 30 °C.
3. After the reaction ceased at 50% conversion, PAS (26 mg, 454 nmol) was added to the mixture without isolation of unreacted (*R*)-**1a** and shaking was continued for another 24 hours.
4. Alcohol (*R*)-**1b** was isolated by extraction of the aqueous phase with *t*-butyl methyl ether (3 × 100 mL). The organic phase was dried with Na_2SO_4, filtered, and carefully evaporated at 220 mbar and ~30 °C to avoid loss of product. (*R*)-**1b** was afforded as a yellow liquid (82% yield, 0.45 g, 3.6 mmol).
5. For determination of the *ee*, a sample of (*R*)-**1b** (5 mg) was acetylated with acetic anhydride (100 μL) and cat. DMAP (1 mg, 8 μmol) in ethyl acetate (1 mL) and shaken in a 1.5 mL Eppendorf tube for 1 hour at 120 rpm and 30 °C. The reaction was quenched with distilled water (300 μL) and dried with Na_2SO_4, and the *ee* was measured via GC-FID. It was found to be 98% [10].

3.2.2 Procedure 2: Two-Step Chemoenzymatic Deracemization Using Inverting Pisa1 and Retaining Acid Hydrolysis (Scheme 3.5)

3.2.2.1 Materials and Equipment

- Pisa1 (6.5 mg, 88 nmol, expressed and purified as previously described) [13]
- Substrate *rac*-**2a** (1 g, 4.4 mmol, synthesized as previously described) [14]
- Tris/HCl buffer (100 mM, pH 8.0, 100 mL)
- Na_2SO_4 anhydrous
- *p*-TsOH monohydrate (1.75 g, 9.2 mmol, *p*-toluenesulfonic acid monohydrate)
- *t*BuOMe (594 mL)
- 1,4-Dioxane (100 μL)
- $NaHCO_3$ solution (saturated, 50 mL)
- Ethyl acetate (1 mL)
- Acetic anhydride (100 μL)
- DMAP (4-dimethylaminopyridine, 1 mg)
- Round-bottom flasks (250 mL)

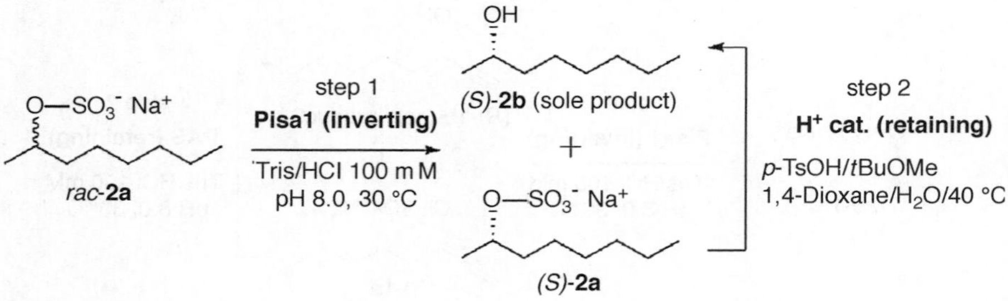

Scheme 3.5 *Stepwise deracemization of an aliphatic sulfate ester (rac-2a) using Pisa1 for inversion (step 1) and H$^+$ catalysis for retention (step 2) to yield (S)-2b as the sole product.*

- Rotary shaker at 30 °C
- Water bath at 40 °C
- Eppendorf tubes

3.2.2.2   Procedure

1. Substrate *rac*-**2a** (1 g, 4.4 mmol) was dissolved in 100 mM Tris/HCl buffer, pH 8.0 (100 mL), in a 250 mL round-bottom flask. Purified Pisa1 (6.5 mg, 88 nmol) was added to the solution.
2. The mixture was shaken for 24 hours at 120 rpm and 30 °C.
3. After the reaction ceased at 50% conversion, (S)-**2b** was extracted from the aqueous phase using *t*-butyl methyl ether (3 × 100 mL). The combined organic phases were dried with Na$_2$SO$_4$, filtered, and evaporated (250 mbar, rt), which afforded the first crop of product (S)-**2b** as a clear yellow liquid (248 mg, 1.9 mmol).
4. Unreacted (S)-**2a** was recovered by overnight lyophilization of the aqueous phase and was placed in a 250 mL round-bottom flask.
5. For the chemical hydrolyis step, *p*-TsOH monohydrate (1.75 g, 9.2 mmol), 1,4-dioxane (100 µL, 1.2 mmol), *t*-BuOMe (194 mL), and distilled water (6 mL) were added to the lyophilized residue and the mixture was heated under reflux at 40 °C for 5 hours.
6. The reaction was cooled to room temperature and saturated NaHCO$_3$ solution (50 mL) was added. The aqueous phase was extracted with *t*-BuOMe (2 × 50 mL) and the organic phases were combined, dried with Na$_2$SO$_4$, and evaporated under reduced pressure at 250 mbar and room temperature to avoid loss of product.
7. The second crop of (S)-**2b** was afforded as a clear yellow liquid (243 mg, 1.9 mmol) from the chemical reaction step. Combining both crops from the individual steps yielded 87% (491 mg, 3.8 mmol) of (S)-**2b** with an *ee* of >99% (after derivatization as acetate as described in Procedure 1) [13].

3.2.3   Analytical Methods

3.2.3.1   Determination of Enantiomeric Excess

The *ee* of alcohols (R)-**1b** and (S)-**2b** was determined after derivatization to the corresponding acetates by GC-FID using a chiral DEX-CB column (25 m × 0.32 mm × 0.25 µm film) and He as carrier gas. Retention times were as shown in Table 3.1.

Table 3.1 *GC-FID retention times for acetylated alcohols 1 and 2. Injector temperature 200°C, flow 2.0 mL.min^{-1} He: 80°C, hold for 1.0 minutes, 15°C.min^{-1} to 110°C, 4°C.min^{-1} to 130°C, 10°C.min^{-1} to 180°C.*

Acetylated alcohol	Retention time (minutes)	
	(R)	(S)
1	5.5	5.9
2	5.9	5.3

3.2.3.2 Optical Rotation

(*R*)-**1b**: $[\alpha]_D{}^{20}$ + 4.68; c = 1, CHCl$_3$ [10]
(*S*)-**2b**: $[\alpha]_D{}^{20}$ + 7.21; c = 2, CHCl$_3$ [13]

3.2.3.3 Nuclear Magnetic Resonance Spectroscopy

(*R*)-**1b**: ^{1}H-NMR (300 MHz; CDCl$_3$) δ 4.39 (dt, *J* = 29 and 9.2 Hz, 1H), 2.48 (d, *J* = 5.2 Hz, 1H), 1.80–1.67 (m, 2H), 1.53–1.25 (m, 6H), 0.92 (t, *J* = 6.1 Hz, 3H); ^{13}C-NMR (75 MHz; CDCl$_3$) δ 85.0, 72.8, 62.3, 37.6, 31.4, 24.7, 22.5, 14.0 [10]
(*S*)-**2b**: ^{1}H-NMR (300 MHz; CDCl$_3$) δ 3.86–3.75 (m, 1H), 1.55–1.17 (m, 13H), 0.93–0.86 (m, 3H); ^{13}C-NMR (75 MHz; CDCl$_3$) δ 68.2, 39.4, 31.8, 29.3, 25.7, 23.5, 22.6, 14.0 [13]

3.2.4 Conclusion

Sulfatases have proven their great potential for the deracemization of *sec*-alcohols in both purely enzymatic and chemoenzymatic strategies, which are based on the stereo-convergent transformation of a pair of enantiomers via a retaining and an inverting step [10,13]. Although the one-pot approach using two enzymes simultaneously is clearly the most elegant version, it is somewhat limited in terms of scope, because both enzymes have to display a double selectivity – opposite enantiopreference with matching inversion and retention of configuration (process **A**, Table 3.2). So far, this reaction has been successfully demonstrated for substrates **A1** and **A2**, yielding alcohol products with *ee*s >98%. In contrast, the one-pot two-step approach (process **B**) shows a broader substrate tolerance, because only one enzyme – used in the first step – has to be enantioselective (substrates **B1** to **B5**, *ee*s 91 to >99%, Table 3.2). The least restrictive strategy is the chemoenzymatic approach (process **C**), in which the first enantioselective inversion is catalyzed by Pisa1, leading to a homochiral mixture of sulfate ester and alcohol. In the second step, the unreacted starting material is converted with retention to the corresponding alcohol via acidic hydrolyis. Hence, the substrate scope is considerably wider, featuring not only unfunctionalized saturated alkyl chains (substrates **C1** to **C8**), but also branched (**C3**), olefinic (**C9** to **C14**), and acetylenic analogs (**C15**, **C16**). The *ee* of product ranges from moderate (60% for **C8**) to perfect (>99% for **C4**). The synthetic utility of this technology has been demonstrated in the chemoenzymatic asymmetric total synthesis of a macrolide antibiotic [17].

Table 3.2 *Substrate scope of the deracemization of sec-alcohols through enantio-convergent hydrolysis of the corresponding sulfate esters via chemoenzymatic protocols **A** through **C**.*

Entry	One-pot two enzymes simultaneous (**A**)		One-pot two enzymes two steps (**B**)		One-pot chemoenzymatic two steps (**C**)	
	R_{small}	R_{large}	R_{small}	R_{large}	R_{small}	R_{large}
1	Me	C≡C-Ph	Et	C≡C-Me	Me	n-Pr
2	Me	m,m-(CF$_3$)$_2$C$_6$H$_3$	Et	C≡C-Et	Me	n-Bu
3			C≡CH	CH$_2$-CHMe$_2$	Me	(CH$_2$)$_2$CHMe$_2$
4			C≡CH	n-Bu	Me	n-hexyl
5			C≡CH	n-pentyl	Me	n-octyl
6					Et	n-pentyl
7					Et	n-Bu
8					n-Pr	n-Bu
9					Me	CH$_2$-CH=CH$_2$
10					Me	(CH$_2$)$_2$-CH=CH$_2$
11					CH=CH$_2$	n-Bu
12					CH=CH$_2$	n-pentyl
13					CH=CH$_2$	n-hexyl
14					CH=CH$_2$	n-heptyl
15					Me	C≡C-Et
16					Me	CH$_2$-C≡C-Et

3.3   Dynamic Kinetic Resolution of a Primary Amine by an Efficient Bifunctional Pd-CALB Hybrid Catalyst. A Metalloenzyme Mimic for Enhanced Cooperative Catalysis

Oscar Verho, Karl P. J. Gustafson, and Jan-E. Bäckvall

Department of Organic Chemistry, Arrhenius Laboratory, Stockholm University, Sweden

Chiral amines are an important class of compounds as they are commonly used as building blocks in the synthesis of a wide range of pharmaceuticals, fragrances, and agrochemicals. Therefore, there exists a need for the development of synthetic protocols that allow for straightforward access to enantiomerically pure amines, and in this respect chemoenzymatic dynamic kinetic resolution (DKR) constitutes a simple and efficient way of preparing these compounds [18–21]. Recently, our group reported on a fully hetero-geneous catalytic system for the DKR of β-amino esters [21], which made use of *Candida antarctica* lipase A (CALA) immobilized on siliceous mesocellular foam (MCF) and a Pd nanocatalyst on alumina (Scheme 3.6) [22]. Following this study, we envisioned that it should be possible to construct a bifunctional hybrid catalyst, in which both of these two catalytic species were co-immobilized onto the MCF material [23]. This would position the Pd nanoparticles and the lipase within the same cavities of the support and bring the two catalysts in close proximity to one another, which was anticipated to lead to an enhanced cooperativity between the metal-catalyzed racemization and the enzyme-catalyzed resolution.

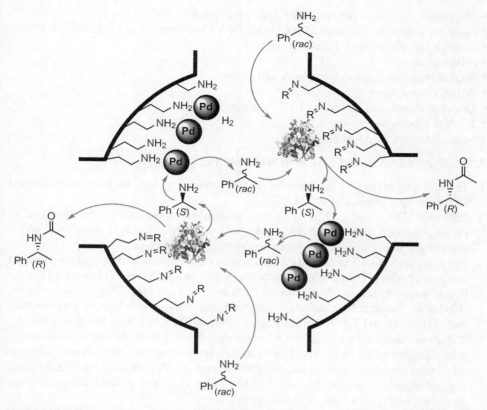

Scheme 3.6 *Dynamic kinetic resolution of an amine performed by a bifunctional biomimetic catalyst, where Pd nanoparticles and a lipase (CALB) have been co-immobilized within the same cavities of a heterogeneous siliceous support.*

3.3.1 Materials and Equipment

- Aqueous solutions of CALB from either Novozymes A/S (CALB-L) or Fermenta Biotech Limited (~5 000 LU.g^{-1} of liquid). A lipase unit (LU) is defined as the quantity of enzyme that will liberate 1 µmol of butyric acid per minute under the conditions of the test.
- Anhydrous toluene
- 0.5 M Solution of pentadecane in anhydrous toluene as internal standard (IS)
- Freshly distilled 1-phenylethylamine stored over molecular sieves
- Freshly distilled ethyl methoxyacetate
- Glutaraldehyde (50% in H$_2$O)
- Dry sodium carbonate
- Potassium phosphate buffer (100 mM; pH 7.0, 7.2, and 8.0)
- Hydrogen gas, balloon
- Bradford assay
- Activated molecular sieves 4 Å
- Newly prepared Pd^0-AmP-MCF

- Centrifuge capable of reaching 4100 rpm while holding 4 °C
- Gas chromatography (GC) with a CP-Chiralsil-DEX CB column (25 m × 0.32 mm × 0.25 μm)
- Amicon Ultra 10 K centrifugal column
- Stirrer plate
- Suitable reaction vial equipped with stirring bar
- 50 mL Falcon tubes
- Desiccator containing saturated LiCl solution
- UV/Vis spectrophotometer
- Manifold coupled to vacuum pump

3.3.2 Synthesis of the Pd-CALB Hybrid Catalyst

1. The synthesis of the Pd-CALB hybrid catalyst commenced from the previously reported Pd^0-AmP-MCF nanocatalyst [24,25].
2. Pd^0-AmP-MCF (0.40 g, 7.91 wt%) was suspended in potassium phosphate buffer (25 mL, 100 mM, pH 8). Glutaraldehyde (50% in H_2O, 0.18 g, 0.78 mmol) was added and the reaction was stirred at room temperature for 24 hours.
3. The glutaraldehyde-functionalized Pd nanocatalyst was isolated by centrifugation (4100 rpm) and washed with potassium phosphate buffer (3 × 45 mL, pH 8) and acetone (3 × 45 mL).
4. Before immobilization of commercial CALB (10 mL, ~5 000 LU.g^{-1}), the enzyme solution was subjected to a buffer exchange and concentration with a potassium phosphate buffer (100 nM, pH 7.0, 150 mL) using an Amicon Ultra 10 K centrifugal column.
5. The glutaraldehyde-functionalized Pd nanocatalyst was stirred overnight (16 hours) in potassium phosphate buffer (1 mL per 100 mg support, 100 mM, pH 7.2) with a 50 mg. mL^{-1} solution of CALB (0.34 mL per 100 mg support). The obtained Pd-CALB hybrid was subsequently isolated by centrifugation (4100 rpm) and washed with potassium phosphate buffer (2 × 5 mL, 100 mM, pH 7.2).
6. The catalyst was dried under reduced pressure and stored over a solution of LiCl to adjust the water activity of the CALB to 0.11, before it was used in catalytic experiments.

3.3.3 DKR of 1-Phenylethylamine

1. Pd-CALB hybrid catalyst (30 mg, 15.6 wt%, 4.80 wt% Pd), dry Na_2CO_3 (50 mg, 0.60 mmol), and molecular sieves 4 Å (300 mg) were added to a flame-dried flask. The flask was evacuated and filled with hydrogen gas.
2. Toluene (2 mL) was added and the reaction mixture was heated to 70 °C.
3. Pentadecane (158 μL, 0.5 M in toluene, IS), ethyl methoxyacetate (142 mg, 1.20 mmol), and 1-phenylethylamine (73 mg, 0.60 mmol) were added to the reaction, and the flask was mounted with a hydrogen replacement balloon.
4. After 16 hours, a liquid aliquot was withdrawn from the reaction mixture and analyzed by chiral GC to determine the yield and *ee* of the reaction [23].

3.3.4 Conclusion

By performing the DKR with the hybrid catalyst, it was possible to obtain amide (*R*)-**2** in quantitative yield and 99% *ee* within 16 hours. This can be compared to the corresponding reaction involving Pd nanoparticles and CALB separately immobilized on to MCF, which afforded the desired product in 89% yield and 99% *ee* after 20 hours. These results demonstrate that the close proximity of the two catalytic species within the same cavity of the MCF leads to an enhanced cooperativity that enables an overall highly efficient DKR. This co-immobilization concept holds great promise for the construction of novel catalytic tandem protocols, which combines the utility of bio- and transition metal catalysts. An interesting aspect of this approach is that it could provide for a simple way of combining enzymes and transition metal catalysts, which are not compatible with one another when they are used together in a homogeneous phase by conventional means. In addition, this strategy is associated with the typical advantages of heterogeneous catalysis, which involves simple separation and recycling of the catalyst.

3.4 Highly Efficient DKR of Secondary 1-Phenylethanol Derivatives Using a Low-Cost Solid Super Acid as Racemization Catalyst

Gang Xu, Jian-Ping Wu, Yong-Jun Chen, Liang Wang, Xiao Meng, and Li-Rong Yang

Department of Chemical and Biological Engineering, Zhejiang University, China

DKR is a widely researched method used to prepare optically pure chiral secondary alcohol. By combining enzyme-catalyzed kinetic resolution with *in situ* racemization normally catalyzed by chemical catalyst, DKR increases the maximum yield of enantiopure product in kinetic resolutions from 50 to 100%. The key part of a successful and pragmatic DKR system is a highly efficient and low-cost racemization catalyst capable of functioning under the mild reaction conditions required by enzymes [26,27]. We developed a new type of acid racemization catalyst, a nano solid super acid, which showed highly efficient racemization capability. When coupling this new racemization catalyst with the lipase Novozym 435, good biocompatibility was demonstrated, and optically pure aromatic enantiomer was obtained (Scheme 3.7) [28,29].

3.4.1 Procedure 1: Preparation of the Solid Super Acid TiO$_2$/SO$_4^{2-}$

3.4.1.1 Materials and Equipment

- TiCl$_4$ (20 mL)
- Deionized water (1000 mL)
- Concentrated ammonia water (weight percentage = 25.28%, 250 mL)
- H$_2$SO$_4$ (0.5 M, 20 mL)

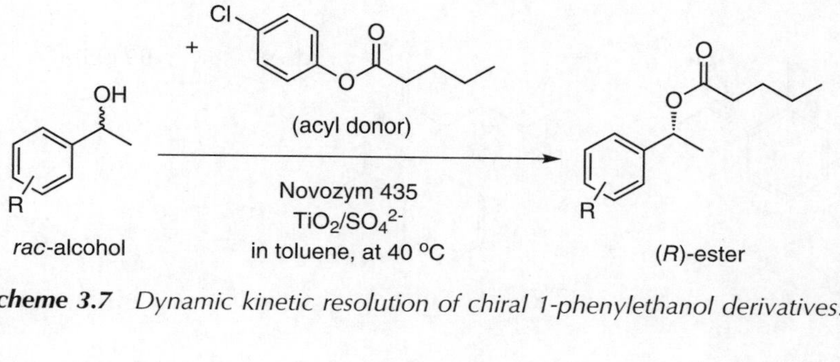

Scheme 3.7 *Dynamic kinetic resolution of chiral 1-phenylethanol derivatives.*

3.4.1.2 Procedure

1. TiCl$_4$ (20 mL) was added to deionized water (500 mL) dropwise at 0 °C and stirred until completely dissolved.
2. The pH of the solution was adjusted to 8.0 with concentrated ammonia water (weight percentage = 25.28%) and the precipitate was filtered and washed several times by deionized water until it reached pH 7.0. The resulting solid, Ti(OH)$_4$, was dried at 60 °C.
3. The dried Ti(OH)$_4$ (2.0 g) was added to H$_2$SO$_4$ (0.5 M, 20 mL) and stirred for 2 hours, and the acid-treated solid was filtered and dried at 60 °C for another 1 hour.
4. The solid was then roasted at 400 °C for 5 hours in a muffle furnace to afford TiO$_2$/SO$_4{}^{2-}$.

3.4.2 Procedure 2: DKR of Chiral Secondary Aromatic Alcohol

3.4.2.1 Materials and Equipment

- Racemic 1-aryl ethanol (200 μmol)
- 4-Chlorophenyl pentanoate (600 μmol)
- Toluene (2 mL)
- Novozym 435 (immobilized *Candida antarctica* lipase B from Novozym, 10 mg)

3.4.2.2 DKR Procedure

Racemic alcohol (1–15, 100 mM), acyl donor (4-chlorophenyl pentanoate, 300 mM), TiO$_2$/SO$_4{}^{2-}$ (50 mg mL^{-1}), and Novozym 435 (10 mg mL^{-1}) were mixed with toluene (2 mL) in a 10 mL tube, shaken at 40 °C at 200 rpm for 10~14 hours. The results for each alcohol are shown in Table 3.3.

Table 3.3 DKR results for chiral secondary aromatic alcohols.

Entry	Substrate	Substrate	ee$_p$ (%)	Conversion[a] (Yield[b])	Time (h)
1			>99	>99 (95)	10
2			>99	>99 (95)	10

Table 3.3 (*Continued*)

Entry	Substrate	Substrate	ee_p (%)	Conversion[a] (Yield[b])	Time (h)
3	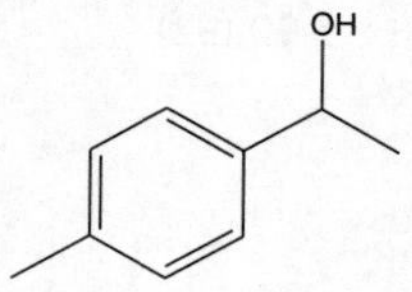	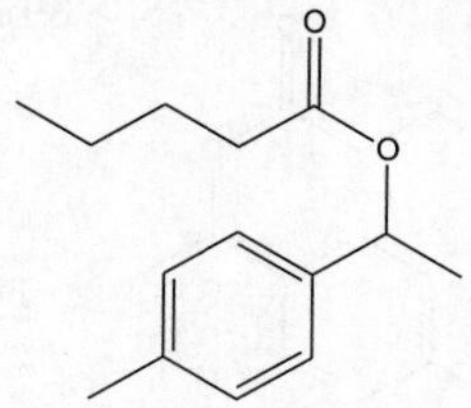	>99	>99 (95)	10
4	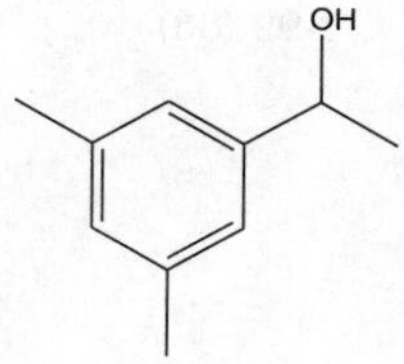	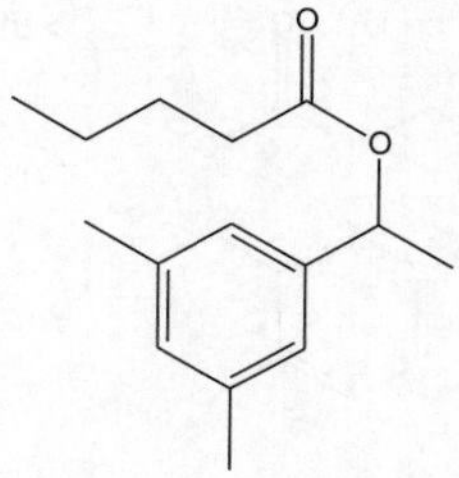	>99	>99 (95)	10
5	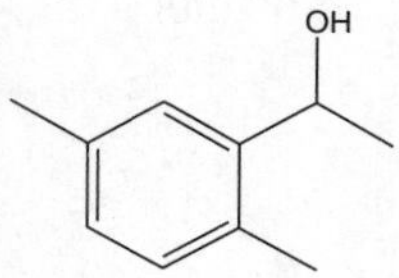	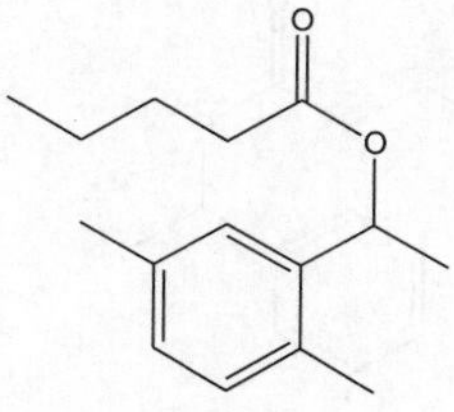	>99	>99 (94)	10
6	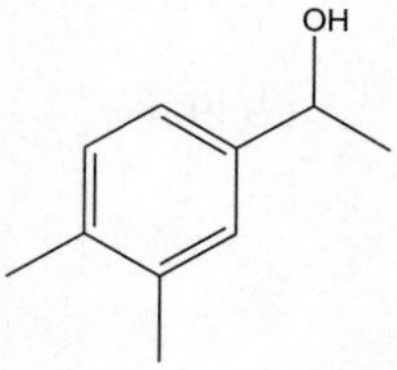	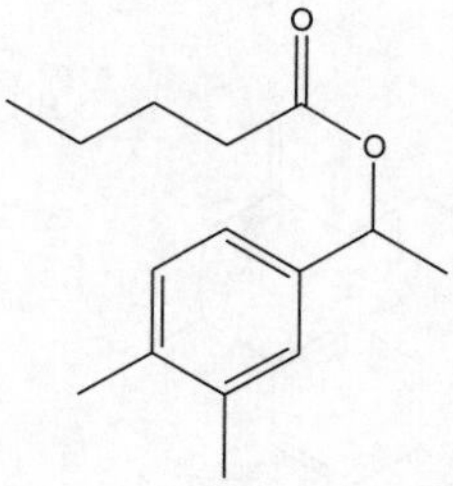	>99	>99 (90)	10
7	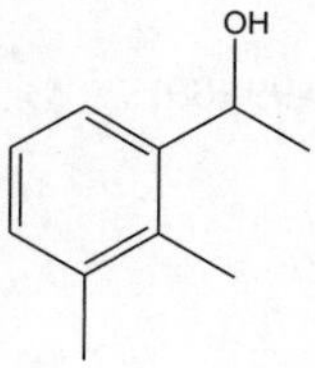	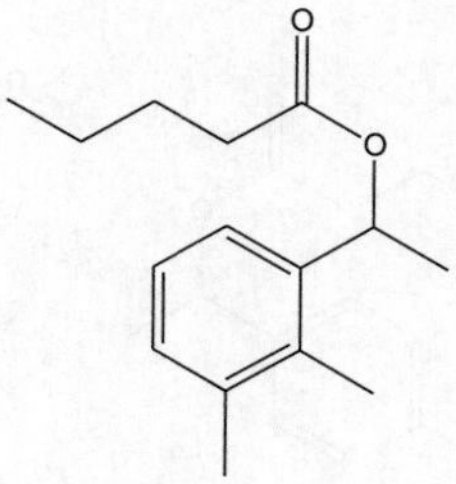	>99	>99 (92)	10

(*continued*)

Table 3.3 *(Continued)*

Entry	Substrate	Substrate	ee_p (%)	Conversion[a] (Yield[b])	Time (h)
8			97.5	>99 (92)	14
9			>99	>99 (88)	14
10			>99	>99 (90)	14
11			89.7	>99 (90)	14
12			>99	>99 (89)	14

Table 3.3 (*Continued*)

Entry	Substrate	Substrate	ee_p (%)	Conversion[a] (Yield[b])	Time (h)
13			>99	>99 (92)	12
14			>99	>99 (88)	14
15			>99	>99 (90)	14

[a] Conversion ratio (%) = [converted substrate (mol)]/[total initial substrate (mol)] × 100%.
[b] Isolated yield (%) = [isolated product (mol)]/[total initial substrate (mol)] × 100%. The enantiomerically pure DKR products were isolated from the reactant by column chromatography using n-hexane: ethyl acetate = 10 : 1 (v/v) as developing agent.

3.4.2.3 Analysis of the DKR Product

All substrates (*rac*-alcohols) and products (enantiopure esters) were analyzed using an HP6890 GC equipped with FID and a Supelco Beta DEX 120 fused silica capillary column (30 m × 0.25 mm × 0.25 μm film thickness). For entries 1–10 in Table 3.3, the column temperature was 150 °C, and for entries 11–16, it was 190 °C. The temperatures of the injector and detector were 240 and 250 °C, respectively. Hydrogen was used as carrier gas.

3.4.2.4 Purification of the DKR Products

The enantiomerically pure DKR products were isolated from the reactant by column chromatography using *n*-hexane: ethyl acetate = 10 : 1 (v/v).

3.4.3   Conclusion

A highly efficient DKR of secondary aromatic alcohols using the solid super acid TiO_2/SO_4^{2-} as racemization catalyst, coupled with enzymatic kinetic resolution, was developed. During the reaction, the solid super acid was found to be competent and sufficiently stable, and the system has achieved high ee_p and conversion ratio with several aromatic secondary alcohols.

3.5   Identification of New Biocatalysts for the Enantioselective Conversion of Tertiary Alcohols

Robert Kourist,[1] Susanne Herter,[2] and Uwe T. Bornscheuer[2]
[1]*Junior Research Group for Microbial Biotechnology, Ruhr-University Bochum, Germany*
[2]*Institute of Biochemistry, Department of Biotechnology and Enzyme Catalysis, University of Greifswald, Germany*

Chiral tertiary alcohols may be synthesized by enantioselective enzymatic hydrolysis of their esters, catalyzed by esterases and lipases (Scheme 3.8) [30,31]. While most esterases and lipases show poor activity toward tertiary alcohol esters, hydrolases bearing a so-called "GGG (A)X" motif in their active centers are mostly active and often have excellent enantiose-lectivity [32]. This makes identification of new esterases bearing this motif straightfor-ward [33]. However, the substrate scope of enzymes haboring the GGG(A)X motif is rather narrow, hence making simple methods (ODER approaches) toward the identification of new biocatalysts highly desirable. Microorganisms bearing hydrolytic enzymes with activity toward tertiary alcohols can be identified by two strategies. Following the first strategy, we screened 47 bacterial strains from a strain collection of crude-oil based hydrocarbon degrading microorganisms for their hydrolytic activity toward the model compound *tert*-butyl acetate. The second approach is based on the ability of bacteria to utilize acetate as the sole source of carbon and energy [34].

3.5.1   Materials and Equipment

- Sodium phosphate buffer (100 mM, pH 7.5)
- MMB mineral salt medium ($NH_4H_2PO_4$ 5 g.L^{-1}, K_2HPO_4 2.5 g.L^{-1}, $MgSO_4 \cdot 7H_2O$ 0.5 g. L^{-1}, NaCl 0.5 g.L^{-1}, K_2SO_4 0.46 g.L^{-1}, $CaCl_2$ 0.07 g.L^{-1}, $FeCl_3 \cdot 6H_2O$ 2 mg.L^{-1}, H_2BO_3 0.5 mg.L^{-1}, $CuSO_4 \cdot 5H_2O$ 0.1 mg.L^{-1}, KI 0.1 mg.L^{-1}, $CoCl_2$ 0.1 mg.L^{-1}, $MnSO_4 \cdot 5H_2O$

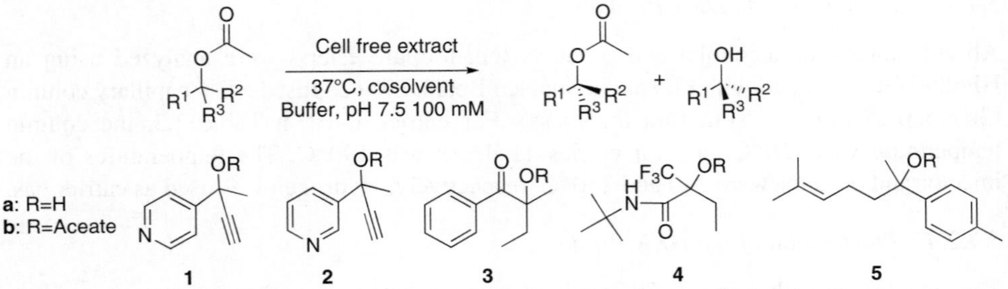

Scheme 3.8   *Esterase-catalyzed kinetic resolution of tertiary alcohols.*

$0.4\,\mathrm{mg.L^{-1}}$, $ZnSO_4\cdot7H_2O$ $0.4\,\mathrm{mg.L^{-1}}$, $NaMoO_3$ $0.2\,\mathrm{mg.L^{-1}}$, adjusted to pH 6.3 using aqueous NaOH)

- Filter-sterilized aqueous solution of *tert*-butyl acetate (0.01%, 0.1%, 0.2%, 0.3% v/v)
- Filter-sterilized aqueous solution of tributyrin
- Bromothymol blue
- Racemic tertiary alcohol acetates (for details of the synthesis, see references [34–36])
- Racemic tertiary alcohols as standards for chiral analytics
- Hydrodex γ-TBDAc Octakis-(2,3-di-O-acetyl-6-O-t-butyldimethyl-silyl)-γ-cyclodextrin column from Macherey-Nagel (Düren, Germany)
- MMB agar plates (1.8% w/v agar concentration, pH 6.8)
- Petri dishes with nutrient agar (pH 7.2)
- Fastprep system (Qbiogene Inc., Carlsbad, USA) for cell disruption
- GC-14A gas chromatograph (Shimadzu, Tokyo, Japan)

3.5.2 Enrichment Selection

1. Samples were taken from activated sewage sludge of a local waste water treatment plant (Greifswald, Germany).
2. Samples (4 mL) were transferred to 500 mL Erlenmeyer flasks containing 100 mL of MMB mineral salt medium. Cultures were supplemented with different concentrations of filter-sterilized *tert*-butyl acetate and sealed with Parafilm to prevent substrate evaporation. Cultures were incubated for 7 days at 30 °C and 130 rpm.
3. Culture supernatants were repeatedly (4x) recultured under the same conditions. They were finally plated out on nutrient agar plates.
4. Isolated pure cultures were retested by streaking them out on MMB agar plates. To prevent evaporation of the model substrate, the plates were incubated in closed glass vessels under a gas phase saturated with *tert*-butyl acetate.
5. Strains were identified using 16S-RNA sequencing [34].

3.5.3 Functional Screening

1. Strains from the enrichment selection and 47 bacterial strains from the strain collection of the Institute of Microbiology, Department of Applied Microbiology, University of Greifswald, Germany with known ability to degrade aromatic compounds and hydrocarbons were cultivated on nutrient agar plates at 30 °C for 24 or 72 hours.
2. For the induction of esterase production, solutions of *tert*-butyl acetate or tributyrin were added. Cells were scraped off the plates and disrupted using the Fastprep system at $4\,\mathrm{m.s^{-1}}$ for 40 seconds with a 5-minute cooling interval. Cell debris was removed by centrifugation (4 °C, 15 minutes, $13\,000\times g$) and the cell-free extracts were used for biocatalysis experiments.
3. Hydrolysis of *tert*-butyl acetate results in a shift of pH that can be detected using bromothymol blue. Cell-free extract (300 μL) was added to a solution of *tert*-butyl acetate (30 mM) in water supplemented with the pH indicator bromothymol blue (0.1% w/v). After 1 hour of incubation at 37 °C and 1100 rpm, color change from blue-green to yellow indicated enzymatic activity due to the release of acetic acid.

3.5.4 Determination of Enantioselectivity

1. Acetates of racemic tertiary alcohols dissolved in DMSO (75 μL, 25 mM) and phosphate buffer (825 μL) were added to cell-free extract in phosphate buffer (600 μL, 100 mM,

pH 7.5). The reaction mixture was shaken in a Thermoshaker (Eppendorf, Germany) at 37 °C for a certain time (1, 4, 8, and 24 hours), after which 300 μL samples were taken. The samples were extracted twice with dichloromethane (400 μL). The combined organic fractions were dried over Na_2SO_4, filtered, and transferred to a GC vial. Enantioselectivity and conversion were calculated according to Chen *et al.* [37]. The *E*-values for biocatalysis with tertiary alcohol esters were calculated based on ee_S and ee_P.

3.5.5 Analytical Methods

The *ee* was determined using GS on a GC-14A gas chromatograph (Shimadzu, Tokyo, Japan) equipped with a Hydrodex γ-TBDAc Octakis-(2,3-di-O-acetyl-6-O-t-butyldimethyl-silyl)-γ-cyclodextrin column from Macherey-Nagel (Düren, Germany), as described previously [34–36].

3.5.6 Conclusion

Functional screening resulted in 10 gram-positive strains with hydrolytic activity toward esters of tertiary alcohols. The enrichment selection from a sewage sludge sample of a waste water treatment plant resulted in 14 active strains, of which 5 were Gram-negative. Four strains showed moderate to high enantioselectivity ($E = 10$–70) toward several substrates, which underlines the usefulness of this method for the identification of enantioselective biocatalysts for the kinetic resolution of tertiary alcohols. Functional screening and enrichment selection using *tert*-butyl acetate thus yielded strains bearing enzymes with activity toward much larger tertiary alcohols. The method is generally applicable and is directly transferable toward other substrates.

3.6 Enzyme-Catalyzed Hydrolysis of Bicycloheptane Diester to Monoester

Zhiwei Guo, Michael Kwok Y. Wong, Matthew R. Hickey, Bharat P. Patel, Xinhua Qian, and Animesh Goswami
Chemical Development, Bristol-Myers Squibb, USA

Diacid formation is a major problem in the conventional chemical hydrolysis of diesters. Hydrolysis of dimethyl bicyclo[2.2.1]heptane-1,4-dicarboxylate **1** by base results in the formation of significant levels of diacid (Scheme 3.9) [38–44]. The downstream isolation

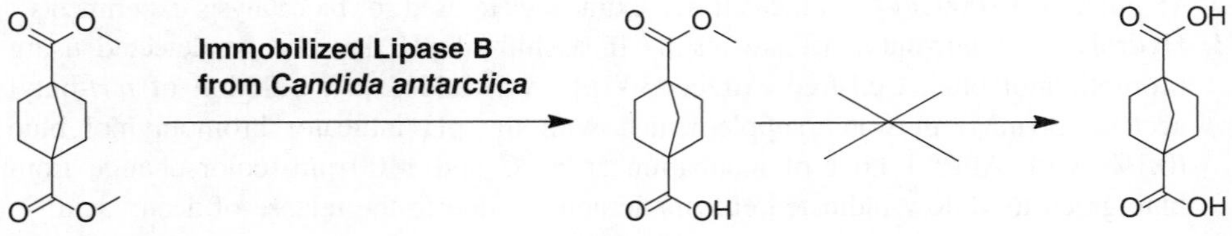

Scheme 3.9 *Hydrolysis of dimethyl bicyclo[2.2.1]heptane-1,4-dicarboxylate.*

process for the separation of remaining diester **1**, desired monoester **2**, and undesired diacid **3** requires careful pH control and is tedious. Monoester **2** is a building block of many potential therapeutic candidates [42–44]. Enzymatic hydrolysis is a powerful tool for the selective hydrolysis of esters [45,46]. Hydrolysis and desymmetrization of diesters to monoesters by porcine liver esterase [47–51], porcine pancreatic lipase [48,49], lipase from *Pseudomonas* species [50], lipase from *Candida antarctica* [49–52], and other lipases [53,54] has been reported extensively in the literature. We have recently reported [55] results of our work on the selective enzyme catalyzed hydrolysis of dimethyl bicyclo[2.2.1] heptane-1,4-dicarboxylate **1**. This section provides experimental details of the hydrolysis of bicyclo[2.2.1]heptane-1,4-dicarboxylate **1** to its corresponding monoester **2** by the immobilized lipase B from *Candida antarctica*.

3.6.1 Materials and Equipment

- Diester **1** purchased from commercial suppliers and prepared on a large scale, as described in the literature [40–44]
- Immobilized lipase B from *Candida antarctica* (Novozym 435) from Novozymes, Inc.
- Other chemicals of reagent grade, purchased from common commercial suppliers
- Sodium phosphate monobasic (NaH_2PO_4)
- Sodium phosphate dibasic (Na_2HPO_4)
- Sodium hydroxide (NaOH)
- Sulfuric acid (H_2SO_4)
- Methyl *t*-butyl ether (MTBE)
- Jacketed glass reactor (1L)
- Overhead mechanical stirrer
- pH Meter
- Water circulator
- Glass Buchner funnel with fritted disc (medium porosity)
- Buchner flask (1 L)
- Separatory funnel (2 L)
- Laboratory-scale rotary evaporator fitted with appropriate vacuum pump and cold-water circulator
- Vacuum oven
- Vacuum pump

3.6.2 Preparation of Monoester 2 by Immobilized Lipase B from *Candida antarctica* (Novozym 435)-Catalyzed Hydrolysis of the Diester 1

1. A 1 L jacketed reactor equipped with an overhead mechanical stirrer, a pH meter, and a water circulator to control heating and cooling was set up.
2. Immobilized lipase B from *Candida antarctica* (Novozym 435) (600 mg) was added.
3. Diester **1** (30 g, 141.4 mmol, HPLC area 41% assay 87%, slightly yellowish color) was added.Note: This was a crude stream and contained several impurities originating from its synthesis. There were a major impurity of 2,6-di-tert-butyl-4-methylphenol (BHT) of 35% area, an unknown impurity of 6% area and many unknown impurities of 2% area or less in HPLC. BHT and, probably, other impurities have stronger UV absorption, causing their amounts to be overestimated from HPLC areas.

4. 0.2 M sodium phosphate, pH 7.0 buffer (600 mL) was added. Note: The buffer was prepared as follows: NaH_2PO_4, 9.36 g (78 mmol, FW 119.98); Na_2HPO_4, 17.32 g (122 mmol, FW 141.96). Distilled water was added to a total volume of 1000 mL. The observed pH was 7.01 at 22 °C.

5. Stirring was maintained at 200 rpm. The batch was slowly heated (to prevent overheating) to 55 °C and was maintained at 55 ± 2 °C (batch temperature) throughout the reaction.

6. The pH was adjusted to 7.0 with aqueous 5 N NaOH every half hour in the first 4 h and then every hour in the following 4 h. The pH was in the range of 6.8–7.1 in the first 8 h. A total of 19.9 g (d = 1.184, 16.8 mL, 84.0 mmol) of 5 N NaOH was added to maintain the constant pH in the first 8 h. Note: The pH dropped during the reaction, due to the generation of acid. The enzyme activity decreases when the pH falls below 7.0, hence the need for this adjustment. Too quick addition of base may result in local non-enzymatic hydrolysis to diacid side product. This was especially important near the end of the reaction, when only a small amount of diester remained. A lower base concentration (e.g., 1 N NaOH) can be used for pH adjustment at high conversion (e.g., ≥80%).

7. After 8 h, the reaction was allowed to continue overnight without pH adjustment. After 24 h, the pH was ~6.13. It was adjusted back to 7.0 by slow addition of 5 N NaOH (12.4 g). After 28 hours, the pH was again adjusted to 7.0 by addition of 5 N NaOH (1.2 g). A total of 33.5 g, 28.3 mL, 141.5 mmol NaOH (1 equivalent) was added to maintain the pH during the entire reaction.

8. In order to monitor the progress of the reaction, samples (200 μL) were taken periodically using a wide-bore pipette tip, mixed with 1 N HCl (200 μL) and acetonitrile (2 mL), filtered, and analyzed by HPLC. The HPLC results and the amount of NaOH added to maintain constant pH showing the reaction progress are given in Table 3.4.

9. When the HPLC analysis indicated no more than 3% area of starting material (conversion no less than 97% area), the reaction was considered complete. Note: The diacid formation was very slow under the reaction conditions. The diacid would not have increased significantly even if the reaction had been allowed to continue for a few more hours.

10. After the reaction was completed (after 32 h), the reactor vessel was disconnected (stirrer, pH meter, and water circulator) and stored in a cold room (4 °C) for 16 h (overnight). Note: This would not have been necessary if the work-up could have been done immediately.

11. The reactor vessel was reconnected (stirrer, pH meter, and water circulator) and the mixture was warmed to 20 °C under mild stirring. The mixture was adjusted to pH 8.0 by the addition of 5 N NaOH (5.5 g).

Table 3.4 Reaction progress.

Time (h)	4	8	24	26	28	30	32	50[a]
5 M NaOH total (g)	8.8	19.9	32.3	—	33.5	—	33.5	39.0
Diester 1%	72.6	46.1	6.2	4.9	3.2	1.7	1.3	0.9
Monoester 2%	27.4	53.9	93.8	94.6	96.2	97.7	98.0	98.3
Diacid 3%	< 0.5	< 0.5	< 0.5	0.5	0.6	0.7	0.7	0.8

[a] Sample from reaction mixture after overnight storage and adjustment to pH 8 before filtration.

12. The mixture was filtered under vacuum through a glass Buchner funnel with fritted disc (medium porosity) fitted with a Buchner flask. The residue was washed with 0.2 M phosphate buffer, pH 8.0 (50 mL), prepared as follows: NaH_2PO_4, 1.27 g (10.6 mmol, FW 119.98); Na_2HPO_4, 26.89 g (189.4 mmol, FW 141.96); distilled water added to a total volume of 1000 mL.

13. The combined filtrate and washing was extracted twice with MTBE (300 mL + 200 mL) to remove unreacted diester **2** and other impurities coming from the starting diester batch. The aqueous phase was collected to isolate the product monoester **1** as described in Step 14 onwards. Note: The unreacted starting material and impurities were obtained from the MTBE phase as follows. The combined MTBE phase was used to wash the immobilized enzyme in the Buchner funnel. The MTBE washing was concentrated to dryness under reduced pressure and further dried in a vacuum oven at 30 °C overnight to give 3.86 g residue, 12.9 wt% of starting diester input. HPLC analysis showed the impurities mostly came from the starting diester, as well as a very small amount of unreacted diester **1** 0.76% area and product monoester **2** 0.14 % area. The immobilized enzyme was collected and dried in a vacuum oven at room temperature overnight to give 600 mg of recovered immobilized enzyme.

14. The aqueous phase from Step 13 was immediately adjusted to pH 2.5 with 5 M H_2SO_4 (35.8 g). Note: Steps 11 to 14 were carried out as quickly as possible to minimize the time at pH 8. Subjecting the monoester **2** to pH 8 for longer time raises the possibility of causing over-hydrolysis of monoester **2** to diacid **3**.

15. The aqueous layer was extracted with MTBE (300 mL) and the MTBE phase was separated from the aqueous phase.

16. The aqueous phase was again extracted with MTBE (300 mL) and the two phases were again separated. Note: HPLC of the aqueous phase showed 2.1% monoester product left after the first extraction and no product left after the second extraction.

17. The MTBE layers obtained in Steps 15 and 16 were combined and washed with water (200 mL).

18. Solvent was removed from the MTBE layer under reduced pressure in a rotary evaporator at 30 °C to provide monoester **2** as a solid.

19. The solid was dried overnight in a vacuum oven at 30 °C. Monoester **2** was obtained as a white solid 23.41 g, (as is yield 83.5%). HPLC (Conc 5 mg mL^{-1} in acetonitrile) showed excellent quality of monoester **2**, 91.9% area, no detectable diester **1** or diacid **3**, no detectable BHT, and an unknown impurity 6.8% area (the same unknown impurity was present with 6% area in the starting diester input). The HPLC assay of the product monoester **2** was 96.3%, compared against an authentic standard. The yield corrected for the purity of both starting material and product was 92.5%. Monoester **2** was used successfully in the next synthetic step.

3.6.3 NMR Data

Solutions for nuclear magnetic resonance spectroscopy (NMR) were prepared in $CDCl_3$ or acetic acid-d4.

NMR spectra were recorded on a Bruker-500 instrument.

3.6.3.1 Diester 1 (CDCl$_3$)

^{1}H δ 3.68 (s, 6H), 2.02 (m, 4H), 1.90 (s, 2H), 1.68 (m, 4H) ppm
^{13}C δ 175.33 (2C), 52.56 (2C), 51.63 (2C), 44.94, 32.94 (4C) ppm

3.6.3.2 Monoester 2 (CDCl₃)

^{1}H δ 12.5-10.5 (COOH), 3.70 (s, 3H), 2.08 (m, 4H), 1.95 (s, 2H), 1.72 (m, 4H) ppm
^{13}C δ 181.32, 175.27, 52.80, 52.43, 51.75, 44.96, 32.95 (2C), 32.85 (2C) ppm

3.6.3.3 Diacid 3 (Acetic Acid-d4)

^{1}H δ 11.54 (COOH), 2.10 (m, 4H), 1.98 (s, 2H), 1.74 (m, 4H) ppm
^{13}C δ 181.70 (2C), 53.70 (2C), 45.59, 33.69 (4C) ppm

3.6.4 HPLC Method

The conversion and the amounts of substrate and product in the reaction mixture were determined by the relative area percentage (% area) of the HPLC peaks, without any correction. HPLC area versus concentration standard curves of authentic standards were used to assay various batches of diesters and monoesters.

The reversed-phase HPLC method was used to monitor the hydrolysis of dimethyl bicyclo[2.2.1]heptane-1,4-dicarboxylate **1** using a Waters XTerra RP-18 column (3.5 μm, 150 × 4.6 mm) at ambient temperature with UV detection at 210 nm and a flow rate of 1 mL. min^{-1} of solvent A (0.05% TFA in water: methanol 80 : 20) and solvent B (0.05% TFA in acetonitrile: methanol 80 : 20), with a gradient from 0 to 100% solvent B over 20 minutes. The retention times were 8.0 minutes for diacid **3**, 10.1 minutes for monoester **2**, and 12.0 minutes for diester **1**.

3.7 Double Mutant Lipase with Enhanced Activity and Enantioselectivity for Bulky Secondary Alcohols

Tadashi Ema, Shusuke Kamata, Yasuko Nakano, Daiki Yoshida, and Takashi Sakai
Division of Chemistry and Biotechnology, Graduate School of Natural Science and Technology, Okayama University, Japan

Secondary alcohols with bulky substituents on both sides of the hydroxy group are poor substrates for most lipases. We altered a *Burkholderia cepacia* lipase on the basis of a transition-state model (Figure 3.1) [56,57]. Among several variants, a I287F/I290A double mutant was the best biocatalyst, showing a high conversion and a high *E* value (>200) for a poor substrate, 1-phenyl-1-hexanol **1a**, for which the wild-type enzyme showed a low conversion and a low *E* value (5). This enhancement of the activity and enantioselectivity of the variant results from the cooperative action of the two mutations: Phe287 makes favorable contact (CH/π interactions) with the alkyl chain of (*R*)-**1a**, while the I290A mutation creates a space to accommodate (*R*)-**1a** (Figure 3.1), both of which accelerate the acylation of (*R*)-**1a**. This double mutant could also be applied to the DKR of **1a.**

3.7.1 Procedure 1: Kinetic Resolution of Secondary Alcohol 1a (Scheme 3.10)

3.7.1.1 Materials and Equipment

• Immobilized I287F/I290A double mutant lipase (0.5% w/w enzyme/Toyonite-200M, 700 mg)

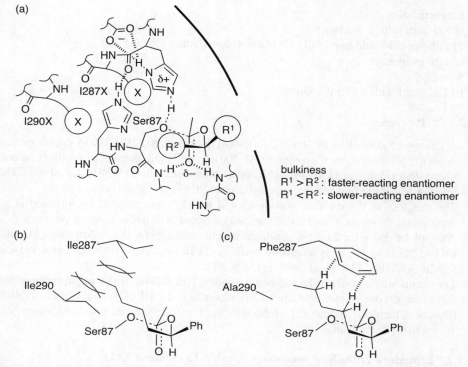

Figure 3.1 *(a) Transition-state model rationalizing enantioselectivity in the lipase-catalyzed kinetic resolution of secondary alcohols (showing mutated residues 287 and 290). (b) Steric repulsion in the transition state for the wild-type enzyme-catalyzed acylation of (R)-**1a**. (c) CH/π interactions in the transition state for the I287F/I290A double mutant-catalyzed acylation of (R)-**1a**.*

- Alcohol **1a** (89.1 mg, 0.50 mmol)
- Vinyl acetate (93 µL, 1.0 mmol)
- Dry *i*-Pr₂O (5 mL)
- Hexane
- Ethyl acetate
- 2-Propanol
- Molecular sieves 3 Å (three pieces)
- Test tube with a rubber septum
- Syringes

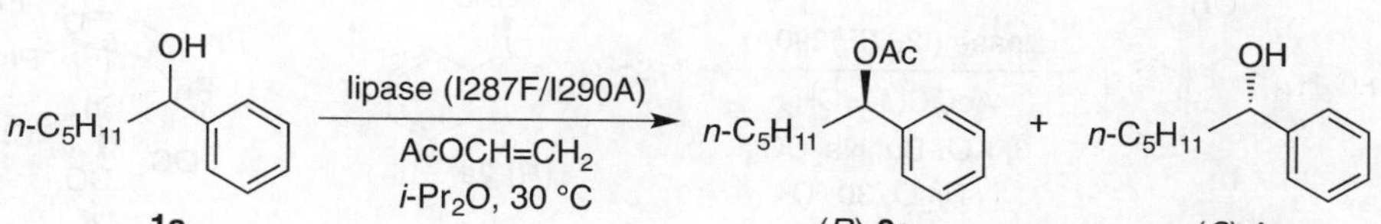

Scheme 3.10 *Kinetic resolution of secondary alcohol **1a**.*

- Magnetic stirrer
- Water bath with a thermostat
- Thin-layer chromatography (TLC) plates (Merck silica gel 60 F$_{254}$)
- Rotary evaporator
- Silica gel
- HPLC fitted with a chiral column

3.7.1.2 Procedure

1. A mixture of alcohol **1a** (89.1 mg, 0.50 mmol), immobilized I287F/I290A double mutant lipase (0.5% w/w enzyme/Toyonite-200M, 700 mg), and molecular sieves 3 Å (three pieces) in dry *i*-Pr$_2$O (5 mL) in a test tube with a rubber septum was stirred at 30 °C for 30 mins. The reaction was started by addition of vinyl acetate (93 µL, 1.0 mmol) via a syringe.
2. The progress of the reaction was monitored by TLC and stopped by filtration at an appropriate conversion, and the filtrate was concentrated under reduced pressure.
3. Alcohol **1a** and ester **2a** were separated by chromatography over silica using hexane/ EtOAc (50 : 1 to 10 : 1) as an eluent to afford (*S*)-**1a** (44.2 mg, 50% yield, 98% ee) and (*R*)-**2a** (47.9 mg, 43% yield, 99% ee). *E* > 200.
4. The enantiomeric purity of **1a** was determined by HPLC fitted with a chiral column, and that of **2a** was determined after chemical hydrolysis to **1a**. HPLC for **1a**: Chiralcel OB-H (Daicel Chemical Industries), hexane/*i*-PrOH = 9 : 1, 0.5 mL.min^{-1}, 254 nm, (*S*) 10.6 mins, (*R*) 12.2 mins.

3.7.2 Procedure 2: DKR of Secondary Alcohol 1a (Scheme 3.11)

3.7.2.1 Materials and Equipment

- Immobilized I287F/I290A double mutant lipase (0.5% w/w enzyme/Toyonite-200M, 400 mg)
- Alcohol **1a** (190 µL, 1.0 mmol)
- Isopropenyl acetate (165 µL, 1.5 mmol)
- Ru catalyst **3** (63.8 mg, 0.100 mmol)
- KO*t*-Bu (14 mg)
- Na$_2$CO$_3$ (106 mg)
- Dry THF (250 µL)
- Hexane
- Ethyl acetate
- 2-Propanol

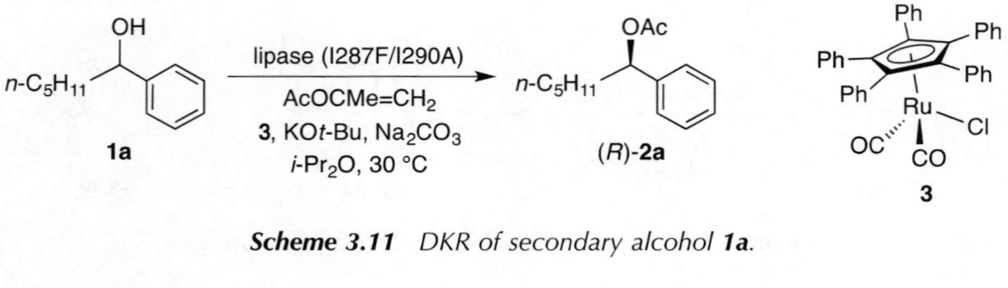

Scheme 3.11 *DKR of secondary alcohol* ***1a***.

- Dry *i*-Pr$_2$O (2 mL)
- Silica gel
- Molecular sieves 4 Å (20 pieces)
- Argon gas
- Schlenk flask
- Syringes
- Magnetic stirrer
- Water bath with a thermostat
- TLC plates
- Rotary evaporator
- Silica gel
- HPLC fitted with a chiral column

3.7.2.2 Procedure

1. A solution of KO*t*-Bu (0.5 M in dry THF, 250 µL) was added to a Schlenk flask, and THF was removed under reduced pressure.
2. Immobilized I287F/I290A double mutant lipase (0.5% w/w enzyme/Toyonite-200M, 400 mg), Ru catalyst **3** (63.8 mg, 0.100 mmol), Na$_2$CO$_3$ (106 mg, 1.00 mmol), and molecular sieves 4 Å (20 pieces) were added to the flask. The flask was evacuated and filled with Ar.
3. Dry *i*-Pr$_2$O (2 mL) was added, and the mixture was stirred at 30 °C for 15 mins. Alcohol **1a** (190 µL, 1.0 mmol) was added, and the mixture was stirred for 15 mins. The reaction was started by addition of isopropenyl acetate (165 µL, 1.5 mmol). The mixture was stirred at 30 °C for 3 days.
4. The reaction was stopped by filtration, and the mixture was concentrated under reduced pressure. Purification by chromatography over silica using hexane/EtOAc (100 : 1) as eluent gave (*R*)-**2a** (196 mg, 88% yield, 95% *ee*).

3.7.3 Conclusion

Secondary alcohols with bulky substituents on both sides of the hydroxy group are inherently poor substrates for most lipases, which has been an unresolved weakness in the application of lipases in resolutions of alcohol substrates. The synergic effects of only two mutations (I287F/I290A) dramatically enhanced both catalytic activity and enantioselectivity for these poor substrates. The I287F/I290A double mutant showed broad substrate scope (Table 3.5), and

Table 3.5 *Substrate scope of the I287F/I290A double mutant and wild-type enzyme.[a]*

Entry	Substrate **1**	Time (h)	I287F/I290A		Wild-type	
			c (%)[b]	E [c]	c (%)[d]	E
1	OH **1b**	7	39	>200	0	–
2	MOMO OH **1c**	22	46	143	6	–

(*continued*)

Table 3.5 (*Continued*)

Entry	Substrate **1**	Time (h)	I287F/I290A c (%)[b]	E[c]	Wild-type c (%)[d]	E
3	**1d**	6.5	41	>200	3	–
4	**1e**	9	35	147	0	–
5	**1f** CF$_3$	54	45	>200	4	–
6	**1g** OMe	7	41	>200	2	–
7	**1h** OMOM	7	42	>200	3	–
8	**1i**	7	41	>200	0	–

[a] Conditions: immobilized lipase (200 mg, 0.5% w/w enzyme/Toyonite-200 M), **1** (0.50 mmol), vinyl acetate (1.0 mmol), molecular sieves 3 Å (three pieces), dry *i*-Pr$_2$O (5 mL), 30 °C.
[b] Conversion calculated from $c = ee(\mathbf{1})/(ee(\mathbf{1}) + ee(\mathbf{2}))$.
[c] Calculated from $E = \ln[1 - c(1 + ee(\mathbf{2}))]/\ln[1 - c(1 - ee(\mathbf{2}))]$.
[d] Conversion calculated from ^{1}H-NMR.

this variant could be applied to the DKR of various bulky secondary alcohols for which the wild-type enzyme showed little or no activity [56,57]. We expect that this variant will be useful in the synthesis of chiral intermediates that cannot be prepared with other lipases.

3.8 Stereoselective Synthesis of β-Amino Acids by Hydrolysis of an Aryl-Substituted Dihydropyrimidine by Hydantoinases

Ulrike Engel, Christoph Syldatk, and Jens Rudat
Institute of Process Engineering in Life Sciences, Section II: Technical Biology, Karlsruhe Institute of Technology, Germany

The hydantoinase process is a well-established DKR for the industrial production of non-canonical α-amino acids via a cascade reaction of two or three enzymes (Figure 3.2a). The process is mainly carried out using whole-cell catalysis with wild-type or recombinant strains containing the particular hydantoinase and carbamoylase, ideally accompanied by an appropriate racemase leading to near quantitative yields and very high *ee* values.

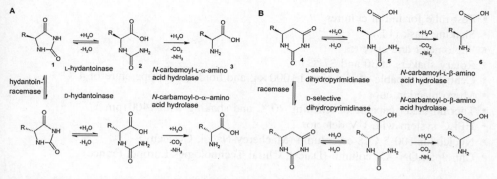

Figure 3.2 *(a) Industrially applied hydantoinase process, starting with racemic hydantoins 1 and leading to chiral α-amino acids 3 via N-carbamoyl-α-amino acids 2. (b) Modified process, starting with racemic dihydropyrimidines 4 and leading to β-amino acids 6 via N-carbamoyl-β-amino acids 5.*

By applying substituted dihydropyrimidines as substrates instead of hydantoins, this process might be useful for the synthesis of β-amino acids (Figure 3.2b), which are important building blocks for pharmaceuticals and fine chemicals [58], and which until now have been hard to access by asymmetric synthesis [59]. Several hydantoinases have been shown to cleave the six-membered dihydropyrimidine ring system as well, thus acting as dihydropyrimidinases [60]. Both stereoselectivities have been documented and the corresponding genes have been identified [61], so the first step in a modified hydantoinase process leading to β-amino acids has been rendered possible.

3.8.1 Materials and Equipment

- Phenyldihydrouracil [60] (PheDU, 50 mg, 0.26 mmol)
- *N*-Carbamoyl-β-phenylalanine [60] (*N*CβPhe, 50 mg, 0.24 mmol)
- LB-broth (4 L; 10 g.L^{-1} bacto-tryptone, 5 g.L^{-1} yeast extract, and 5 g.L^{-1} NaCl, pH 7.2)
- Ampicillin stock solution (100 mg.mL^{-1})
- D, L-5-(3-Indolylmethyl)-3-*N*-methylhydantoin (CH_3-IMH; 0.2 g, 0.82 mmol)
- $ZnSO_4 \cdot 7H_2O$ (0.29 g)
- Rhamnose stock solution (100 g.L^{-1})
- Growth medium (1 L; 10 g.L^{-1} glucose, 6.5 g.L^{-1} $(NH_4)_2SO_4$, 0.2 g.L^{-1} $MgSO_4 \cdot 7 H_2O$, 0.02 g.L^{-1} $MnCl_2$ $4 \cdot H_2O$, 0.02 g.L^{-1} $FeSO_4$ $7 \cdot H_2O$, 0.02 g.L^{-1} $CaCl_2$ $2 \cdot H_2O$, 0.28 g.L^{-1} citrate, 335 mL potassium sodium phosphate buffer (with K_2HPO_4 and NaH_2PO_4, 0.2 M; pH 6.8)
- Bacterial strains:
   - *Escherichia coli* JM109/pMW11
   - *Arthrobacter polychromogenes* DSM20136
   - *Bacillus* sp. DSM25052, *Aminobacter* sp. DSM24755
   - *Rhizobium* sp. DSM24917
- Media
- Sterile loops
- Petri dishes

- Test tube for initial cultures
- Shaking flask (1 L)
- Photometer and cuvettes
- Rotary shakers at 30 and 37 °C
- Centrifuge capable of reaching $13\,000 \times g$ and holding a temperature of 4 °C
- Micro reaction cups
- Thermoshaker capable of reaching 70 °C and shaking with 1400 rpm
- HPLC system with UV detector
- Nucleodur 100-5 C18 ec column (Macherey-Nagel, Germany)
- Chiralpak QN-AX column (Daicel, Chiral Technologies Europe, France)

3.8.2 Media and Solution Preparation

1. LB-Amp-broth (1 L LB-broth + 1 mL Ampicillin stock solution).
2. LB$_{Amp/Zn}$-broth (1 L LB-Amp-broth + 0.29 g ZnSO$_4$·7H$_2$O).
3. LBi-broth (1 L LB-broth + 0.1 g CH$_3$IMH).
4. GMi-broth (1 L GM broth + 0.2 g CH$_3$IMH).
5. A 4 mM PheDU stock solution was prepared by dissolving PheDU (25 mg, 0.13 mmol) in potassium phosphate buffer (0.1 M, pH 8, 32.85 mL), assisted by heating for 30 mins at 70 °C.

3.8.3 Initial Cultures

1. A single colony of *E. coli* JM109/pMW11 was inoculated in LB-Amp (5 mL) at 30 °C and 150 rpm overnight.
2. Single colonies of wild-type strains (*Arthrobacter polychromogenes* DSM20136, *Bacillus* sp. DSM25052, *Aminobacter* sp. DSM24755, *Rhizobium* sp. DSM24917) were each inoculated in LB (5 mL) at 30 °C and 140 rpm overnight, or until they reached an optical density at 600 nm (OD$_{600}$) of 1.

3.8.4 Main Cultures

1. The initial culture of *E. coli* JM109/pMW11 (2 mL) was added to LB$_{Amp/Zn}$ (100 mL) in a 1 L shaking flask and incubated at 37 °C and 150 rpm until the OD$_{600}$ was 0.4. Subsequently, the culture was induced by adding rhamnose stock solution (0.2 mL) and incubated for 6 h at 30 °C and 140 rpm.
2. The initial cultures of the wild-type strains of *A. polychromogenes* DSM20136 (4 mL) and *Bacillus* sp. DSM25052 (4 mL) were added to GMi (100 mL) in a 1 L shaking flask and incubated at 30 °C and 140 rpm.
3. The initial culture of the wild-type strain of *Aminobacter* sp. DSM24755 (4 mL) was added to LBi (100 mL) in a 1 L shaking flask and incubated at 30 °C and 140 rpm.
4. The initial culture of the wild-type strain of *Rhizobium* sp. DSM24917 (4 mL) was added to 100 mL LB (100 mL) in a 1 L shaking flask and incubated at 30 °C and 140 rpm.

3.8.5 Preparation of Resting Cells

1. The cells were harvested by centrifugation ($8000 \times g$, 10 mins, 12 °C) after the appropriate induction time (*E. coli* JM109/pMW11) or in the late exponential growth phase (wild-type strains).

2. The supernatant was discarded and the cells were washed twice with potassium phosphate buffer (0.1 M, pH 8), followed by centrifugation.
3. Resting cells were obtained by resuspending the cells in potassium phosphate buffer (0.1 M, pH 8, 10 mL).

3.8.6 Biotransformation Reactions

1. Biotransformation reactions with PheDU were started by adding PheDU stock solution (500 μL, 4 mM) to a resting cell suspension (500 μL).
2. For the PheDU control reaction, PheDU stock solution (4 mM, 500 μL) was added to a potassium phosphate buffer (0.1 M, pH 8, 500 μL).
3. For resting cell control reactions, the appropriate resting cell suspensions (500 μL) were added to a potassium phosphate buffer (0.1 M, pH 8, 500 μL).
4. Reactions were incubated at 40 °C and 1400 rpm for 24 h.
5. Reactions were stopped by centrifugation (13 000 × g, 1 mins). The supernatants were stored at −20 °C until analysis.

3.8.7 Analysis

All substrate and product concentrations were analyzed by HPLC on an Agilent 1100 system (Agilent Technologies, Santa Clara, USA). For non-chiral analysis, a Nucleodur 100-5 C18 ec column (Macherey-Nagel, Germany) was used. The mobile phase for the analysis of PheDU, *N*CβPhe, consisted of 20% MeOH/80% [0.1% v/v H_3PO_4 (pH 3.0, NaOH)]. The flow rate was 0.8 mL.min^{-1}, the temperature was 30 °C, and the detection wavelength was 210 nm. The retention times were 21.1 minutes (*N*CβPhe) and 26.2 minutes (PheDU). For chiral analysis, a Chiralpak QN-AX column (Daicel, Chiral Technologies Europe, France) was applied. The mobile phase was prepared according to the manufacturer's instructions and consisted of 90% MeOH/10% acetic acid 0.2 M with an apparent pH of 6, adjusted with concentrated aqueous ammonia. The flow rate was 0.8 mL.min^{-1}, the temperature was 25 °C, and the detection wavelength was 210 nm. The retention times for the *N*CβPhe enantiomers were 14.3 mins (*S*) and 18.2 mins (*R*).

3.8.8 Conclusion

The recombinantly expressed D-hydantoinase (pMW11; originating from *Arthrobacter crystallopoietes* DSM20117 [62]) and the wild-type strains showed activity toward the tested aryl-substituted dihydropyrimidine PheDU. With a resting cell suspension of *E. coli* JM109/pMW11 (cell dry weight (cdw) = 6 g.L^{-1}), 27% of the PheDU was converted (cv) to *N*CβPhe with an ee_L of 61% within 24 hours of biotransformation reaction. Under the same conditions, the wild-type strains also showed an enantioselective hydrolysis of the dihydropyrimidine substrate: *A. polychromogenes* DSM20136 (cdw = 6 g.L^{-1}; cv = 12%; ee_D = 90%), *Bacillus* sp. DSM25052 (cdw = 15 g.L^{-1}; cv = 29%; ee_D = 51%), *Aminobacter* sp. DSM24755 (cdw = 9 g.L^{-1}; cv = 44%; ee_D = 61%), *Rhizobium* sp. DSM24917 (cdw = 9 g.L^{-1}; cv = 30%; ee_D = 20%). In previous studies, all of the tested wild-type strains were shown to possess hydrolysis activity toward 5′-monosubstitued hydantoins [63]. Thus, these results confirm the hypothesis that hydantoinases are not only able to hydrolyze 5′-monosubstituted heterocyclic hydantoins, five-ring systems, into α-amino acids, but can also hydrolyze 6′-monosubstituted dihydropyrimidines, heterocyclic six-ring systems, into β-amino acids.

References

1. Westheimer, F.H. (1987) *Science*, **235**, 1173–1178.
2. Jones, S. and Smanmoo, C. (2004) *Tetrahedron Letters*, **45**, 1585–1588.
3. Van Herk, T., Hartog, A.F., Van der Burg, A.M., and Wever, R. (2005) *Advanced Synthesis & Catalysis*, **347**, 1155–1162.
4. Van Herk, T., Hartog, A.F., Schoemaker, H.E., and Wever, R. (2006) *The Journal of Organic Chemistry*, **71**, 6244–6247.
5. Babich, L., Hartog, A.F., Van Hemert, L.J.C. *et al.* (2012) *ChemSusChem*, **5**, 2348–2353.
6. Tanaka, N., Hasan, Z., Hartog, A.F. *et al.* (2003) *Organic & Biomolecular Chemistry*, **1**, 2833–2839.
7. Hanson, S.R., Best, M.D., and Wong, C.-H. (2004) *Angewandte Chemie (International Edition)*, **43**, 5736–5763.
8. Knaus, T., Schober, M., Kepplinger, B. *et al.* (2012) *FEBS Journal*, **279**, 4374–4384.
9. Schober, M. and Faber, K. (2013) *Trends in Biotechnology*, **31**, 468–478.
10. Schober, M., Toesch, M., Knaus, T. *et al.* (2013) *Angewandte Chemie (International Edition)*, **52**, 3277–3279.
11. Lukatela, G., Krauss, N., Theis, K. *et al.* (1998) *Biochemistry*, **37**, 3654–3664.
12. Boltes, I., Czapinska, H., Kahnert, A. *et al.* (2001) *Structure*, **9**, 483–491.
13. Schober, M., Gadler, P., Knaus, T. *et al.* (2011) *Organic Letters*, **13**, 4296–4299.
14. Wallner, S.R., Nestl, B., and Faber, K. (2005) *Tetrahedron*, **61**, 1517–1521.
15. Guthrie, J.P. (1978) *Canadian Journal of Chemistry*, **56**, 2342–2354.
16. Schober, M., Knaus, T., Toesch, M. *et al.* (2012) *Advanced Synthesis and Catalysis*, **354**, 1737–1742.
17. Fuchs, M., Toesch, M., Schober, M. *et al.* (2013) *European Journal of Organic Chemistry*, 356–361.
18. Paetzold, J. and Bäckvall, J.-E. (2005) *Journal of the American Chemical Society*, **127**, 17620.
19. Parvulescu, A.N., Jacobs, P.A., and De Vos, D.E. (2007) *Chemistry - A European Journal*, **13**, 2034.
20. Thalén, L.K., Zhao, D., Sortais, J.–B. *et al.* (2009) *Chemistry - A European Journal*, **15**, 3403.
21. Engström, K. and Shakeri, M. (2011) *European Journal of Organic Chemistry*, **10**, 1827.
22. Han, Y., Lee, S.S., and Ying, J. (2006) *Chemistry of Materials*, **18**, 643.
23. Engström, K., Johnston, E.V., Verho, O. *et al.* (2013) *Angewandte Chemie-International Edition*, **52**, 14006.
24. Shakeri, M., Tai, C.–W., Göthelid, E. *et al.* (2011) *Chemistry - A European Journal*, **17**, 13269.
25. Johnston, E.V., Verho, O., Kärkäs, M.D. *et al.* (2012) *Chemistry - A European Journal*, **18**, 12202.
26. Cheng, Y.-M., Xu, G., Wu, J.-P. *et al.* (2010) *Tetrahedron Letters*, **51**, 2366.
27. Xu, G., Chen, Y.-J., Wu, J.-P. *et al.* (2011) *Tetrahedron: Asymmetry*, **22**, 1373.
28. Xu, G., Wang, L., Chen, Y.-J. *et al.* (2013) *Tetrahedron Letters*, **54**, 5026.
29. Wu, Jian-Ping, Meng, Xiao, Wang, Liang *et al.* (2014) *Tetrahedron Letters*, **55**, 5129.
30. Kourist, R., Dominguez de Maria, P., and Bornscheuer, U.T. (2008) *ChemBioChem*, **9**, 491–498.
31. Kourist, R. and Bornscheuer, U.T. (2011) *Applied Microbiology and Biotechnology*, **91**, 505–517.
32. Henke, E., Bornscheuer, U.T., Schmid, R.D., and Pleiss, J. (2003) *ChemBioChem*, **4**, 485–493.
33. Kourist, R., Hari Krishna, S., Patel, J.S. *et al.* (2007) *Organic and Biomolecular Chemistry*, **5**, 3310–3313.

34. Herter, S., Nguyen, G.-S., Thompson, M.L. *et al.* (2011) *Applied Microbiology and Biotechnology*, **90**, 929–939.
35. Kourist, R., Bartsch, S., and Bornscheuer, U.T. (2007) *Advanced Synthesis and Catalysis*, **349**, 1393–1398.
36. Nguyen, G.-S., Kourist, R., Paravidino, M. *et al.* (2010) *European Journal of Organic Chemistry*, 2753–2758.
37. Chen, C., Fujimoto, Y., Girdaukas, G., and Sih, C. (1982) *Journal of the American Chemical Society*, **104**, 7294–7299.
38. Wilcox, C.F. and Leung, C. (1968) *The Journal of Organic Chemistry*, **33**, 877–880.
39. Adcock, W. and Gangodawila, H. (1989) *The Journal of Organic Chemistry*, **54**, 6040–6047.
40. Della, E.W. and Tsanaktsidis, J. (1986) *Australian Journal of Chemistry*, **39**, 2061–2066.
41. Qian, X., Ramirez, A., Wong, M.K.Y. *et al.* (2010) 240th ACS National Meeting, Boston, MA, August 22–26, 2010.
42. Al Hussainy, R., Verbeek, J., van der Born, D. *et al.* (2010) *Journal of Medicinal Chemistry*, **54**, 3480–3491.
43. Peddi, S., Patel, M.V., and Rohde, J.J. (October 21, 2010) US Patent Application US2010/0267738A1.
44. Bennett, B.L., Elsner, J., Erdman, P. *et al.* (October 26, 2012) PCT International Patent Application WO2012/145569A1.
45. Bornscheuer, U.T. and Kazlauskas, R.J. (2006) *Hydrolases in Organic Synthesis: Regio- and Stereoselective Biotransformations*, 2nd edn, Wiley-VCH Verlag GmbH, Weinheim, pp. 61–184.
46. Paravidino, M., Bohm., P., Gröger, H., and Hanefeld, U. (2012) *Enzyme Catalysis in Organic Synthesis*, 3rd edn (eds K. Drauz, H. Gröger, and O. May), Wiley-VCH Verlag GmbH, Weinheim, pp. 249–362.
47. Sabbioni, G. and Jones, J.B. (1987) *The Journal of Organic Chemistry*, **52**, 4565–4570.
48. Hultin, P.G., Mueseler, F.-J., and Jones, J.B. (1991) *The Journal of Organic Chemistry*, **56**, 5375–5380.
49. Chenevert, R., Ngatcha, B.T., Rose, Y.S., and Goupil, D. (1998) *Tetrahedron, Asymmetry*, **9**, 4325–4329.
50. Eycken, J.V.E., Vandewalle, M., Heinemann, G. *et al.* (1989) *Journal of the Chemical Society. Chemical Communications*, 306–308.
51. Torres, C., Bernabe, M., and Otero, C. (1999) *Enzyme and Microbial Technology*, **25**, 753–761.
52. Goswami, A. and Kissick, T.P. (2009) *Organic Process Research & Development*, **13**, 483–488.
53. Kashima, Y., Liu, J., Takenami, S., and Niwayama, S. (2002) *Tetrahedron, Asymmetry*, **13**, 953–956.
54. Cabrera, Z. and Palomo, J.M. (2011) *Tetrahedron, Asymmetry*, **22**, 2080–2084.
55. Guo, Z., Wong, M.K.Y., Hickey, M.R. *et al.* (2014) *Organic Process Research & Development*, **18**, 774–780.
56. Ema, T., Kamata, S., Takeda, M. *et al.* (2010) *Chemical Communications*, **46**, 5440–5442.
57. Ema, T., Nakano, Y., Yoshida, D. *et al.* (2012) *Organic and Biomolecular Chemistry*, **10**, 6299–6308.
58. Seebach, D. and Gardiner, J. (2008) *Accounts of Chemical Research*, **41**, 1366–1375.
59. Weiner, B., Szymanski, W., Janssen, D.B. *et al.* (2010) *Chemical Society Reviews*, **39**, 1656–1691.
60. Engel, U., Syldatk, C., and Rudat, J. (2012) *Applied Microbiology and Biotechnology*, **94**, 1221–1231.
61. Engel, U., Syldatk, C., and Rudat, J. (2012) *AMB Express C7-33*, **2**, 1–13.
62. Werner, M., Las Heras-Vazques, F.J., Fritz, C. *et al.* (2004) *Engineering in Life Sciences*, **4**, 563–572.
63. Dürr, R. (2007) Screening and description of novel hydantoinases from distinct environmental sources. Fakultät für Chemieingenieurswesen und Verfahrenstechnik. Karlsruhe, Doctoral Thesis, Universität Karlsruhe (TH) Dr.-Ing.

4

Non-Redox Lyases and Transferases for C−C, C−O, C−S, and C−N Bond Formation

4.1 Regioselective Enzymatic Carboxylation of Phenols and Hydroxystyrenes Employing Co-Factor-Independent Decarboxylases

Johannes Gross,[1] Tamara Reiter,[1] Christiane Wuensch,[1] Silvia M. Glueck,[1] and Kurt Faber[2]

[1]*Austrian Centre of Industrial Biotechnology, Department of Chemistry, Organic and Bioorganic Chemistry, University of Graz, Austria*
[2]*Department of Chemistry, Organic and Bioorganic Chemistry, University of Graz, Austria*

The use of CO_2 as a versatile readily available C_1 carbon source for the production of valuable chemicals has become attractive in recent years [1]. However, so far only a very limited number of carboxylation processes are performed on industrial scale due to the low energy level of carbon dioxide, which strongly impedes its synthetic application [2]. In order to circumvent this limitation, the use of transition metal [3–5] or organo-catalysts [6] has been explored. These are powerful methods. An attractive and eco-friendly alternative for the carboxylation of aromatic compounds using a biocatalyst has recently been developed [7–10]. The enzymatic carboxylation of phenol- and styrene-type substrates catalyzed by various co-factor-independent decarboxylases proceeds in a highly regioselective fashion, depending on the nature of the enzyme:

Benzoic acid decarboxylases (BDCs) selectively catalyze the *o*-carboxylation of phenols to yield the corresponding *o*-hydroxybenzoic acid derivatives in up to 80% conversion. In a regiocomplementary fashion, **phenolic acid decarboxylases** (PADs)

Practical Methods for Biocatalysis and Biotransformations 3, First Edition.
Edited by John Whittall, Peter W. Sutton, and Wolfgang Kroutil.
© 2016 John Wiley & Sons, Ltd. Published 2016 by John Wiley & Sons, Ltd.

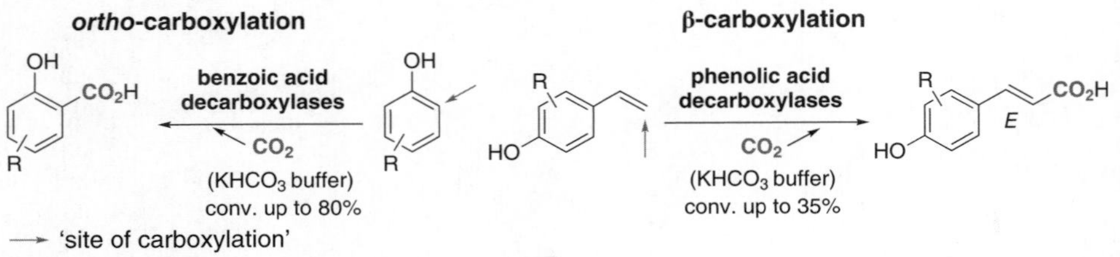

ortho-carboxylation　　　　　　　　　　　　　　**β-carboxylation**

'site of carboxylation'

Scheme 4.1 *Regiocomplementary carboxylation of phenol- and styrene-type substrates employing decarboxylases.*

exclusively act at the β-carbon atom of the side chain of *p*-hydroxystyrenes, forming (*E*)-cinnamic acid derivatives in up to 35% conversion. The reversible, endothermic reaction is favored at elevated concentrations (3 M) [8,9] of bicarbonate as CO_2 source. The equilibrium can be further shifted toward carboxylation by solvent engineering employing organic cosolvents or bicarbonate-based ionic liquids [11].

The *p*-carboxylation of phenols, which are catalyzed by 4- [12] and 3,4-(di)hydrox-ybenzoate decarboxylases [13] or ATP-dependent phenylphosphate carboxylases, is less explored [14]. The regioselective carboxylation of non-phenolic (hetero)aromatics, such as pyrrole and indole, is catalyzed by pyrrole-2- and indole-3-carboxylate decarboxylases, respectively. However, these enzymes are highly substrate specific and thus are of limited practical use [15]. This section focuses on the *o*- and β-carboxylation of phenolic compounds (Scheme 4.1).

4.1.1 Regioselective Ortho-Carboxylation of Phenols Employing BDCs

The enzyme catalyzed *o*-carboxylation of phenols represents a biocatalytic equivalent to the Kolbe–Schmitt reaction [16], which is applied on large scale for the production of salicylic acid. Although the regioselectivity for the *o*- or *p*-carboxylation product can be controlled by starting from Na^+- or K^+-phenolate [17], regio-isomeric product mixtures are often obtained, and the procedure requires harsh reaction conditions; that is, high CO_2 pressure and elevated temperatures. For the biocatalytic variant, several BDCs have been identified which catalyze the *o*-carboxylation of phenol and derivatives thereof in a highly regiose-lective manner (Scheme 4.2). BDCs are metal-dependent enzymes that require a catalytic Zn^{2+} in the active site for catalysis, and their mechanism bears a strong resemblance to the Kolbe–Schmitt reaction [18].

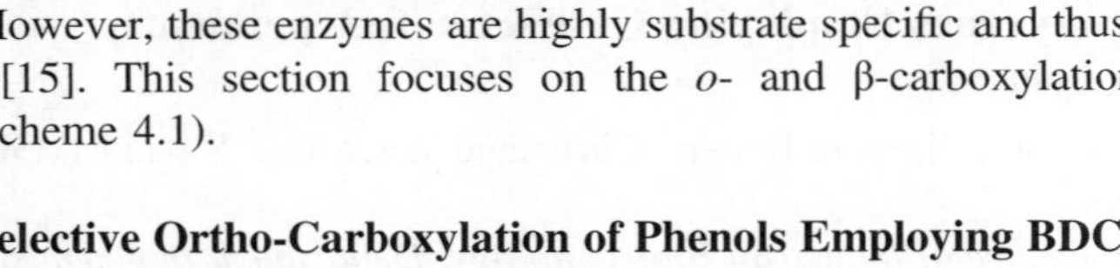

Scheme 4.2 *Regioselective o-carboxylation of 3-aminophenol (1a) [10,19] to yield 4-amino-salicylic acid (1b), employing salicylic acid decarboxylases from* Trichosporon moniliiforme.

4.1.1.1 Enzymes

- 2,3-Dihydroxybenzoic acid decarboxylase from *Aspergillus oryzae* (GI: 94730373)
- Salicylic acid decarboxylase from *Trichosporon moniliiforme* (GI: 225887918)
- 2,6-Dihydroxybenzoic acid decarboxylase from *Rhizobium* species (GI: 116667102)

The genes were synthesized and ligated into a pET 21a (+) vector at *Life Technologies* (Germany). *E. coli* BL21(DE3) was transformed with the respective plasmid using isopropyl β-D-1-thiogalactopyranoside (IPTG) for induction [10]. As biocatalyst, lyophilized *E. coli* cells harboring the corresponding BDC were employed.

4.1.1.2 Materials and Equipment

- Recombinant salicylic acid decarboxylase from *Trichosporon moniliiforme* (30 mg lyophilized *E. coli* cells containing the overexpressed enzyme)
- Substrate **1a** (10 mM final concentration)
- Phosphate buffer (1 mL final volume, 100 mM, pH 5.5)
- Potassium bicarbonate (300 mg, 3 M final concentration)
- Glass vials with screw cap and septum (2 mL)
- Vortex shaker
- Rotary shaker at 30 °C and 120 rpm
- Desktop centrifuge (13 000 rpm)
- Acetonitrile (500 μL)
- Trifluoroacetic acid (30 μL)

4.1.1.3 Screening Procedure for the Carboxylation of Phenols, Exemplified for 1a

Lyophilized cells (30 mg *E. coli* cells containing the overexpressed enzyme) were resuspended in phosphate buffer (1 mL, initial pH 5.5, 100 mM) and rehydrated for 30 minutes by shaking at 30 °C and 120 rpm in a horizontal position, for better mixing. Afterwards, the substrate (10 mM final concentration) and $KHCO_3$ as CO_2 source (300 mg, 3 M final concentration, final pH 8.5) were added to the cell suspension. In order to avoid any loss of CO_2, the reaction vials were immediately tightly closed using a septum and the mixture was shaken for 24 hours at 30 °C and 120 rpm in a horizontal position. Thereafter, the reaction mixture was centrifuged (13 000 rpm, 15 minutes) and an aliquot of 100 μL was diluted with an H_2O/acetonitrile mixture (1 mL, 50% v/v) supplemented with trifluoroacetic acid (3% v/v, 30 μL). The samples were recentrifuged (13 000 rpm, 15 minutes) after incubation at room temperature for 5 minutes and analyzed on reverse-phase high-performance liquid chromatography (HPLC) to determine the conversion [10].

For upscaling, the standard screening procedure was employed, with lyophilized whole cells (100 mg) at 50 mM substrate concentration in a 3 mL final volume of phosphate buffer [10,11].

4.1.1.4 Determination of Conversion

HPLC analysis was performed on a Shimadzu HPLC system equipped with a diode array detector (SPD-M20A) and a reversed-phase column (Phenomenex Luna, C_{18} (2) 100 Å, 250×4.6 mm, 5 μm). Compounds were spectrophotometrically detected at 254 and 280 nm.

The conversion was calculated based on calibration curves for substrate **1a** and product **1b**, prepared with authentic reference material. The method for **1a/1b** was run over 55 minutes with phosphate buffer (0.12 M, pH 2.4) as the mobile phase at a flow rate of 0.5 mL.min^{-1}. The column temperature was set to 30 °C. Retention times: **1a** = 5.3 minutes, **1b** = 45.9 minutes.

4.1.1.5 Substrate Spectrum of BDCs

BDCs displayed a remarkably broad substrate tolerance toward non-natural substrates (Scheme 4.3). A range of structurally and electronically diverse phenol-type substrates could be applied to the biocatalytic *o*-carboxylation. BDCs exclusively acted on the aromatic ring system in the *o*-position of the "directing" phenolic group, and no trace of regio-isomeric *p*-carboxylation products could be detected. Substrates lacking a free *o*-position and non-phenolic substrates were unreactive. In general, steric effects seemed to

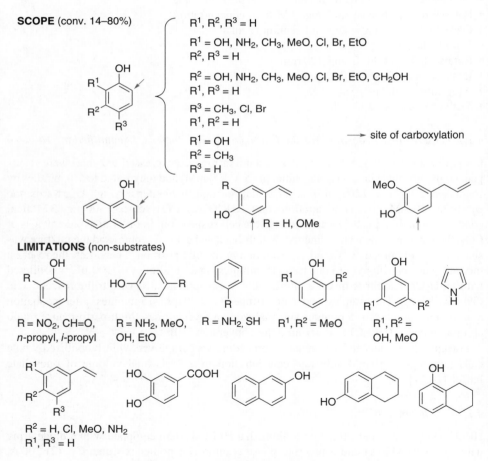

Scheme 4.3 *Substrate tolerance of BDCs.*

play a certain role; however, at this point, general rules for substrate acceptance as a function of the substitution pattern cannot be delineated. In contrast, the electronic effects of substituents were weak, and similar conversions have been obtained for substrates bearing alkyl, alkoxy, halo, or amino groups, the only exception being strong electron-withdrawing substituents (e.g., NO_2 or carbonyl groups), which were not tolerated.

Depending on the type of substrate and the biocatalyst, the corresponding carboxylated products were obtained with 14–80% conversion using the standard screening conditions, already described. Furthermore, the *o*-carboxylation was scaled up using orcinol (5-methylbenzene-1,3-diol) as the model substrate. Identical conversions (67–68%) were obtained when the amount of substrate was raised from 10 mM (4.3 mg) to 50 mM (21.3 mg), but a further increase (100 mM (42.7 mg) and 235 mM (100 mg)) led to a significant drop in conversion (23 and 4%, respectively).

4.1.2 Regioselective β-Carboxylation of Styrene-Type Substrates Employing PADs

In a regiocomplementary fashion, the β-carboxylation of styrene derivatives occurred exclusively on the β-C atom of the side chain of *p*-hydroxystyrene-type substrates, leaving the aromatic system untouched (Scheme 4.4). PADs from bacterial sources have been identified that are able to catalyze the energetically "uphill" carboxylation reaction of non-natural styrene-type substrates [10,11] to yield the corresponding (*E*)-cinnamic acid derivatives in up to 35% conversion. This "side-chain" carboxylation is especially remarkable since no chemical counterpart exists in traditional chemistry. The closest chemical analogs involve a detour via organoboronic esters employing Rh or Cu complexes or Ni-catalyzed hydrocarboxylation of styrenes to obtain the corresponding α- or β-carboxylated products [4,5]. Recombinant PADs are robust biocatalysts that do not require any co-factor. Their mechanism is assumed to proceed through metal-independent acid–base catalysis involving a central quinone methide intermediate [20–22].

4.1.2.1 Enzymes

- PAD from *Lactobacillus plantarum* (GI: 300769086)
- PAD from *Bacillus amyloliquefaciens* (GI: 308175189)
- PAD from *Mycobacterium colombiense* (GI: 342860341)
- PAD from *Methylobacterium* sp. (GI: 168197631)
- PAD from *Pantoea* sp. (GI: 304396594)
- PAD from *Lactococcus lactis* (GI: 15673912)
- Ferulic acid decarboxylase from *Enterobacter* sp. (GI: 212525355)

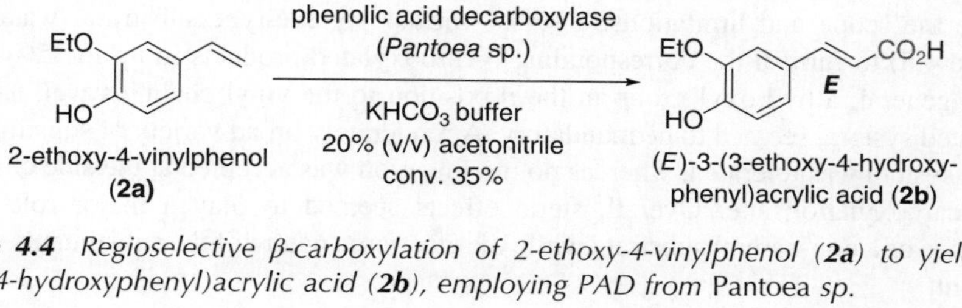

Scheme 4.4 *Regioselective β-carboxylation of 2-ethoxy-4-vinylphenol (2a) to yield (E)-3-ethoxy-4-hydroxyphenyl)acrylic acid (2b), employing PAD from* Pantoea *sp.*

Cloning and overexpression were performed as previously reported [22] in the same way as described for the BDCs, using a pET 21a (+) or a pET 28a (+) vector for subcloning of the corresponding genes.

4.1.2.2 Materials and Equipment

- Recombinant PAD from *Pantoea* sp. (30 mg lyophilized *E. coli* cells containing the overexpressed enzyme)
- Substrate **2a** (10 mM final concentration)
- Phosphate buffer (1 mL final volume, 100 mM, pH 5.5)
- Potassium bicarbonate (300 mg, 3 M final concentration)
- Acetonitrile (1×200 µL, 1×500 µL)
- Glass vials with screw cap and septum (2 mL)
- Vortex shaker
- Rotary shaker at 30 °C and 120 rpm
- Desktop centrifuge (13 000 rpm)
- Trifluoroacetic acid (30 µL)

4.1.2.3 Screening Procedure for the Carboxylation of Hydroxystyrenes, Exemplified for 2a

The procedure was the same as described for the *o*-carboxylation, except that biotransformations were performed in the presence of acetonitrile (20% v/v, 200 µL). As biocatalysts, the corresponding recombinant PADs were applied (30 mg lyophilized *E. coli* cells) [10,11].

4.1.2.4 Determination of Conversion

HPLC analysis was performed on a Shimadzu HPLC system equipped with a diode array detector (SPD-M20A) and a reversed-phase column (Phenomenex Luna, C$_{18}$ (2) 100 Å, 250×4.6 mm, 5 µm). The column temperature was set at 24 °C. The conversion was calculated using calibration curves for substrate **2a** and product **2b** prepared with authentic reference material. Compounds were spectrophotometrically detected at 270 and 280 nm. The method for **2a/2b** was run over 17 minutes at a flow rate of 1 mL.min^{-1} with water (+ trifluoroacetic acid (0.1%), solvent **A**) and acetonitrile (+ trifluoroacetic acid (0.1%), solvent **B**), with the following gradient: 0–2 minutes, 0% **B**; 2–15 minutes, 0–100% **B**; 15–17 minutes, 100% **B**. Retention times: **2a** = 15.4 minutes, **2b** = 13.0 minutes.

4.1.2.5 Substrate Spectrum of PADs

In comparison to BDCs, PADs showed a more restricted substrate tolerance. In order to evaluate the scope and limitations, a set of various styrene-type substrates was applied (Scheme 4.5) to furnish the corresponding β-carboxylated products in up to 35% conversion. In general, a hydroxyl group in the *p*-position to the vinyl chain, as well as a fully conjugated system, seemed to be mandatory. A moderately broad variety of substituents on the ring system was tolerated, whereas no modification was accepted at the side chain; that is, the carboxylation site. Overall, steric effects seemed to play a major role for the acceptance of a non-natural substrate, while the electronic nature of the substituents was less important.

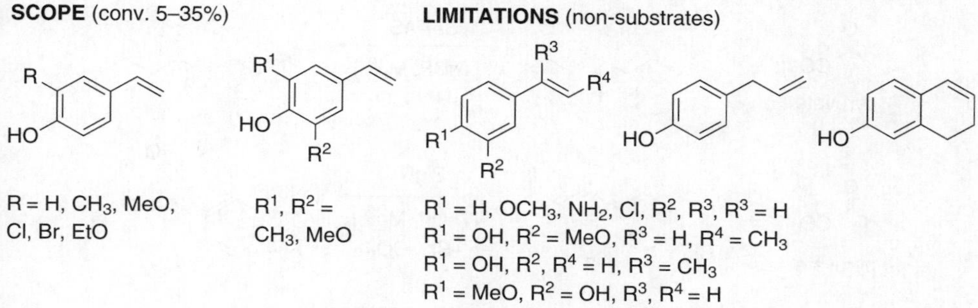

SCOPE (conv. 5–35%)　　　　　　　　LIMITATIONS (non-substrates)

R = H, CH$_3$, MeO,
Cl, Br, EtO

R^1, R^2 =
CH$_3$, MeO

R^1 = H, OCH$_3$, NH$_2$, Cl, R^2, R^3, R^3 = H
R^1 = OH, R^2 = MeO, R^3 = H, R^4 = CH$_3$
R^1 = OH, R^2, R^4 = H, R^3 = CH$_3$
R^1 = MeO, R^2 = OH, R^3, R^4 = H

Scheme 4.5　Substrate tolerance of PADs.

4.1.3　Conclusion

Due to the growing demand for alternative carbon sources for the synthesis of valuable chemicals, the development of CO_2-fixation strategies is a current challenge in synthetic organic chemistry. The enzyme-catalyzed carboxylation of aromatic compounds using recombinant BDCs and PADs represents a promising "green" alternative to classical chemical methods, such as the Kolbe–Schmitt reaction. The biocarboxylation proceeded with excellent regioselectivity: BDCs selectively catalyzed the *o*-carboxylation of phenols to yield salicylic acid derivatives, whereas PADs exclusively enabled the β-carboxylation of *p*-hydroxystyrenes to afford cinnamic acid derivatives. A remarkably broad substrate scope, including industrially relevant targets, mild reaction conditions, and the easy handling of biocatalysts, represents important aspects, which promote the biocatalytic strategy as an interesting biosynthetic tool.

4.2　Stetter Reactions Catalyzed by Thiamine Diphosphate-Dependent Enzymes

Carola Dresen, Elena Kasparyan, Lydia S. Walter, Fabrizio Bonina, Simon Waltzer, and Michael Müller

Institute of Pharmaceutical Sciences, Albert Ludwigs University of Freiburg, Germany

The Stetter reaction represents the conjugate 1,4-addition of an "umpoled" aldehyde derivative to an α,β-unsaturated carbonyl compound. The synthetic value of this name reaction is enormous, but catalytic asymmetric versions have rarely been identified. We have succeeded in identifying the first stetterase enzyme, PigD, and the conditions resulting in catalytic asymmetric versions of the Stetter reaction [23]. Starting with the amino acid sequence of PigD from *Serratia marcescens*, other putative stetterases have been identified, such as *Se*AAS from *Saccharopolyspora erythraea* and HapD from *Hahella chejuensis*. The substrate range of the three enzymes is distinctly different; still, all of them give almost identical stereochemistry in the enzymatic products [24]. A comprehensive overview of thiamine diphosphate (ThDP)-dependent enzymes is given in the literature [25–28], and

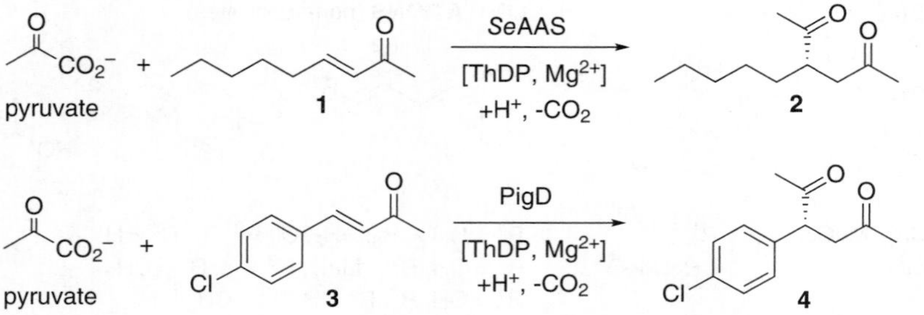

Scheme 4.6 *Enzymatic Stetter reaction of enones.*

experimental details in this section are taken primarily from references [23] and [24] (Scheme 4.6).

4.2.1 Procedure 1: SeAAS-Catalyzed Formation of Aliphatic Products

4.2.1.1 Materials and Equipment

- Sodium pyruvate (33 mg, 0.3 mmol)
- Non-3-en-2-one (33.7 mg; 39.7 μL, 0.24 mmol)
- Dimethyl sulfoxide (DMSO ≥99%; 2.4 mL)
- 50 mM HEPES Buffer pH 7.5 (9.4 mL, 2.5 mM $MgCl_2$, 100 mM NaCl, 0.1 mM ThDP)
- SeAAS (12 mg, purified protein)

4.2.1.2 Synthesis of 1,4-Addition Product 2

1. The acceptor substrate **1** (non-3-en-2one, 20 mM) and sodium pyruvate (25 mM) were dissolved in DMSO (2.4 mL), and 50 mM HEPES buffer (9.6 mL) was added.
2. The reaction was started with the addition of SeAAS ($c_{end} = 1\,mg \times mL^{-1}$) and the reaction mixture was incubated at 30 °C and 300 rpm for 24 hours.
3. The reaction mixture was extracted with ethyl acetate ($3 \times 20\,mL$) and the combined organic layers were washed with brine and dried over Na_2SO_4. The solvent was evaporated at 28 °C and 150 mbar.
4. The crude product was purified by silica gel column chromatography (SiO_2, cyclohexane/ethyl acetate, 5 : 1, $R_f = 0.34$).
5. GC-MS of the crude product showed a conversion of 60% (ee >99%). Column chromatography afforded 17 mg (S)-3-pentylhexane-2,5-dione (**2**, 38%, ee >99%).

4.2.1.3 Analytical Data

¹H-NMR (400 MHz; $CDCl_3$): δ = 3.07–2.98 (m, 1 H, CHCO), 2.97 (dd, 1 H, J = 17.3, 9.9 Hz, CH_xH_yCO), 2.44 (dd, J = 17.3, 3.0 Hz, 1 H, CH_xH_yCO), 2.25 (s, 3 H, CH_3COCH), 2.15 (s, 3 H, CH_3COCH_2), 1.62–1.52 (m, 2 H, $CHCH_2CH_2$), 1.35–1.20 (m, 6 H, $CH_2CH_2CH_2CH_3$), 0.89 (t, J = 6.9 Hz, 3 H, CH_2CH_3)

¹³C-NMR (100.6 MHz; $CDCl_3$): δ = 211.6 (C_q, CH_3COCH), 207.5 (C_q, CH_3COCH_2), 46.8 (CH), 44.7 (CH_2CO), 31.7 ($CH_2CH_2CH_3$), 31.1 ($CHCH_2CH_2$), 29.9 (CH_3COCH_2), 29.7 (CH_3COCH), 26.7 ($CHCH_2CH_2$), 22.4 ($CH_2CH_2CH_3$), 13.9 (CH_2CH_3)

GC-MS (column HP-5MS, 30 m × 0.250 mm × 0.250 μm; T_{GC} (injector) = 250 °C, T_{MS}(ion source) = 200 °C, time program (oven): T_0 min = 60 °C, T_3 min = 60 °C, T_{14} min = 280 °C (heating rate 20 °C.min^{-1}), T_{19} min = 280 °C): t_R = 8.78 minutes; EI-MS (70 eV, EI); m/z (%): 184 (1) [M$^+$], 169 (3) [M$^+$ – CH$_3$], 141 (10) [M$^+$ – C$_2$H$_3$O], 114 (80) [C$_6$H$_{10}$O$_2$$^+$], 71 (100) [C$_5H_{11}$$^+$]

Product *ee* was determined by chiral phase GC (column FS-Lipodex D, 50 m × 0.25 mm × 1.00 μm, 90 °C, isocratic, carrier gas: He, constant linear velocity 30 cm × sec^{-1}, injector 190 °C): t_R = 94.27 minutes (main enantiomer), 97.99 minutes (minor enantiomer).

4.2.2 Procedure 2: PigD-Catalyzed Formation of Aromatic Products

4.2.2.1 *Materials and Equipment*

- Sodium pyruvate (33 mg, 0.3 mmol)
- 4-(4-Chlorophenyl)bute-3-en-2one (43 mg, 0.24 mmol)
- Methyl *tert*-butyl ether (MTBE; 0.6 mL)
- 46 mM KP$_i$ Buffer pH 7.0 (11.4 mL, 11.5 mM MgSO$_4$, 0.5 mM ThDP)
- PigD (11.3 mg, purified protein)

4.2.2.2 *Procedure*

1. The acceptor substrate **3** (4-(4-chlorophenyl)bute-3-en-2one, 20 mM) and sodium pyruvate (25 mM) were dissolved in MTBE (0.6 mL) and 46 mM KP$_i$ buffer (11.4 mL) was added.
2. The reaction was started with the addition of PigD (c_{end} = 1 mg × mL^{-1}) and the reaction mixture was incubated at 30 °C and 300 rpm for 20 hours.
3. The reaction mixture was extracted with ethyl acetate (3 × 20 mL) and the combined organic layers were washed with brine and dried over Na$_2$SO$_4$. The solvent was evaporated at 28 °C and 150 mbar.
4. The crude product was purified by silica gel column chromatography (SiO$_2$, cyclohexane/ethyl acetate, 5: 1, R_f = 0.20).
5. ^{1}H-NMR analysis of the crude product showed a conversion of 39%. Column chromatography afforded 7 mg (*R*)-3-(4-chlorophenyl)hexane-2,5-dione (**4**, 13%, *ee* 75%).

4.2.2.3 *Analytical Data*

^{1}H-NMR (400 MHz; CDCl$_3$): δ = 7.35–7.30 (m, 2 H, CH$_{Ar}$), 7.19–7.13 (m, 2 H, CH$_{Ar}$), 4.22 (dd, J = 10.0, 4.1 Hz, 1 H, CH), 3.42 (dd, J = 18.0, 10.0 Hz, 1 H, C*H$_x$*H$_y$), 2.57 (dd, J = 18.0, J = 4.1 Hz, 1 H, CH$_x$*H$_y$*), 2.18 (s, 3 H, C*H$_3$*COCH$_2$), 2.14 (s, 3 H, C*H$_3$*COCH)
^{13}C-NMR (100.6 MHz; CDCl$_3$): δ = 206.6 (C$_q$, CH$_3$COCH), 206.3 (C$_q$, CH$_3$COCH$_2$), 136.2 (C$_q$, CAr), 133.6 (C$_q$, CAr), 129.5 (C$_{Ar}$, 2C), 129.3 (C$_{Ar}$, 2C), 53.1 (CH), 46.3 (CH$_2$), 29.9 (*C*H$_3$COCH$_2$), 28.9 (*C*H$_3$COCH)
GC-MS (column HP-5MS, 30 m × 0.250 mm × 0.250 μm; T_{GC} (injector) = 250 °C, T_{MS}(ion source) = 200 °C, time program (oven): T_0 min = 60 °C, T_3 min = 60 °C, T_{14} min = 280 °C (heating rate 20 °C.min^{-1}), T_{19} min = 280 °C): t_R = 10.16 minutes; EI-MS (70 eV, EI);

m/z (%): 226 (33), 224 (100) [M^+], 206 (2) [$M^+ - H_2O$], 183 (26), 182 (84), 181 (56), 167 (21) [$M^+ - C_3H_5O$], 147 (44) [$M^+ - C_2H_3O - Cl$], 139 (30), 138 (42), 125 (31) [$C_7H_6Cl^+$], 115 (9), 112 (12), 103 (64), 77 (37) [$C_6H_5^+$]

$[\alpha]^D_{25} = -314$ ($c = 0.5\,g \times 100\,mL^{-1}$, $CHCl_3$)

Product *ee* was determined by chiral phase HPLC-DAD (Chiralpak AS-H, 25 °C, $0.75\,mL \times min^{-1}$, *n*-hexane/2-propanol $= 97:3$ (v/v)): $t_R = 10.5$ minutes (*R*), $t_R = 11.3$ minutes (*S*).

4.2.3 Conclusion

Catalytic asymmetric C−C bond formations are among the most important synthetic transformations in organic chemistry, if not the most. With the newly identified stetterase group of enzymes, this type of transformation has been introduced to synthetic chemistry. Organocatalytic methods have been developed in parallel [29]. The examples described in this section and in recent publications [23,24] demonstrate the already broad substrate range of some of the known stetterases (Table 4.1). To broaden the synthetic value of this group of enzymes, enantiocomplementary transformations will have to be developed.

Table 4.1 *1,4-carboligation products obtained with PigD, SeAAS, and HapD.*

1,4-carboligation product	Conversion after 24 hours (%) (ee)		
	PigD	SeAAS	HapD
	8%	>99%	7.5%
	37% (>99%) (S)	53% (>99%) (S)	56% (>99%) (S)
	10% (82–95%) (R)	traces (n.d.) (R)	4.5% (83%) (R)
	39% (75%) (R)	3% (84%) (R)	14% (84%) (R)
	11% (<10%)	9% (<10%)	9% (<10%)

4.3 Asymmetric Michael-Type Additions of Acetaldehyde to Nitroolefins Catalyzed by 4-Oxalocrotonate Tautomerase (4-OT) Yielding Valuable γ-Nitroaldehydes

Edzard M. Geertsema, Yufeng Miao, and Gerrit J. Poelarends

Department of Pharmaceutical Biology, Groningen Research Institute of Pharmacy, University of Groningen, The Netherlands

γ-Nitroaldehydes are useful precursors for γ-aminobutyric acid (GABA) derivatives, several of which represent marketed pharmaceuticals [30]. The Michael-type addition of acetaldehyde to nitroolefins gives convenient access to relevant chiral γ-nitroaldehydes (Scheme 4.7). The first successful methodology for this type of reaction was developed in the field of organocatalysis [31,32] and documented in the literature in 2008. We recently presented a biocatalytic procedure for the Michael-type addition of acetaldehyde to various nitroolefins, which involves the homohexameric enzyme 4-oxalocrotonate tautomerase (4-OT) [33–35]. 4-OT is a member of the tautomerase superfamily and is part of a catabolic pathway for aromatic hydrocarbons in the soil bacterium *Pseudomonas putida* mt-2. It is a small enzyme (only 62 amino acid residues per monomer) and is characterized by a catalytic N-terminal proline residue (Pro-1) that resides in the active site [36–39]. We here describe the production and purification of the enzyme 4-OT, as well as its use as a catalyst in the Michael-type addition of acetaldehyde to nitroolefins **1a–b** to produce γ-nitroaldehydes **2a–b** (Scheme 4.7) [33–35].

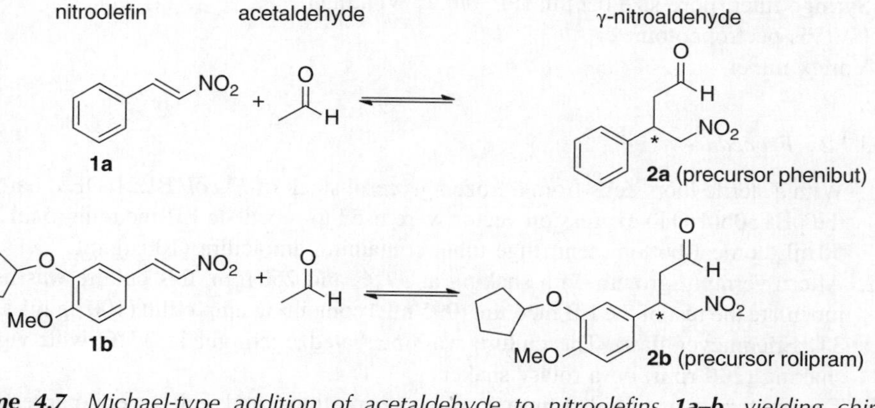

Scheme 4.7 *Michael-type addition of acetaldehyde to nitroolefins **1a–b**, yielding chiral γ-nitroaldehydes **2a–b**. * = chiral center.*

4.3.1 Expression and Purification of 4-OT

4.3.1.1 Materials and Equipment

- *E. coli* BL21(DE3) containing pET20b(4-OT) expression vector (frozen glycerol stock at − 80 °C)
- Ammonium sulfate (3.2 M), 50 mL

- Ampicillin (100 mg.mL^{-1} stock solution in water, filter-sterilized)
- Sodium dihydrogenphosphate (NaH$_2$PO$_4$, 20 mM, pH 7.3), 100 mL
- Sodium dihydrogenphosphate (NaH$_2$PO$_4$, 10 mM, pH 8.0) containing ammonium sulfate ((NH$_4$)$_2$SO$_4$, 1.6 M), 100 mL
- Sodium dihydrogenphosphate (NaH$_2$PO$_4$, 10 mM, pH 8.0) containing sodium sulfate (Na$_2$SO$_4$, 0.5 M), 100 mL
- Sodium dihydrogenphosphate (NaH$_2$PO$_4$, 10 mM, pH 8.0), (buffer A), 250 mL
- Sodium dodecyl sulfate (SDS) gels containing polyacrylamide (10%)
- LB medium (1 L, sterilized)
- (Conical-bottom) centrifuge tube (50 mL, CELLSTAR)
- Centrifuge capable of reaching 15 000 rpm with temperature control
- DEAE-sepharose (CL-6B, Sigma-Aldrich) column (10 × 1.0 cm filled with ~8 mL DEAE-sepharose resin)
- Dialysis membrane with MWCO 2000 Da (Spectrum labs)
- Erlenmeyer flask (3 L)
- Eppendorf tubes (1.5/2.0 mL)
- Magnetic stirrer
- PD-10 Sephadex G-25 gel filtration column (GE Healthcare Bio-Sciences AB)
- Phenyl-sepharose (CL-4B, Sigma-Aldrich) column (10 × 1.0 cm filled with ~8 mL phenyl-sepharose resin)
- Quartz cuvette with 1 or 10 mm light path (Hellma Analytics)
- Rotary shaker with temperature control
- Sterile loop
- Syringe filter, pore size 0.2 µm (FP30/0.2, Whatman)
- UV/Vis pectrophotometer
- Vortex mixer

4.3.1.2   Procedure

1. With a sterile loop, cells from a frozen glycerol stock of *E. coli* BL21(DE3) harboring the pET20b(4-OT) expression vector were used to inoculate LB medium (5 mL), in a 50 mL conical-bottom centrifuge tube, containing ampicillin (100 µg.mL^{-1}).
2. After overnight growth with shaking at 37 °C and 250 rpm, this culture was used to inoculate the rest of the LB medium (995 mL) containing ampicillin (100 µg.mL^{-1}) in a 3 L Erlenmeyer flask. The culture was incubated overnight at 37 °C with vigorous shaking (200 rpm) on a rotary shaker.
3. The culture was divided into two equal portions and cells were harvested by centrifugation (20 minutes at 4500 rpm) after an OD$_{600}$ value of ~4.5 was reached [40].
4. Cells harvested from 0.5 L of culture were resuspended in buffer A (~10 mL) using a vortex mixer. Optional: In order to store the cells at −20 °C, they should be left in 10 mM NaH$_2$PO$_4$ buffer (pH 8.0, buffer A).
5. The cell suspension was transferred to sonication tubes (centrifuge tubes) and sonicated at high power output for 8–10 minutes, in order to disrupt the cells. After centrifugation (45 minutes at 15 000 rpm), the soluble fraction (supernatant) containing 4-OT was separated from the insoluble part (pellet). Note: The cell suspension was kept in a cooling bath (water/ice) during the sonication process to avoid heat inactivation of the 4-OT.

6. The supernatant containing 4-OT was then loaded on to a DEAE-sepharose column, which had previously been equilibrated with buffer A using gravity flow. The flow-through was discarded.
7. The column was washed with buffer A (3×10 mL), and then the protein was eluted by gravity flow using buffer A containing 0.5 M Na_2SO_4 (12 mL). Fractions (1.5 mL $\times 8$) from the elution step were collected in Eppendorf tubes (1.5 mL) and the presence of 4-OT in each fraction was examined by SDS-PAGE.
8. The fractions containing 4-OT were combined and mixed with $(NH_4)_2SO_4$ (final concentration: 1.6 M). After gentle stirring with a magnetic stir bar at 4 °C for 2 hours, the precipitate was pelleted by centrifugation (20 minutes at 13 300 rpm, 4 °C).
9. The supernatant was filtered (syringe filter with 0.2 μm pore size) and loaded on to a phenyl-sepharose column that had previously been equilibrated with buffer A containing 1.6 M $(NH_4)_2SO_4$.
10. The supernatant was left to flow through the column and the flow-through was discarded. The column was then washed with buffer A containing 1.6 M $(NH_4)_2SO_4$ (3×10 mL), after which the protein was eluted by gravity flow using buffer A (12 mL). Fractions of the wash step (~30 mL) and the elution step (~1.5 mL $\times 8$) were analyzed by SDS-PAGE. Note: Wild-type 4-OT has little interaction with the phenyl-sepharose column and should therefore elute as homogenous protein (>95% purity as assessed by SDS-PAGE) in the wash step.
11. Fractions containing pure 4-OT were combined and the buffer was exchanged for 20 mM NaH_2PO_4 buffer (pH 7.3) using a pre-packed PD-10 sephadex G-25 gel filtration column. Optional: An additional dialysis step is recommended to reduce the high salt concentration (1.6 M $(NH_4)_2SO_4$) in the sample before exchanging the buffer on a PD-10 column. In a typical dialysis procedure, the fractions containing pure 4-OT were combined and transferred into a dialysis membrane (MWCO 2000 Da). The dialysis membrane containing the sample was then placed in a 5 L beaker containing ~4 L 20 mM NaH_2PO_4 buffer (pH 7.3) with gentle stirring on a magnetic stirrer overnight.
12. The concentration of purified 4-OT in 20 mM NaH_2PO_4 buffer (pH 7.3) was determined with a spectrophotometer using the method of Waddell [41]. The typical yield of purified 4-OT isolated from 0.5 L of cell culture following this protocol was 25–50 mg.
13. The purified protein was divided into aliquots (100–200 μL) and stored at 4 °C until use. Optional: Purified 4-OT can be stored for at least 1 month without significant loss of catalytic activity upon freezing with liquid nitrogen and when kept at −80 °C in concentrations of 10–20 mg.mL^{-1} (in 20 mM NaH_2PO_4 buffer, pH 7.3).

4.3.2 Synthesis of 2a

4.3.2.1 Materials and Equipment

- *trans*-β-Nitrostyrene **1a** (18.0 mg, 0.12 mmol; commercially available, Sigma-Aldrich)
- Acetaldehyde (132 mg, ~0.17 mL, 3.0 mmol)
- EtOH (6 mL)
- 20 mM NaH_2PO_4 buffer, pH 5.5 (~54.0 mL)
- 4-OT (11.3 mg, 1.7×10^{-3} mmol) from aliquots with 4-OT concentrations of 10–20 mg. mL^{-1} (in 20 mM NaH_2PO_4 buffer, pH 7.3) (see Step 13 of the Procedure for the Expression and Purification of 4-OT)

- Diethyl ether (120 mL)
- MgSO$_4$ (~0.5 g)
- Deuterated chloroform (CDCl$_3$, 0.65 mL for ^{1}H-NMR analysis)
- Balance
- Measuring cylinder (50 mL)
- Glass flask (150 mL)
- Standard pipette for 4-OT addition
- 1 mL Syringe (NORM-JECT) for acetaldehyde addition
- UV/Vis spectrophotometer
- Cuvette (preferably, path length = 1 mm, volume = 300 μL; see Analytical Methods)
- Two vivaspin columns (Sartorius Stedim Biotech SA, France) with a 5000 Da molecular weight cut-off filter
- Separatory funnel (250 mL)
- Two Erlenmeyer flasks (100 and 250 mL)
- Magnetic stirrer
- Magnetic stir bar (for drying with MgSO$_4$ during work-up procedure)
- Funnel with filter paper
- Round-bottom glass flask (250 mL)
- Rotary evaporator connected to a vacuum pump
- NMR tube
- NMR spectrometer

4.3.2.2   Procedure (from Reference [34])

1. In a glass flask (150 mL), a solution of nitroolefin **1a** (18.0 mg, 0.12 mmol) in EtOH (6.0 mL) was added to a mixture of acetaldehyde (132 mg, 0.17 mL, 3.0 mmol) and 4-OT (11.3 mg, ~0.6–1.2 mL of aliquots with 4-OT concentrations of 10–20 mg.mL^{-1}, 1.7 × 10^{-3} mmol 4-OT, 1.4 mol% compared to **1a**) in 20 mM NaH$_2$PO$_4$ buffer, pH 5.5 (~53.2–52.6 mL). The change of pH caused by the addition of aliquots containing 4-OT (in buffer, pH 7.3) was ignorable. Final volume of reaction mixture: 60 mL.

2. The mixture was incubated at room temperature and the reaction progress was monitored by recording the UV spectra of aliquots taken from the reaction mixture after regular time intervals. Note: To avoid inactivation of the enzyme, it is recommended not to stir the reaction mixture with a stirring magnet, but rather to shake gently every 30 minutes.

3. After 2 hours, all **1a** was converted. The reaction mixture was divided into two equal portions, transferred to two Vivaspin columns, and centrifuged (4000 rpm, RT) to remove 4-OT.

4. The flow-throughs were collected, combined, and extracted with diethyl ether (3 × 40 mL). The combined organic layers were dried with MgSO$_4$, filtered, and concentrated *in vacuo* with a rotary evaporator to yield **2a** with high purity (16.2 mg, 8.4 × 10^{-2} mmol, 70%) as a colorless oil.

5. The ^{1}H-NMR spectroscopic data for **2a** were in agreement with published data [42,43]. *ee* was determined by derivatization of the aldehyde moiety of **2a** into the corresponding cyclic acetal [33]. Normal phase HPLC analysis of derivatized **2a** using a Chiracel OD column revealed an *ee* of 81% in favor of the (*S*)-enantiomer. HPLC parameters: eluent = (*n*-heptane/*i*-PrOH 90:10, 40 °C), flow rate = 1 mL.min^{-1}, UV detection at

210 nm, t_R (minor) = 13.0 minutes, (major) = 15.4 minutes. The HPLC data were in accordance with the literature [43].

4.3.3 Synthesis of 2b

4.3.3.1 Materials and Equipment

- Nitroolefin **1b**: (*E*)-2-(cyclopentyloxy)-1-methoxy-4-(2-nitrovinyl)benzene (31.6 mg, 0.12 mmol) [44]
- Acetaldehyde (132 mg, ~0.17 mL, 3.0 mmol)
- DMSO (24 mL)
- 20 mM NaH_2PO_4 Buffer, pH 5.5 (~36.0 mL)
- 4-OT (30 mg, 4.4×10^{-3} mmol) from aliquots with 4-OT concentrations of 10–20 mg. mL^{-1} (in 20 mM NaH_2PO_4 buffer, pH 7.3) (see Step 13 of the Procedure for the Expression and Purification of 4-OT)
- Water (18 mL)
- Chloroform (48 mL)
- $MgSO_4$ (~0.5 g)
- Deuterated chloroform ($CDCl_3$, 0.65 mL for ^1H-NMR analysis)
- Balance
- Measuring cylinder (50 mL)
- Glass flask (150 mL)
- Standard pipette for 4-OT addition
- 1 mL Syringe (NORM-JECT) for acetaldehyde addition
- UV/Vis spectrophotometer
- Cuvette (preferably, path length = 1 mm, volume = 300 µL; see Analytical Methods)
- Six CELLSTAR tubes (15 mL, polypropylene (PP), conical bottom)
- Acid-resistant CentriVap vacuum concentrator (Labconco, 78100 series) connected to a cold trap and a vacuum pump
- Separatory funnel (100 mL)
- Two Erlenmeyer flasks (25 and 100 mL)
- Magnetic stirrer
- Magnetic stir bar (for drying with $MgSO_4$ during work-up procedure)
- Funnel with filter paper
- Round-bottom glass flask (100 mL)
- Rotary evaporator connected to a vacuum pump
- NMR tube
- NMR spectrometer

4.3.3.2 Procedure (from Reference [35])

1. In a glass flask (150 mL), a solution of nitroolefin **1b** (31.6 mg, 0.12 mmol) in DMSO (6.0 mL) was added to a mixture of acetaldehyde (132 mg, 0.17 mL, 3.0 mmol) and 4-OT (30 mg, 1.5–3.0 mL of aliquots with 4-OT concentrations of 10–20 mg.mL^{-1}, 4.4×10^{-3} mmol 4-OT, 3.7 mol% compared to **1b**) in 20 mM NaH_2PO_4 buffer, pH 5.5 (~34.5–33.0 mL) and DMSO (18 mL). The change of pH caused by the addition of aliquots containing 4-OT (in buffer, pH 7.3) was ignorable. Final volume

of reaction mixture: 60 mL. Note: DMSO was required as a co-solvent to achieve sufficient solubility of nitroolefin **1b** and product **2b** in aqueous buffer. DMSO enhances the solubility of **1b** and **2b**, does not impede the catalytic activity of 4-OT, and does not chemically react with any of the reagents (see [35] for a full explanation).

2. The mixture was incubated at room temperature and the reaction progress was monitored by recording the UV spectra of aliquots taken from the reaction mixture after regular time intervals. Note: To avoid inactivation of the enzyme, it is recommended not to stir the reaction mixture with a stirring magnet, but rather to gently shake every 30 minutes.

3. After 2.5 hours, all **1b** was converted and the reaction mixture was divided over six polypropylene tubes (15 mL, CELLSTAR). The solvents were evaporated using an acid-resistant CentriVap vacuum concentrator (55 °C, overnight).

4. Water (3 mL) and chloroform (3 mL) were added to each tube. The dry residues were dissolved/suspended by vigorous stirring (additional scraping with a spatula may be required). The combined water and chloroform layers were separated in a separatory funnel. The water layer was extracted with chloroform (3 × 10 mL). The combined organic layers were dried with MgSO$_4$, filtered, and concentrated *in vacuo* with a rotary evaporator to yield **2b** with high purity (23.6 mg, 7.7×10^{-2} mmol, 64%) as a colorless oil.

5. The ^{1}H-NMR spectroscopic data for **2b** were in agreement with published data [31]. The *ee* was determined by derivatization of the aldehyde moiety of **2b** into its corresponding methyl ester [45]. Normal phase HPLC analysis of derivatized **2b** using a Chiralpak IB column revealed an *ee* of 96% in favor of the (*S*)-enantiomer [45]. HPLC parameters: eluent = (*n*-heptane/*i*-PrOH 95:5, 25 °C), flow rate = 1 mL.min^{-1}, UV detection at 220 nm, t$_R$ (minor) = 32.2 minutes, (major) = 34.9 minutes.

4.3.4 Analytical Methods

The biocatalytic conversions of nitroolefins **1a–b** into **2a–b** were monitored by UV spectroscopy. After regular time intervals (typically 30 minutes), 300 μL of reaction mixture was transferred into a 300 μL cuvette with a path length of 1 mm and a UV spectrum was recorded (200–500 nm). A decrease of absorbance at λ_{max} of **1a–b** indicates the depletion of **1a–b** ($\lambda_{max,1a} = 320$ nm, $\varepsilon_{max} = 14.4$ mM^{-1}.cm^{-1}; $\lambda_{max,1b} = 378$ nm, $\varepsilon_{max} = 13.5$ mM^{-1}. cm^{-1}). After the UV spectrum had been recorded, the contents of the cuvette were recombined with the reaction mixture (with a syringe). The work-up procedure was initiated when the absorbance at λ_{max} of the nitroolefin (**1a–b**) had totally vanished, which indicated full conversion of the nitroolefin (**1a–b**).

4.3.5 Conclusion

The enzyme 4-OT can be readily produced and purified. About 50–100 mg of 4-OT can be obtained from 1 L of cell culture. The enzyme can be stored at −80 °C in concentrations of 10–20 mg.mL^{-1} (in 20 mM NaH$_2$PO$_4$ buffer, pH 7.3) for at least 1 month without significant loss of catalytic activity. 4-OT can easily be tested for full catalytic activity by assaying its natural tautomerase activity with the substrate 2-hydroxymuconate [46] or phenyl(enol)pyruvate [47] and comparing with literature data.

The enzyme 4-OT promiscuously catalyzes the Michael-type addition of acetaldehyde to nitroolefins **1a–b**, yielding γ-nitroaldehydes **2a–b** (Scheme 4.7). The methodology is

Table 4.2 *Conditions for, and results of, 4-OT-catalyzed Michael-type additions of acetaldehyde to nitroolefins 1a and 1b in 20 mM NaH$_2$PO$_4$ buffer (pH 5.5) and co-solvent yielding 2a and 2b (Scheme 4.7).*

Nitroolefin	4-OT (mol%)	Co-solvent (v/v)	Reaction time (h)	Product	yield (%)	ee (%)
1a	1.4	EtOH 10%	2	2a	70	81
1b	3.7	DMSO 40%	2.5	2b	64	96

characterized by relatively low catalyst loading (compared to 10–20 mol% in organo-catalysis) (reagents: acetaldehyde and **1a**; mol% catalyst: 20%; reaction time: 15 hours; yield: 51%; *ee* 92% [31]; reagents: acetaldehyde and **1a**; mol% catalyst: 10%; reaction time: 18 hours; yield: 75%; *ee* 96% [32]; reagents: acetaldehyde and **1a**; mol% catalyst: 10%; reaction time: 72 hours; yield: 55%; *ee* 96% [48]; reagents: acetaldehyde and **1a**; mol% catalyst: 10%; reaction time: 48 hours; yield: 58%; *ee* 0% [49]; reagents: acetaldehyde and **1a**; mol% catalyst: 20%; reaction time: 3 hours; yield: 75%; *ee* 95% [50]; reagents: acetaldehyde and **1a**; mol% catalyst: 10%; reaction time: 22 hours; yield: 57%; *ee* 91% [51]), high stereoselectivities, and predominantly aqueous reaction media (Table 4.2). Reaction times of 2.5 hours or less are sufficient for full conversion of nitroolefins **1a** and **1b**, respectively ($c_{1a,b} = 2.0$ mM; $c_{4\text{-}OT} = 2.8 \times 10^{-3}$ mM for **1a**; $c_{4\text{-}OT} = 7.4 \times 10^{-3}$ mM for **1b**). Side products are hardly formed, as revealed by straightforward work-up procedures without column chromatography, which yielded **2a–b** with high purity (^{1}H-NMR spectroscopy). Respective yields of 70 and 64% were established for **2a** and **2b**, while *ee* values of 81 and 96% were found, respectively. Catalysis of the Michael-type additions takes place in the active site of 4-OT, and its characteristic N-terminal proline is a key catalytic residue, as demonstrated by various control experiments [33–35]. 4-OT fully retains its "Michaelase" activity in aqueous solvent systems containing up to 50% DMSO (v/v) [35]. This methodology is therefore not only restricted to water-soluble substrates, but also allows application of poorly water-soluble chemicals such as **1b**, since its solubility in aqueous solvent systems is greatly enhanced in the presence of DMSO. In addition to acetaldehyde and aromatic nitroolefins, 4-OT also accepts linear aldehydes up to octanal as donors [34] and aliphatic nitroolefins as acceptors [35] for Michael-type addition reactions.

4.4 Michael-Type Addition of Aldehydes to β-Nitrostyrenes by Whole Cells of *Escherichia coli* Expressing 4-Oxalocrotonate Tautomerase (4-OT)

Tanja Narancic,[1] Gordana Minovska,[1] Predrag Jovanovic,[2] Jelena Radivojevic,[1] and Jasmina Nikodinovic-Runic[1]

[1] *Institute of Molecular Genetics and Genetic Engineering, University of Belgrade, Serbia*

[2] *Department of Organic Chemistry, Faculty of Pharmacy, University of Belgrade, Serbia*

4-OT encoded by the *xylH* gene is a part of the aromatic compound degradation pathway in *Pseudomonas putida* mt-2 [52]. It naturally catalyzes the conversion of

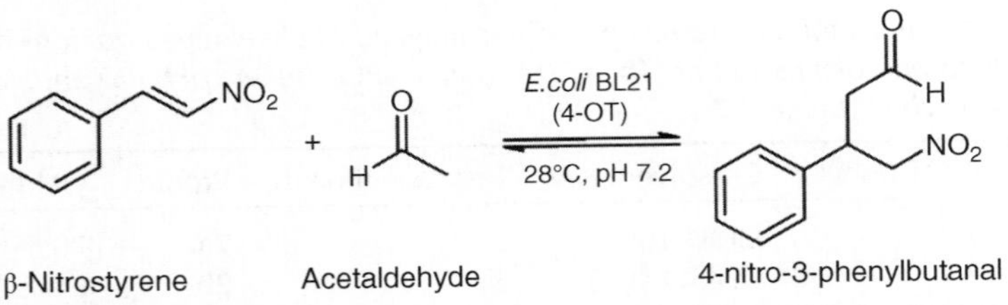

Scheme 4.8 *Biocatalytic synthesis of 4-nitro-3-phenylbutanal using* E. coli *whole cells expressing the* xylH *gene (4-OT).*

2-hydroxyhexa-2,4-dienedioate into 2-oxo-3-hexendioate, with terminal proline functioning as a catalytic base during proton transfer [52,53]. This particular enzyme in purified form has recently been described catalyzing Michael-type addition of acetaldehyde to β-nitrostyrene [54], the isomerization of *cis*-nitrostyrene to *trans*-nitrostyrene [55], and aldol condensation and dehydratation [56] in aqueous media with good enantioselectivity. The *xylH* gene (192 nt) has been cloned and expressed in *Escherichia coli* BL21(DE3), allowing an efficient and scalable biocatalytic system based on resting whole cells [57] (Scheme 4.8). A range of nitrostyrenes has been easily biotransformed using this system, affording valuable nitroaldehydes suitable for further modifications. Experimental details have previously been described [57].

4.4.1 Materials and Equipment

- *E. coli* BL21(4-OT) Harboring plasmid pRSET-TAUT (Figure 4.1) (frozen glycerol stocks)
- M9 Medium [58] supplemented with:
  - US trace element solution 1 mL.L^{-1} (1 M hydrochloric acid and the following salts, per liter: 1.50 g of MnCl$_2$·4 H$_2$O, 1.05 g of ZnSO$_4$, 0.30 g of H$_3$BO$_3$, 0.25 g of Na$_2$MoO$_4$·2 H$_2$O, 0.15 g of CuCl$_2$·2 H$_2$O, and 0.84 g of Na$_2$EDTA·2 H$_2$O)
  - casamino acids (NZ-Amine) 5 g.L^{-1}
  - glucose 10 g.L^{-1} (50% w/v stock solution in water, filter-sterilized)
  - ampicillin 50 µg.mL^{-1} (50 mg.mL^{-1} stock solution in water, filter-sterilized)
- M9 Agar plates (M9 medium supplemented with bacteriological agar 1.5 g.L^{-1})
- IPTG (238.3 mg.mL^{-1} stock solution in water)
- 20 mM Sodium phosphate buffer (pH 7.2) [58]
- HPLC-grade ethanol (5 mL)
- HPLC-grade ethyl acetate (500 mL)
- HPLC-grade isopropanol
- HPLC-grade heptane
- HPLC-grade hexanes or petroleum ether (500 mL)
- HPLC-grade water (200 mL)
- Acetaldehyde (HPLC-grade; 1 mL)
- β-Nitrostyrene stock solution in ethanol (29.8 mg.mL^{-1})
- Saturated solution of NaCl in water (brine)
- Magnesium sulfate anhydrous (10 g)
- 0.22 µm Polyvinylidene difluoride syringe filters

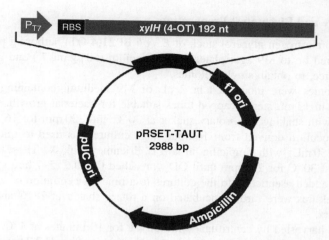

Figure 4.1 *4-OT expression construct based on pRSET-B (Invitrogen) vector.*

- Filter paper (Whatman No.1)
- 1–100 and 100–1000 µL Automatic pipettes
- Sterile loop
- Petri dish
- Sterile tubes (15 mL) for bacterial growth
- Two glass Erlenmeyer flasks (1 L)
- Duran Schott glass bottle (250 mL) with autoclavable screw-type plastic lid
- Nalgene solvent-resistant centrifuge bottles (400 mL)
- Rotary shaker at 28 and 30 °C
- Centrifuge capable of reaching $5000 \times g$ while holding 4 °C
- Centrifuge capable of reaching $10\,000 \times g$, suitable for holding 1.5–2.0 mL plastic tubes with lids
- UV/Vis spectrophotometer and suitable quartz or plastic cuvettes
- Set up for dry-flash column chromatography
- Silica gel 60 (230–400 mesh)

4.4.1.1 Optional

- Silica gel TLC plates (silica gel 60 F_{254}, Merck)
- 50% v/v H_2SO_4 in water (100 mL)
- Reagent spray bottle
- Heat gun
- Glass separating funnel (500 mL)
- CHIRALPAK IA 250 column (5 µm, 150 × 3.2 mm; Chiral Technologies Europe, Cedex, France)
- HPLC system and UV detection
- NMR system

4.4.2 Culture and Biocatalyst Preparation

1. Crystals from a frozen glycerol stock of *E. coli* BL21(4-OT) harboring pRSET-TAUT were streaked on to M9 agar plates with ampicillin (50 µg.mL^{-1}) and glucose as the carbon source, to obtain single colonies.
2. Single colonies were inoculated in 2 mL of M9 medium (containing 50 µg.mL^{-1} ampicillin) in 13 mL sterile capped tubes suitable for bacterial growth. Cultures were incubated with shaking on a rotary shaker at 30 °C and 200 rpm for 16 hours.
3. A 0.1% inoculum derived from 16-hour stage culture was used to initiate fresh M9 cultures (250 mL) with ampicilin in a 1 L Erlenmeyer flask. These cultures were incubated at 30 °C for 200 rpm until OD$_{600}$ reached 0.5–0.6 (5–7 hours).
4. IPTG was added aseptically to the cultures to a final concentration of 23.83 mg.L^{-1}.
5. Induced cultures were further incubated on a rotary shaker at 28 °C and 200 rpm for 12–14 hours.
6. Cells were harvested by centrifugation 5000 × *g* for 10 minutes at 4 °C.
7. Cell pellets were resuspended in cold 20 mM phosphate buffer pH 7.2 to a cell dry weight (cdw) concentration of 5 g.L^{-1} (OD$_{600}$ = 20).

4.4.3 Whole-Cell Biotransformation

1. Prepared cell suspension (60 mL) was transferred to a Duran Schott glass bottle.
2. A β-nitrostyrene stock solution was prepared in ethanol (29.8 mg.mL^{-1}) and 600 µL was added to the biotransformation sequentially (3 × 200 µL) upon complete utilization of β-nitrostyrene (usually time points: 0, 30, and 70 minutes upon reaction start). Concurrently with β-nitrostyrene addition, acetaldehyde was added directly to the cell suspension (3 × 23 µL). Using these amounts, final concentrations of 2 mM β-nitrostyrene and 20 mM acetaldehyde were used.
3. Reactions were incubated at 28 °C with shaking at 150 rpm and samples (800 µL) were withdrawn from the reaction over time and centrifuged at 10 000 × *g* for 3 minutes. The supernatants were analyzed spectrophotometrically and the depletion of β-nitrostyrene was followed by reduction of absorbance at 320 nm (ε = 14.4 mM^{-1}.cm^{-1}). Alternatively, biotransformations could be monitored by TLC (see Analytical Methods).
4. Upon complete depletion of β-nitrostyrene (approximately 30–40 minutes for the amounts given), the next batch of β-nitrostyrene and acetaldehyde was added to the reaction and Steps 3 and 4 were repeated.

4.4.4 Product Extraction and Purification

1. The reaction mixture was extracted with ethyl acetate (2 × 100 mL). Extraction was carried out in Nalgene solvent-resistant centrifuge bottles (400 mL volume), with centrifugation at 5000 × *g* for 5 minutes at 4 °C between extractions. Classical glass-separating funnels could be used instead.
2. The combined organic extract was washed with brine (100 mL), dried over anhydrous magnesium sulfate, and filtered.
3. Solvent was removed under reduced pressure and the residue was purified by dry flash column chromatography (silica gel), eluting with petroleum ether/ethyl acetate mixture in gradient (from 7 : 3 to 9 : 1) to afford pure product (4-nitro-3-phenylbutanal) in a yield of 60% (Figure 4.1).

4.4.5 Analytical Methods

Biotransformation was monitored by UV/Vis spectrophotometry (spectra from 200 to 500 nm were recorded, while depletion of β-nitrostyrene corresponded to a reduction of absorbance at 320 nm). Standard solutions were prepared by dissolving known volumes of β-nitrostyrene stock solution in 20 mM sodium phosphate buffer (pH 7.2). Aliquots of 0.8 mL of biotransformation samples were withdrawn and microcentrifuged at $10\,000 \times g$ for 3 minutes. The supernatants were analyzed by UV-Vis spectrometry, using 20 mM sodium phosphate buffer (pH 7.2) as blank.

TLC monitoring of biotransformation progress was carried out after quick, small-scale ethyl acetate extraction. A sample (0.5 mL) was placed in a plastic microfuge tube and about 100–200 mg of NaCl was added, followed by 200 μL of ethylacetate. The contents were vigorously mixed (vortex 20 seconds) and microcentrifuged at $10\,000 \times g$ for 3 minutes. The organic layer was spotted on silica gel plates (10–20 μL), along with 10–20 μg of standard compounds. Plates were developed with petroleum ether/ethylacetate 9 : 1 v/v solvent and visualized by spraying with a 50% H_2SO_4 spray reagent, followed by gentle heating. R_f values of standards were: β-nitrostyrene, 0.8; 4-nitro-3-phenylbutanal, 0.2.

Identity of the purified product was achieved by NMR analysis. ^{1}H- and ^{13}C-NMR spectra were recorded on a Varian Gemini 200 at 200/50 MHz in deuterated chloroform (CDCl$_3$) [57].

The HPLC system used a mobile phase consisting of heptane/iPrOH (80 : 20 v/v). Chiral separation of purified product and ee was achieved by isocratic elution over CHIRALPAK IA column (Chiral Technologies Europe) at 210 nm and flow rate 1 mL.min^{-1} (ee > 99%).

Alternative nitrostyrenes, such as 4-chloro-, 2-chloro, and 4-fluoro-β-nitrostyrene, are also suitable for this biotransformation, as are other aldehydes (propanal, butanal, pentanal, etc.).

4.4.6 Conclusion

Whole resting cells of recombinant *E. coli* expressing 4-OT are an excellent biocatalyst for Michael-type addition of acetaldehyde to β-nitrostyrenes. High activity and excellent enantioselectivity were achieved using this economically viable route. Furthermore, the successful application of the whole-cell catalyst in repetitive batch conversions was demonstrated. This bioprocess may be an efficient alternative for practical synthesis of 4-nitro-3-phenylbutanal and other aldehydes that are demonstrated synthons for desirable amino alcohols.

4.5 Norcoclaurine Synthases for the Biocatalytic Synthesis of Tetrahydroisoquinolines

Benjamin Lichman,[1] Eleanor D. Lamming,[2] John M. Ward,[1] and Helen C. Hailes[2]
[1] Department of Biochemical Engineering, University College London, UK
[2] Department of Chemistry, Christopher Ingold Laboratories, University College London, UK

The benzylisoquinoline alkaloids (BIAs) are a large and structurally diverse group of natural products. Many BIAs have significant biological activities, including morphine (analgesic), papaverine (antispasmodic), and berberine (antibiotic and antineoplastic). The tetrahydroisoquinoline (THIQ) moiety found in BIAs is also present in a number of other natural products (e.g., saframycin) and some synthetic pharmaceuticals. In the BIA pathway, the THIQ skeleton is formed by a Pictet–Spengler condensation between dopamine and 4-hydroxyphenylacetaldehyde (4-HPAA), catalyzed by the enzyme (*S*)-norcoclaurine synthase (NCS) [59]. Recent studies have shown that NCS can accept a variety of aldehyde acceptors, thus demonstrating its potential for use as a general biocatalytic tool for the synthesis of chiral THIQs [60,61]. A key challenge in developing facile biotransformations with NCS is the presence of a high background reaction in cell lysate. A biomimetic synthesis of racemic THIQs by phosphate catalysis has previously been developed, and phosphates present in the cell lysate are likely to be the cause of the high background reaction [62]. Here, we show two representative milligram-scale stereoselective syntheses of chiral THIQs: the first utilizes a purified mutant enzyme (L76V-*Tf*NCS), while the second uses cell lysate (containing *Cj*NCS2). The selection of different methods, enzymes, and substrates in these two reactions demonstrates the diversity of biotransformations possible with NCS.

4.5.1 Procedure 1: Purified Enzyme (Enzyme: L76V-*Tf*NCS, Aldehyde: Hexanal) (Scheme 4.9)

4.5.1.1 Materials and Equipment

- Glycerol stock of *E. coli* BL21(DE3)/pQR1045 (pJ411 with L76V-*Tf*NCS gene)
- Terrific broth
- Kanamycin
- IPTG
- BugBuster Protein Extraction Reagent (Merck)
- PD-10 desalting column (GE Healthcare)
- Syringe filter (0.2 µm cellulose acetate, glass fiber prefilter; Minisart plus)
- Ni-NTA Agarose (Qiagen)
- Binding buffer (0.1 M HEPES, 20 mM imidazole, 100 mM NaCl, pH 7.5)
- Wash buffer (0.1 M HEPES, 40 mM imidazole, 100 mM NaCl, pH 7.5)
- Elution buffer (0.1 M HEPES, 500 mM imidazole, 100 mM NaCl, pH 7.5)
- 0.1 M HEPES (pH 7.5)
- Glycerol
- Dopamine-HCl (38 mg, 0.20 mmol)
- Hexanal (30 mg, 0.30 mmol)

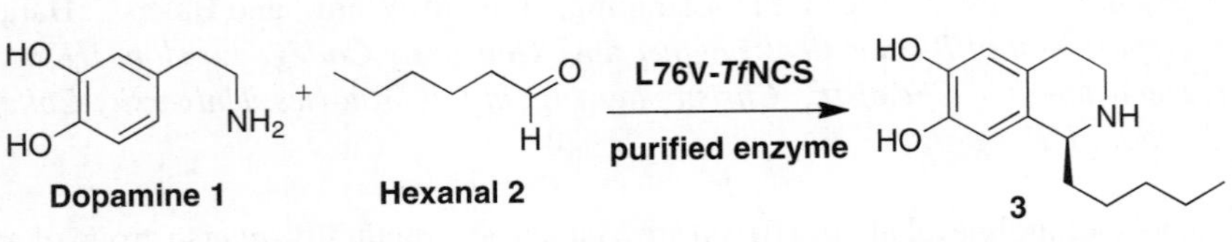

Scheme 4.9 *Purified enzyme (enzyme: L76V-TfNCS; aldehyde: hexanal).*

- MeCN
- 1 M HCl
- HPLC system and UV detection
- Ascentis Supelco C18–10 (10 × 2.12 cm) column
- Supelco Astec Chirobiotic T2 column
- HiChrom ACE C18-5 (150 × 4.6 mm) column

4.5.1.2   Selection and Preparation of Mutant NCS

Mutations of *Tf*NCS were designed based on analysis of the enzyme crystal structure, together with mechanistic predictions made by docking calculations [61,63]. Selected N-terminal His-tagged *Tf*NCS genes were synthesized and cloned into pJ411 vectors by DNA2.0. The plasmids obtained were transformed into *E. coli* BL21(DE3) cells by a standard heat-shock protocol. The transformed cells were stored as glycerol stocks at −80 °C. Screening assays of mutants with various substrates showed L76 V had good activity toward aliphatic substrates, and thus it was selected for use in this biotransformation.

4.5.1.3   Culture and Expression

1. An aliquot from a frozen glycerol stock of *E. coli* BL21(DE3)/pQR1045 (pJ411 with the N-terminal His-tagged L76V-*Tf*NCS gene) was inoculated into 20 mL of TB medium (containing 50 μg.mL^{-1} kanamycin). Starter cultures were incubated overnight at 37 °C, shaking at 250 rpm.
2. A 4 mL sample of overnight starter cultures was added to 100 mL of TB medium (containing 50 μg.mL^{-1} kanamycin). The expression cultures were incubated for 2 hours at 37 °C, followed by 1 hour at 25 °C, while shaking at 250 rpm.
3. Expression was induced by the addition of 100 μL of 500 mM IPTG. Cultures were incubated for a further 3 hours at 25 °C prior to harvesting, while shaking at 250 rpm.
4. Cells were harvested by centrifugation at 10 000 × *g* for 10 minutess at 4 °C. Supernatant was removed and the cells were stored at −20 °C until purification.

4.5.1.4   Protein Purification

1. Cell pellets were thawed and suspended in BugBuster (10 mL).
2. The insoluble portion of the lysate was pelleted by centrifugation at 10 000 × *g* for 30 minutes at 4 °C. The supernatant was removed and filtered through a glass fiber prefilter and 0.2 μm cellulose acetate syringe filter.
3. An empty and clean PD-10 column was charged with Ni-NTA (2 mL). The column was washed with distilled water (10 mL), followed by binding buffer (10 mL).
4. The filtered supernatant (from Step 2) was passed through the Ni-NTA column, and the column was then washed with binding buffer (10 mL), followed by wash buffer (20 mL, or until no protein could be detected coming off the column by a standard Bradford's assay). The bound protein was then eluted with elution buffer (5 mL).
5. The eluant containing pure enzyme was buffer-exchanged into 0.1 M HEPES, using a PD-10 column. Glycerol was added (10% v/v) and the concentration of the protein was

determined by absorbance at 280 nm (typical yield was 1.0 mg.mL^{-1}). The protein was divided into 0.5 mL samples and frozen in liquid nitrogen. The purified protein was stored at −80 °C.

4.5.1.5 Biotransformation

1. Substrate solutions of dopamine.HCl (38 mg, 0.20 mmol) in water (2 mL) and hexanal (30 mg, 0.30 mmol) in MeCN (2 mL) were prepared.
2. The reaction mixtures were prepared by the addition of the dopamine solution (2 mL, 100 mM, 0.20 mmol), the hexanal solution (2 mL, 150 mM, 0.30 mmol), and purified enzyme (1.0 mg.mL^{-1}, 2 mL) to water (14 mL). The final reaction mixtures (20 mL) contained 10 mM dopamine, 15 mM hexanal, and approximately 0.1 mg.mL^{-1} purified enzyme.
3. The reaction was incubated at 37 °C for 3 hours, before being quenched by the addition of 1 M HCl (2 mL). The mixture was centrifuged (10 000 × g for 30 minutes) to remove protein precipitate and then filtered. Samples for analysis (see Product Analysis) were taken at this point, and indicated a conversion yield of 70%. The reaction mixture was then concentrated under reduced pressure.

4.5.1.6 Product Purification

1. Product **3** was purified on an HPLC system using an Ascentis Supelco C18-10 (10 × 2.12 cm) column with an isocratic gradient of 15% MeCN (0.1% TFA) and 85% H$_2$O (0.1% TFA) at a flow rate of 8 mL.min^{-1}, with detection at 280 nm. Compound **3** eluted after 32.0 minutes.
2. Solvent was removed under reduced pressure and the product was washed with MeOH to remove excess TFA, yielding (1S)-1-pentyl-1,2,3,4-tetrahydro-isoquinoline-6,7-diol as the TFA salt [(S)-**3**.TFA] (40 mg, 57%, ee > 98%), with spectroscopic data consistent with those previously reported [64].

4.5.1.7 (1S)-1-Pentyl-1,2,3,4-tetrahydroisoquinoline-6,7-diol (TFA salt) [(S)-3.TFA] [64]

^{1}H-NMR (500 MHz; CD$_3$OD): δ = 0.93 (3 H, t, J = 7.1 Hz, (CH$_2$)$_3$CH$_3$), 1.38–1.50 (6 H, m, CH$_2$)$_3$CH$_3$), 1.87 (1 H, m, CHCHH), 2.01 (1 H, m, CHCHH), 2.86–3.00 (2 H, m, 4-H$_2$), 3.29–3.34 (1 H, m, 3-HH), 3.50 (1 H, m, 3-HH), 4.34 (1 H, dd, J = 8.3 and 5.0 Hz, 1-H), 6.60 (1 H, s, 5-H), 6.64 (1 H, s, 8-H); ^{13}C-NMR (125 MHz; CD$_3$OD): δ = 14.3, 23.5, 25.7, 26.1, 32.7, 35.1, 41.0, 56.7, 113.9, 116.2, 118.3 (q, $^1J_{CF}$ 295 Hz, CF$_3$), 123.7, 124.3, 145.9, 146.7, 163.2 (br, CF$_3$CO$_2$); m/z [HRMS ES+] found MH$^+$ 236.1639. C$_{14}$H$_{22}$NO$_2$ requires 236.1651

4.5.1.8 Product Analysis

Conversion yield was determined by analytical normal phase HPLC using an ACE C18 (150 × 4.6 mm) column and a 1 mL.min^{-1} gradient of H$_2$O (0.1% TFA)/MeCN from 90/10% to 30/70% over 6 minutes. The column was used at 30 °C, and compound detection was at 280 nm. Product **3** eluted after 5.9 minutes, and showed a conversion yield of 70%.

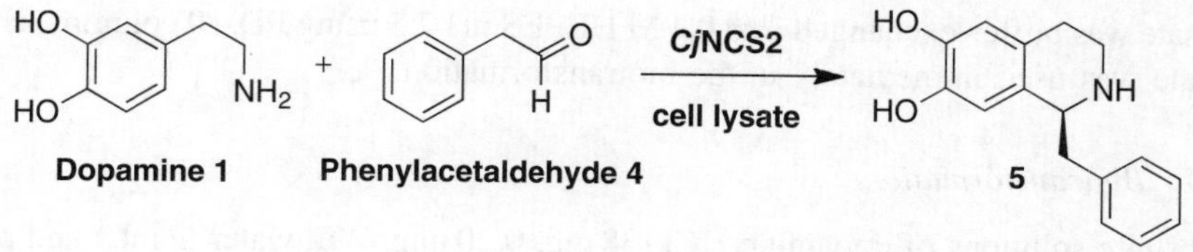

Scheme 4.10 *Cell lysate (enzyme: CjNCS2; aldehyde: phenylacetaldehyde).*

Product *ee* was determined by HPLC analysis using a Supelco Astec Chirobiotic T2 column and an isocratic MeOH (0.1% TFA, 0.2% TEA) mobile phase at 1 mL.min^{-1} and 30 °C. Compounds were detected by UV absorbance at 230 nm. The retention times of the *S* and *R* isomers were identified from a preparation of the racemate using the biomimetic phosphate reaction; they were 12.7 and 15.9 minutes, respectively [62]. The procedure provided **3** (*S*-isomer as for other NCSs with aliphatic aldehydes) [61] with an *ee* > 98%.

4.5.2 Procedure 2: Cell lysate (Enzyme: *Cj*NCS2, Aldehyde: Phenylacetaldehyde) (Scheme 4.10)

4.5.2.1 Materials and Equipment

- Glycerol stock of *E. coli* BL21(DE3)/pQR1025 (pET29a with *Cj*NCS2 gene[3])
- Terrific broth
- Kanamycin
- IPTG
- BugBuster protein extraction reagent (Merck)
- PD-10 Desalting column (GE Healthcare)
- 0.1 M HEPES (pH 7.5)
- Dopamine-HCl (38 mg, 0.30 mmol)
- Phenylacetaldehyde (36 mg, 0.20 mmol)
- MeCN
- 1 M HCl
- HPLC system and UV detection
- Ascentis Supelco C18-10 (10 × 2.12 cm) column
- Supelco Astec Chirobiotic T2 column
- HiChrom ACE C18-5 (150 × 4.6 mm) column

4.5.2.2 Culture and Expression

See Procedure 1 for details.

4.5.2.3 Lysate Preparation

1. Cell pellets were thawed and suspended in BugBuster (10 mL). If necessary, lysate was sonicated until the mixture became homogenous.
2. The insoluble portion of the lysate was pelleted by centrifugation at 10 000 × *g* for 45 minutes at 4 °C.

3. Lysate was buffer-exchanged into 0.1 M HEPES pH 7.5 using PD-10 columns. HEPES-lysate was used immediately in the biotransformation.

4.5.2.4 Biotransformation

1. Substrate solutions of dopamine-HCl (38 mg, 0.20 mmol) in water (2 mL) and phenyl-acetaldehyde (36 mg, 0.30 mmol) in MeCN (2 mL) were prepared.
2. The reaction mixtures were prepared by the addition of the dopamine solution (2 mL, 0.20 mmol), the phenylacetaldehyde solution (2 mL, 0.30 mmol), and HEPES-lysate (7 mL) to water (9 mL). The final reaction mixtures (20 mL) contained 10 mM dopamine and 15 mM phenylacetaldehyde.
3. The reaction was incubated at 37 °C for 1 hour, before being quenched by the addition of 1 M HCl (2 mL). The mixture was centrifuged (10 000 × g for 30 minutes) to remove protein precipitate and then filtered. Samples for analysis (see Product Analysis) were taken at this point, and indicated a conversion yield of 81%. The reaction mixture was then concentrated under reduced pressure.

4.5.2.5 Product Purification

1. Product **5** was purified on an HPLC system using a C18 Ascentis Supelco column with an isocratic gradient of 15% MeCN (0.1% TFA) and 85% H_2O (0.1% TFA) at a flow rate of 8 mL.min^{-1}, with UV detection at 280 nm. Compound **5** eluted after 18.4 minutes.
2. Solvent was removed under reduced pressure and the product was washed with MeOH to remove excess TFA, yielding (1S)-1-benzyl-1,2,3,4-tetrahydroisoquinoline-6,7-diol as a TFA salt [(S)-**5**.TFA] (33 mg, 45%, *ee* > 97%), with spectroscopic data consistent with those previously reported [62].

4.5.2.6 Product Analysis

Conversion yield was determined by analytical normal phase HPLC using an ACE C18 (150 × 4.6 mm) column and a 1 mL.min^{-1} gradient of H_2O (0.1% TFA)/MeCN from 90/10% to 30/70% over 6 minutes. The column was used at 30 °C, and compound detection was at 280 nm. Product **5** eluted after 5.5 minutes, and showed a conversion yield of 81%.

Product *ee* was determined by isocratic normal phase HPLC using a Supelco Astec Chirobiotic T2 column and a MeOH (0.1% TFA, 0.2% TEA) mobile phase at 1 mL.min^{-1} and 30 °C. Compounds were detected by UV absorbance at 230 nm. The retention times of the S and R isomers were identified from a preparation of the racemate using the biomimetic phosphate reaction; they were 12.0 and 16.8 minutes, respectively [62]. The procedure provided **5** (*S*-isomer as previously reported for *Cj*NCS [61]) with an *ee* > 97%.

4.5.3 Conclusion

The development of two methods of biotransformation using NCS allows the reaction to be tailored to the particular substrates being used. In general, reactions with purified enzymes are more time-intensive and cost more to perform, but can give products in greater yields. Here, the method of lysate preparation, involving a desalting step, removes the background reaction that generates racemic product, and the THIQ is formed in very high stereoselectivities. Furthermore, we have shown here that it is possible to use mutant NCS enzymes in biotransformations, enabling optimization of reactions with unnatural substrates. The methods shown here can be used for the facile preparation of many different THIQs.

4.5.4 Acknowledgments

We gratefully acknowledge the Wellcome Trust for studentship funding to B.L. and EPSRC and GlaxoSmithKline for studentship funding to E.L.

4.6 Streptavidin-Based Artificial Metallo-Annulase for the Enantioselective Synthesis of Dihydroisoquinolones

Todd K. Hyster,[1,2] Livia Knörr,[1] Tomislav Rovis,[2] and Thomas R. Ward[1]

[1] *Department of Inorganic Chemistry, University of Basel, Switzerland*
[2] *Department of Chemistry, Colorado State University, USA*

The development of artificial metalloenzymes capable of catalyzing transformations currently inaccessible using biocatalysis is a growing area of interest [65,66]. We prepared an artificial metalloenzyme derived from a streptavidin, a protein with a well-established affinity for biotinylated complexes [67,68] (Scheme 4.11). When coupled with a biotinylated Rh(III) catalyst ([RhCp*[biotin]Cl$_2$]$_2$), this artificial metalloenzyme catalyzed the coupling of acrylates and amides to furnish dihydroisoquinolones [69]. During our studies, we found two mutations to be essential for improved results: introduction of a glutamic acid residue at position 121 (K121E) provided improved activity, while tyrosine at position 112 (S112Y) furnished superior enantioselectivity. The experimental details are taken primarily from reference [69].

4.6.1 Materials and Equipment

- HPLC-grade MeOH
- Distilled water
- ACS-grade DMSO
- ACS-grade EtOAc
- MOPS buffer (pH 5.9, M = 0.7 M)
- Methyl acrylate (99% purity)
- Streptavidin S112Y-K121E mutant (66 nmols, 4.3 mg)

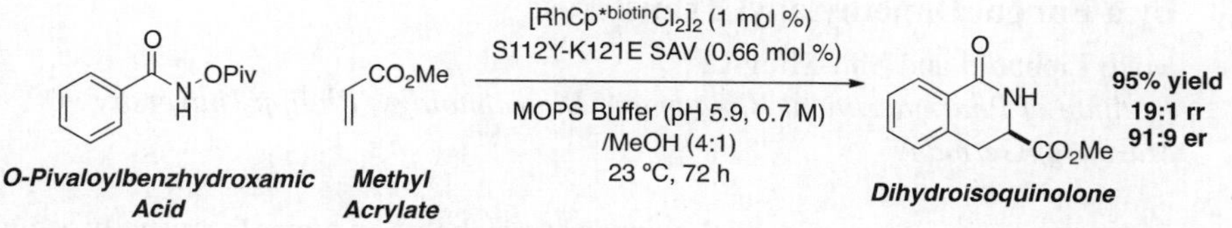

Scheme 4.11 Artificial metalloenzyme-catalyzed synthesis of dihydroisoquinolones.

- [RhCp*$^{*\text{biotin}}$Cl$_2$]$_2$ (0.1 µmols, 0.11 mg) [69]
- *O*-pivaloyl benzhydroxamic acid (0.01 mmols, 2.2 mg) [70]

A detailed description of protein expression and purification can be found in reference [69].

4.6.2 Streptavidin-Based Artificial Metalloenzyme-Catalyzed C–H Activation-Mediated Coupling Reaction

1. Stock solutions were prepared of:
   a. *O*-pivaloyl benzhydroxamic acid in MeOH (250 mM)
   b. [RhCp*$^{*\text{biotin}}$Cl$_2$]$_2$ in DMSO (100 mM)
   c. S112Y-K121E variant in MOPS Buffer (4.13 mM)
   d. 1,3,5-trimethoxybenzene in EtOAc (200 mM)
2. A 1 mL vial was charged with a stir bar and protein solution (160 µL), followed by [RhCp*$^{*\text{biotin}}$Cl$_2$]$_2$ in DMSO (1 µL), and allowed to stir (500 rpm) for 5 minutes.
3. *O*-Pivaloyl benzhydroxamic acid solution (40 µL, 0.01 mmol) was added to the vial, followed by methyl methylacrylate (1.1 µL 0.013 mmol). The vial was sealed and the reaction was allowed to stir for 72 hours at 23 °C.
4. After completion of the reaction, 1,3,5-trimethoxybenzene solution (50 µL) (as internal standard) and EtOAc (400 µL) were added. The biphasic mixture was stirred for 30 minutes and the organic phase was analyzed by GC-MS and chiral HPLC.

4.6.3 Analytical Methods

The yield was determined by integration relative to the 1,3,5-trimethoxybenzene internal standard using GC-MS. The enantiomeric ratio was determined by chiral HPLC using a CHIRALPAK IA column. The mobile phase was a hexane/*i*PrOH (90 : 10) mixture with a flow rate of 1.0 mL·min^{-1}. The retention times and values were as follows: minor (*S*)-enantiomer, 18.12 minutes; major (*R*)-enantiomer, 19.76 minutes.

4.6.4 Conclusion

O-Pivaloylbenzyhydroxamic acid and methyl acrylates can be coupled under the aegis of a streptavidin-based artificial metalloenzyme to provide the desired dihydroisoquinolone in 95% yield and 91 : 9 enantiomeric ratio. The reaction can be applied to a broad range of substrates with comparable yield and enantioselectivity. Essential for the observed reactivity and selectivity are the biotinylated rhodium co-factor and two mutations (S112Y-K121E).

4.7 Regiospecific Benzylation of Tryptophan and Derivatives Catalyzed by a Fungal Dimethylallyl Transferase

Mike Liebhold and Shu-Ming Li

Institute of Pharmaceutical Biology and Biotechnology, Philipp University of Marburg, Germany

Fungal prenyltransferases are microbial secondary metabolite enzymes that usually catalyze highly regiospecific chemical reactions [71]. The acceptance of structurally modified

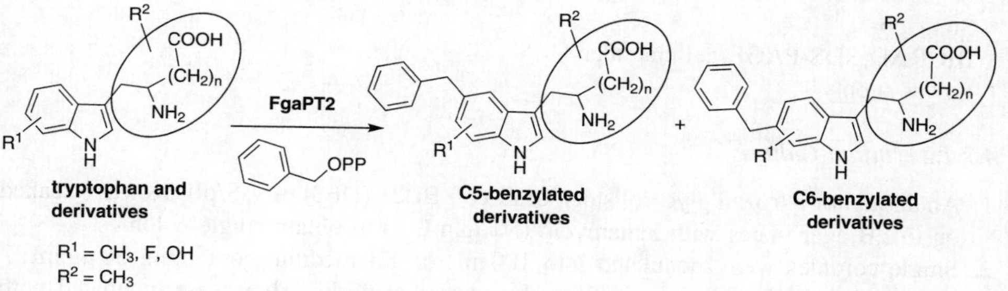

Scheme 4.12 *Enzymatic synthesis of benzylated tryptophan and derivatives.*

non-natural substrates by a number of these enzymes makes them useful as biocatalysts for chemoenzymatic synthesis [72,73]. The dimethylallyl transferase FgaPT2, from *Aspergillus fumigatus*, which catalyzes the C4-prenylation of L-tryptophan, can also use aliphatic dimethylallyl pyrophosphate (DMAPP) analogs as alkyl donors [74,75]. Incubation of L-tryptophan and different tryptophan derivatives with the DMAPP analog benzyl diphosphate leads to regiospecific benzylation of these compounds, mainly at C-5 of the indole ring, with product yields of up to 58% [76]. In two cases, C6-benzylated derivatives were obtained as an additional or sole product (Scheme 4.12). The experimental details described here are taken from reference [76].

4.7.1 Procedure 1: Preparation of the Prenyltransferase FgaPT2

4.7.1.1 Materials and Equipment

- *E. coli* BL21(DE3) pLysS harboring plasmid pIU18 (*FgaPT2* in pHis8) as frozen glycerol stock
- LB medium (per liter: 10 g tryptone, 5 g yeast extract, 10 g NaCl)
- Agar
- TB medium (per liter: 12 g tryptone, 24 g yeast extract, 4 mL glycerol)
- $10 \times$ TB salts (per liter: 23.1 g KH_2PO_4, 125.4 g K_2HPO_4)
- Kanamycin (50 $\mu g.mL^{-1}$ stock solution in water, filter-sterilized)
- IPTG (1 M stock solution in water, filter-sterilized)
- Lysozyme
- Lysis buffer pH 8.0 (10 mM imidazole, 50 mM NaH_2PO_4, 300 mM NaCl)
- Wash buffer pH 8.0 (20 mM imidazole, 50 mM NaH_2PO_4, 300 mM NaCl)
- Elution buffer pH 8.0 (250 mM imidazole, 50 mM NaH_2PO_4, 300 mM NaCl)
- Protein storage buffer pH 7.5 (50 mM Tris-HCl, 15% v/v glycerol)
- Protino Ni-NTA Agarose
- SDS-PAGE
- Low-molecular-weight calibration kit for SDS electrophoresis
- Sephadex G-25 NAP-10
- Shaker flask
- Incubator at 37 °C
- Centrifuge

- Sonicator
- BIORAD SDS-PAGE gel chamber
- Power supply

4.7.1.2   Initial Culture

1. An aliquot of a frozen glycerol stock of *E. coli* BL21 (DE3) pLysS/pIU18 was streaked on to LB agar plates with kanamycin ($50 \, \mu g.mL^{-1}$) to obtain single colonies.
2. Single colonies were inoculated into 100 mL of TB medium containing $50 \, \mu g.mL^{-1}$ kanamycin and $1 \times$ TB salts in 250 mL cylindrical flasks. They were incubated with shaking at 220 rpm on a rotary shaker at 37 °C for 16 hours.
3. A 2% inoculum obtained from 16-hour cultures was transferred to 1 L fresh TB media with $1 \times$ TB salts and $50 \, \mu g.mL^{-1}$ kanamycin in a 2.5 L cylindrical flask. These cultures were incubated at 37 °C with shaking at 220 rpm until an absorption at 600 nm of 0.7 was reached. Cultures were cooled to 22 °C and IPTG was added to a final concentration of 0.5 mM for induction. The cells were cultured at 22 °C with shaking at 220 rpm for a further 6 hours.
4. *E. coli* cells were pelleted by centrifugation at $6750 \times g$ for 10 minutes at 4 °C. The pellets were suspended in 1 mL lysis buffer pH 8.0 per gram wet weight and stored at −80 °C until use.

4.7.1.3   Protein Extraction and Purification

1. After the suspended *E. coli* cells were thawed at 4 °C, $1 \, mg.mL^{-1}$ lysozyme was added and the suspension was incubated for 30 minutes on ice. Cells were sonicated six times for 10 seconds, each at 200 W. To separate the cellular debris from the soluble proteins, the lysate was centrifuged at $20\,000 \times g$ for 30 minutes at 4 °C.
2. One-step purification of the recombinant His_8-FgaPT2 fusion proteins by affinity chromatography with Ni-NTA agarose resin was carried out according to the manufacturer's instructions.
3. The proteins were eluted with elution buffer pH 8.0. In order to remove imidazole, the protein fraction was passed through a NAP-10 column, which had been equilibrated with protein storage buffer pH 7.5 and eluted with the same buffer.
4. The purified protein was analyzed on SDS-PAGE for purity proof and concentration determination, then stored at −80 °C.

4.7.2   Procedure 2: Preparative Synthesis and Structural Elucidation of Benzylated Tryptophan Derivatives

4.7.2.1   Materials and Equipment

- L-Tryptophan and derivatives
- Benzyl diphosphate
- $CaCl_2$
- DMSO
- Tris-HCl
- Glycerol
- HPLC-grade methanol
- HPLC-grade water
- Incubator at 37 °C

- Rotary evaporator
- Agilent HPLC 1200 system and UV detection
- Multospher 120 RP-18 column (250 × 4 or 250 × 10 mm, 5 μm, CS Chromatographie Service, Langenfeld, Germany)

4.7.2.2 Synthesis of Benzyl Diphosphate

1. Benzyl diphosphate was prepared according to the method described for geranyl diphosphate by Woodside *et al.* [77], with a product yield of 58% and a purity of over 90%. ^{1}H-NMR (500 MHz; D$_2$O): $\delta_H = 7.58$ (d, $J = 7.1$, 2 H), 7.53 (t, $J = 7.4$, 2 H), 7.47 (t, $J = 7.2$, 1 H), 5.08 (d, $J = 6.8$, 2 H).

4.7.2.3 Enzymatic Synthesis, Analysis, and Isolation of Benzylated L-Tryptophan and Derivatives

1. The reaction mixtures contained L-tryptophan or derivatives (1 mM), benzyl diphosphate (2 mM), CaCl$_2$ (10 mM), DMSO 0.0–2.5% v/v, Tris-HCl (50 mM, pH 7.5), glycerol 1.5% v/v, and FgaPT2 10 μg.100 μL^{-1}. They were incubated at 37 °C for 16 hours and then terminated with one volume of methanol. Isolation assays were upscaled to a total volume of 10 mL with 0.2 or 0.4 mg.mL^{-1} FgaPT2.
2. The enzymatic products were analyzed on an Agilent HPLC series 1200 using a Multospher 120 RP-18 column (250 × 4 mm) at a flow rate of 1 mL.min^{-1}. A linear gradient of 40–100% solvent B (methanol) in solvent A (water) over 15 minutes was used. The column was then washed with 100% solvent B for 5 minutes and equilibrated with 40% solvent B for 5 minutes. Detection was carried out by a photo diode array detector at 277 nm. For isolation, the same HPLC equipment with a Multospher 120 RP-18 (250 × 10 mm) was used. A linear gradient of 60–80% solvent B in A over 15 minutes, 70–90% over 20 minutes, or 70–90% over 15 minutes at 2.5 mL.min^{-1} was used. The column was washed for 5 minutes with the initial ratio of both components prior to the gradient, then washed with 100% solvent B for 5 minutes and equilibrated with the initial condition for 5 minutes.

4.7.2.4 Structural Elucidation of Benzylated L-Tryptophan and Derivatives

High conversions were obtained for the substrates L-tryptophan (**1a**), L-abrine (**2a**), 4-methyl-DL-tryptophan (**3a**), L-β-homotryptophan (**4a**), 6-fluoro-DL-tryptophan (**5a**), 7-methyl-DL-tryptophan (**6a**), α-methyl-DL-tryptophan (**7a**), 6-methyl-DL-tryptophan (**8a**), and 5-hydroxy-L-tryptophan (**9a**), with yields of 12–58% with 10 μg of protein per 100 μL mixture (see Figure 4.2 for structures with benzylation positions). For structure elucidation, the isolated products were analyzed by ^{1}H-NMR, including H-H COSY and NOESY (JEOL ECA-500 500 MHz spectrometer), as well as high-resolution electron impact (HR-EI) mass spectrometry (Auto SPEC).

Mass spectrometry analysis revealed that the isolated products had masses 90 Da larger than those of the respective substrates.

Detailed interpretation of the NMR spectra of the isolated products and comparison with those of known prenylated tryptophan derivatives revealed the benzylation at C-5 of the indole ring for eight of the nine tested substrates (**1a–8a**) [76]. C6-benzylkated derivatives

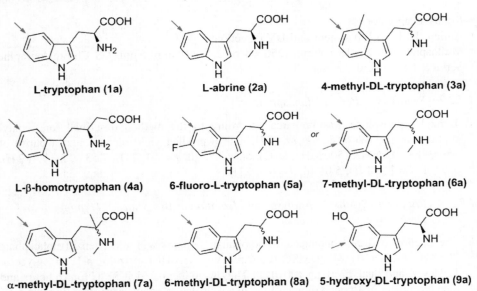

Figure 4.2 Substrates *1a–9a*; arrows indicate the benzylation positions.

were isolated from the incubation mixtures of **6a** and **9a**. By way of example, the ^{1}H-NMR data for 5-benzyl-L-tryptophan are as follows:

^{1}H-NMR (500 MHz; methanol-d_4): δ_H = 7.59 (dd, J = 0.8, 1 H), 7.26 (d, J = 8.3, 1 H), 7.21 (m, 4 H), 7.16 (s, 1 H), 7.11 (m, 1 H), 6.96 (dd, J = 8.3, 1.6, 1 H), 4.04 (s, 2 H), 3.84 (dd, J = 9.8, 4.0, 1 H), 3.49 (dd, J = 15.2, 4.0, 1 H), 3.08 (dd, J = 15.2, 9.8, 1 H)

4.7.3 Conclusion

By using the prenyltransferase FgaPT2 as a biocatalyst, it was possible to synthesize regiospecifically C5-benzylated L-tryptophan and derivatives thereof in moderate to high yields in one-step reactions. Furthermore, benzylation expanded the reaction potential of prenyltransferases for DMAPP analogs from allylic to aromatic donors [75].

4.8 Enantioselective Michael Addition of Water Using *Rhodococcus Rhodochrous* ATCC 17895

Bi-Shuang Chen,[1] Verena Resch,[2] and Ulf Hanefeld[1]
[1] Department of Biotechnology, Delft University of Technology,
The Netherlands
[2] Department of Chemistry, Organic and Bioorganic Chemistry,
University of Graz, Austria

The Michael addition of water to conjugated carbonyl compounds is a major challenge in organic chemistry [78]. Use of water as a nucleophile has rarely been investigated, but it is advantageous for multiple reasons: it represents an abundant substrate and serves as a solvent, and it creates a benign reaction system. Adding water in a Michael-addition fashion also provides a highly atom-efficient route to alcohols [79–81].

Although the addition of water looks straightforward, to date very few procedures have been described. The biocatalytic hydration of α,β-unsaturated carbonyl compounds represents a potent and new route to their respective β-hydroxy ketones or aldehydes, replacing indirect multiple chemical processes [82,83]. Those methods often use complex or expensive catalysts, furnishing racemic products. Even though nature seems to be more capable of performing the addition of water, investigated enzymes often suffer from high substrate specificity and poor stability [78].

The biotransformation of 3-methyl- or 3-ethyl-2-butenolide giving the corresponding (R)-3-hydroxy-3-alkylbutanolides using *Rhodococcus rhodochrous* strain ATCC 17895 as a biocatalyst represents a biocatalytic enantiospecific Michael addition of water (Scheme 4.13). The reaction has been shown to be reliable, and it can readily be employed on gram scale, with good yield and excellent enantioselectivity. A comprehensive genomic study of *R. rhodochrous* strain ATCC 17895 is given in reference [84]. The experimental details in this section are reproduced primarily from reference [85].

4.8.1 Procedure 1: Substrates Synthesis

4.8.1.1 *Materials and Equipment*

- Hydroxyacetone (2 g, 27.02 mmol)
- 1-Hydroxy-2-butanone (2 g, 22.72 mmol)
- (Carbethoxymethylene)triphenylphosphorane (16 g, 2 × 22.98 mmol)
- Toluene (anhydrous, 100 mL)
- Methanol (40 mL)
- 5% Aqueous sodium hydroxide (40 mL)
- Hydrochloric acid (conc., 10 mL)
- Ether (80 mL)
- Ethyl acetate (3 L)
- Petroleum ether (3 L)
- Potassium permanganate staining solution (1.5 g KMnO$_4$, 10 g K$_2$CO$_3$, and 1.25 mL 10% NaOH in 200 mL water)
- Liquid–liquid continuous extractor (500 mL capacity)
- Silica gel TLC plates (silica gel 60 F254, Merck)
- Silica column chromatography
- Bruker Advance 400 NMR instrument

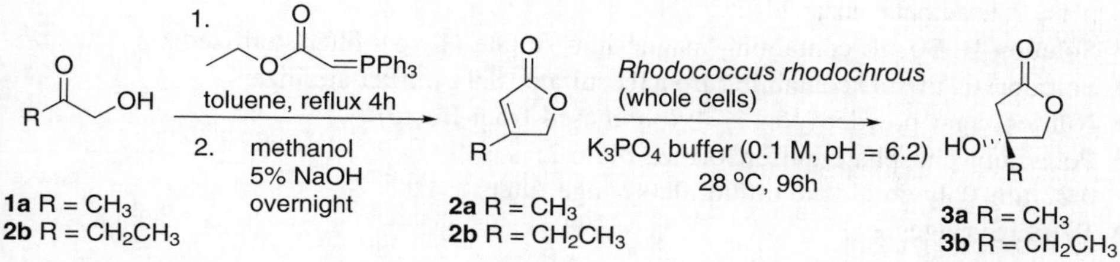

Scheme 4.13 *Substrate synthesis and biocatalytic Michael addition of water.*

- PerkinElmer 241 polarimeter (sodium D line at 20 °C)
- UV lamp/viewing box

4.8.1.2 Procedure

1. Hydroxyacetone (2 g, 27.02 mmol) or 1-hydroxy-2-butanone (2 g, 22.72 mmol) was added to a solution of (carbethoxymethylene)triphenylphosphorane (8 g, 22.98 mmol) in toluene (50 mL) and the mixture was heated at reflux for 4 hours.
2. The resultant mixture was concentrated under reduced pressure and purified by chromatography over silica (eluent: PE/EtOAc 1:1) to afford the hydroxyester intermediate.
3. The hydroxyester was dissolved in methanol (20 mL), treated with 5% aqueous NaOH (20 mL) at 0 °C, and stirred overnight at room temperature.
4. The resultant mixture was diluted with water (40 mL) and washed with ether (2 × 20 mL). The aqueous layer was acidified to pH 1 with conc. HCl, saturated with sodium chloride, and extracted with ethyl acetate (3 × 50 mL).
5. The organic phases were combined and dried with Na_2SO_4 and evaporated, and the product was purified by chromatography over silica (eluent: PE/EtOAc 1:2) to afford **2a** as a colorless solid (mp 52–53 °C) and **2b** as a colorless oil, respectively.

4.8.1.3 3-Methyl-2-Butenolide (2a)

Yield 1.92 g (19.53 mmol), 85%; ^{1}H-NMR (400 MHz, DMSO-d6): δ = 1.95 (s, 3 H), 3.92 (s, 2 H), 5.81–5.88 (s, 1 H); ^{13}C-NMR (100 MHz, DMSO-d6): δ = 15.25, 65.31, 113.12, 158.45, 167.69 (in accordance with reference [85])

4.8.1.4 3-Ethyl-2-Butenolide (2b)

Yield 2.16 g (19.27 mmol), 85%; ^{1}H-NMR (400 MHz, CDCl$_3$): δ = 1.05 (t, J = 7.4 Hz, 3 H), 2.50 (q, J = 7.6 Hz, 2 H), 4.18 (s, 2 H), 5.93 (s, 1 H); ^{13}C-NMR (100 MHz, CDCl$_3$): δ = 12.94, 22.58, 64.71, 112.25, 165.69, 171.76 (in accordance with reference [85])

4.8.2 Procedure 2: Biotransformations

4.8.2.1 Materials and Equipment

- *Rhodococcus rhodochrous* strain ATCC 17895 (purchased from ATCC)
- Nutrient Broth medium (8 g.L^{-1}) (purchased from BD: Becton, Dickinson and Company)
- Solution A: 2.94 L containing potassium dihydrogen phosphate (1.2 g), dipotassium hydrogen phosphate (3.6 g), peptone (15 g), yeast extract (3 g), and glucose (45 g). Final pH = 7.2, autoclaved at 110 °C
- Solution B: 30 mL containing magnesium sulfate (1.5 g), filter-sterilized
- Solution C: 30 mL containing iron(II) sulfate (0.9 g), filter-sterilized
- Nutrient agar powder (15 g.L^{-1}) (purchased from Bacto)
- Potassium phosphate buffer (0.1 M, pH 6.2)
- 0.22 µm Polyvinylidene difluoride syringe filters
- 10 mL Syringes
- Sterile loop
- 2 L and 500 mL Erlenmeyer flasks
- Rotary shakers at 28 °C

- Centrifuge capable of reaching 10 000 rpm while holding 4 °C
- Petri dish

4.8.2.2 Microorganism and Culture Conditions

1. A freeze-dried pellet of *Rhodococcus rhodochrous* strain ATCC 17895, as received from ATCC, was rehydrated with 1.0 mL nutrient broth medium. The suspension was streaked on to nutrient agar plates and incubated at 28 °C to obtain single colonies.
2. A single colony was used to inoculate 3 L of cultivation medium obtained by mixing solutions A, B, and C (potassium dihydrogen phosphate 0.4 g.L^{-1}, dipotassium hydrogen phosphate 1.2 g.L^{-1}, peptone 5 g.L^{-1}, yeast extract 1 g.L^{-1}, glucose 15 g.L^{-1}, magnesium sulfate 0.5 g.L^{-1}, and iron(II) sulfate 0.3 g.L^{-1}).
3. Cultures were incubated at 28 °C using a rotary shaker at 180 rpm for 3 days to OD$_{600} \approx 6.3$. Resting cells were harvested by centrifugation at 10 000 rpm for 20 minutes at 4 °C.

4.8.2.3 Whole-Cell Biotransformations

1. Pelleted cells from the 3 L medium were resuspended in 300 mL of potassium phosphate buffer (0.1 M, pH 6.2), and substrate **2a** (1 g, 10.19 mmol) or **2b** (1 g, 8.92 mmol) was added.
2. Reactions were incubated at 28 °C and shaken at 180 rpm for 96 hours.
3. The cells were removed by centrifugation and the supernatant was saturated with NaCl. Due to the high solubility of the resulting alcohols in water, continuous extraction with ethyl acetate was performed overnight. The extract was then concentrated under reduced pressure and purified by chromatography over silica (eluent: PE/EtOAc 1 : 1).

4.8.2.4 3-Hydroxy-3-Methyl-2-Butanolide (3a)

Yield colorless oil 0.65 g (5.60 mmol), 55%; $[\alpha]_D^{20}$ +46.6 (*c* 0.96, CHCl$_3$) ($[\alpha]_D$ +53.92 (*c* 0.96, CHCl$_3$) [85]); ^{1}H-NMR (400 MHz, CDCl$_3$): δ = 1.47 (s, 3 H), 2.51–2.63 (ABq, 2 H), 2.72 (s, 1 H), 4.09–4.24 (ABq, 2 H); ^{13}C-NMR (100 MHz, CDCl$_3$): δ = 24.94, 43.06, 74.70, 79.82, 176.27 (in accordance with reference [85,88])

4.8.2.5 3-Hydroxy-3-Ethyl-2-Butanolide (3b)

Yield colorless oil 0.58 g (4.46 mmol), 50%; $[\alpha]_D^{20}$ +49.6 (*c* 0.75, CHCl$_3$), ($[\alpha]_D$ +48.9 (*c* 0.72, CHCl$_3$) [85]); ^{1}H-NMR (400 MHz, CDCl$_3$): δ = 0.98 (t, *J* = 7.6 Hz, 3 H), 1.72 (q, *J* = 7.4 Hz, 2 H), 2.49 (s, 1 H), 2.53 (s, 2 H), 4.11–4.23 (ABq, 2 H); ^{13}C-NMR (100 MHz, CDCl$_3$): δ = 8.70, 31.69, 42.36, 78.05, 79.26, 176.82 (in accordance with reference [85])

4.8.3 Procedure 3: *ee* Determination (Scheme 4.14)

4.8.3.1 Materials and Equipment

- (*R*)-α-Methyl-α-trifluoromethylphenylacetic acid (500 mg) (Mosher acid)
- Dicyclohexylcarbodiimide (DCC) (500 mg)
- 4-Dimethylaminopyridine (17 mg)
- Dichloromethane (anhydrous, 30 mL)

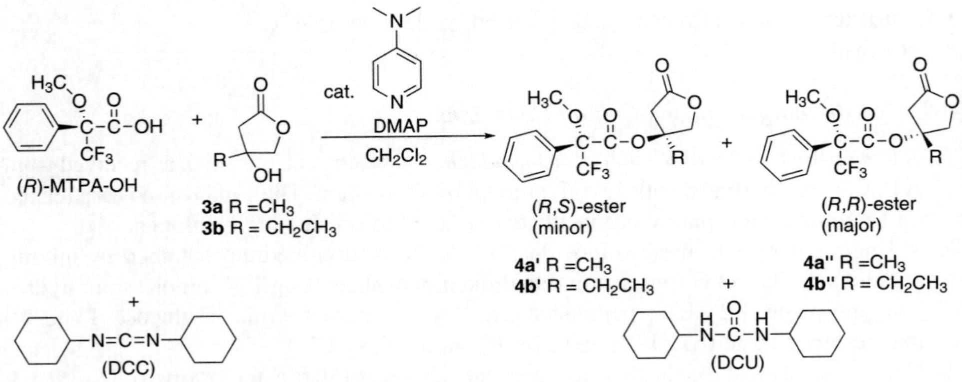

Scheme 4.14 *Synthesis of Mosher esters for ee determination.*

- Pyridine (1 mL)
- Screw-cap drum vials
- Pasteur pipets
- Cotton wool

4.8.3.2 Procedure

1. Solution **1** containing (R)-α-methyl-α-trifluoromethylphenylacetic acid [(R)-MTPA-OH] (20 mL) in anhydrous dichloromethane (50 mg.mL^{-1} = 0.214 mol.L^{-1}) and solution **2** containing 4-dimethylaminopyridine (DMAP) (5 mL) in anhydrous dichloromethane (3.4 g.mL^{-1} = 0.028 mol.L^{-1}) were prepared separately in screw-cap drum vials.

2. 70 mg of Michael product **3a** (0.60 mmol) or **3b** (0.54 mmol) was placed into a 25 mL screw-cap drum vial. DCC (247 mg, 1.19 mmol, 2 equiv.) and pyridine (72 µL, 1.0 equiv.) were added. Solution **1** (6.44 mL, 1.5 equiv.) was added via a syringe, followed by solution **2** (1.08 mL, 0.05 equiv.). The vials were closed tightly, mixed briefly, and left standing in the dark at room temperature under an atmosphere of nitrogen for around 2 days.

3. Upon complete conversion of the alcohols to the corresponding Mosher ester products (checked by TLC analysis), the reaction mixture was diluted with ethyl acetate and filtered through a cotton wool plug in a Pasteur pipet to remove dicyclohexylurea (DCU). Solvents were removed under reduced pressure and the dilution, filtration, and evaporation cycle was repeated twice to remove traces of DCU. The filtrate was again concentrated under reduced pressure for ^{1}H-NMR and ^{19}F-NMR measurement. Yields were quantitative and no further purification was carried out.

4.8.3.3 3-Methyl-5-Oxotetrahydrofuran-3-yl 3,3,3-Trifluoro-2-Methoxy-2-Phenylpropanoate (4a)

87% *ee* was calculated by integration of the proton peak areas of group –OCH$_3$ and fluorine peak areas of group –CF$_3$; ^{1}H-NMR (400 MHz, CDCl$_3$): δ = 1.75 (s, 3 H), 2.70 (d, J = 18 Hz, 1 H), 2.98 (d, J = 18 Hz, 1 H), 3.53 (dd, J_{HF} = 1.2 Hz, J_{HF} = 2.3 Hz, 3 H), 4.28 (d, J = 10.8 Hz, 1 H), 4.65 (d, J = 10.8 Hz, 1 H), 7.37–7.47 (m, 5 H); ^{13}C-NMR (100 MHz, CDCl$_3$): δ = 21.71, 41.28, 55.58 (d, J_{CF} = 1.4 Hz), 76.06, 84.73, 123.32

(d, $J_{CF} = 287\,Hz$), 127.24, 128.87, 130.09, 131.81, 166.04, 173.27. ^{19}F-NMR (376.58 MHz, CDCl$_3$): $\delta = -71.54$, -72.14

4.8.3.4   *3-Ethyl-5-Oxotetrahydrofuran-3-yl 3,3,3-Trifluoro-2-Methoxy-2-Phenylpropanoate (4b)*

94% *ee* was calculated by integration of the proton peak areas of group –OCH$_3$ and fluorine peak areas of group –CF$_3$; ^1H-NMR (400 MHz, CDCl$_3$): $\delta = 3.54$ (dd, $J_{HF} = 1.2\,Hz$, $J_{HF} = 2.3\,Hz$ OMe); ^{19}F-NMR (376.58 MHz, CDCl$_3$): $\delta = -71.31$, -72.59

4.8.4   Conclusion

Enzymes that effectively catalyze the hydration of α,β-unsaturated carbonyl compounds (i.e., the Michael addition of water) have rarely been investigated [89,90]. The hydrations of **2a** and **2b** represent a novel mode of action for *R. rhodochrous* as a biocatalyst, and are the first examples of biocatalytic enantiospecific Michael additions of water to conjugated carbonyl compounds outside primary metabolism. Biocatalytic hydration using *R. rhodochrous* opens up a new route to 3-hydroxy carbonyl compounds, which are essential 1,3-difunctional compounds. The data presented in Table 4.3 clearly show the scalability of the process, which has been demonstrated in operations generating grams of products.

Table 4.3   *Enantioselective Michael addition of water using* Rhodococcus rhodochrous ATCC 17895.

Substrate	Michael product	Isolated yield (%)	ee	Reaction scale
3-methyl-2-butenolide	(R)-3-hydroxy-3-methyl-2-butanolide	55	87	1 g substrates in 300 mL buffer
3-ethyl-2-butenolide	(R)-3-hydroxy-3-ethyl-2-butanolide	50	94	1 g substrates in 300 mL buffer

4.9   Sulfation of Various Compounds by an Arylsulfotransferase from *Desulfitobacterium hafniense* and Synthesis of 17β-Estradiol-3-Sulfate

Michael A. van der Horst, Johan F. T. van Lieshout, Aleksandra Bury, Aloysius F. Hartog, and Ron Wever
Van 't Hoff Institute for Molecular Sciences, University of Amsterdam, The Netherlands

Most sulfated compounds are synthesized using complexes of sulfur trioxide (SO$_3$) with tertiary amines or amides, the use of which suffers from numerous disadvantages, including a lack of selectivity and the need to protect side-group functionalities [91]. Enzymatic methods of sulfating compounds under mild conditions may have advantages, and eukaryotic sulfotransferase using adenosine-3′-phospho-5′-phosphosulfate (PAPS) has been studied in this context [92]. However, PAPS is very expensive and has to be regenerated, introducing complexity. In addition, most PAPS enzymes are very substrate-specific.

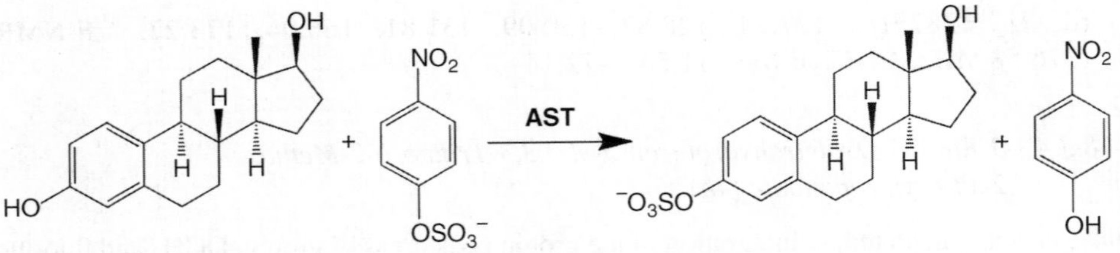

Scheme 4.15 *Enzymatic synthesis of 17β-estradiol 3-sulfate.*

For this reason, we use a recombinant bacterial arylsulfotransferase (AST) from *Desulfitobacterium hafniense*, which catalyzes the efficient transfer of a sulfate group from *p*-nitrophenyl sulfate to phenolic and non-phenolic acceptor molecules [93,94]. We have used this enzyme to prepare a variety of pure sulfated molecules and synthesized 17β-estradiol-3-sulfate (Scheme 4.15).

4.9.1 Materials and Equipment

- 17β-Estadiol (100 mg, 0.36 mmol)
- Acetone (10 mL)
- Tris/glycine buffer (100 mL, 400 mM), pH 8.5
- *p*-Nitrophenylsulfate potassium salt (103 mg, 0.40 mmol)
- Recombinant AST [93]
- CH_2Cl_2 (100 mL)
- *n*-Butanol (100 mL)
- NaCl (100 mL, 5%)
- Ethyl acetate (100 mL)
- 1 N NaOH
- Round-bottom flask 100 mL
- Magnetic/heater stirrer
- Water bath
- HPLC system and UV detection
- Tabletop centrifuge
- pH meter

4.9.2 Procedure

1. A 37 mL reaction solution was prepared, containing acetone (3.7 mL), water (24 mL), and 400 mM Tris/glycine buffer pH 8.5 (9.25 mL), in which *p*-nitrophenylsulfate potassium salt (103 mg, 0.40 mmol) was dissolved. 17β-estradiol (100 mg, 0.36 mmol) was added and the reaction was started by the addition of AST (1.4 mg, 14 U).
2. The solution was stirred at 30 °C in a round-bottom vessel.
3. After 24 hours, when maximal conversion was observed by HPLC [93], the solution was washed with CH_2Cl_2 (3 × 20 mL) to remove acetone, traces of 17β-estradiol, and part of the *p*-nitrophenol, which was discarded. The aqueous phase was extracted with *n*-butanol (3 × 20 mL). This resulted in an organic phase containing 95% of the product and

some *p*-nitrophenol and *p*-nitrophenylsulfate. The organic phase was washed with 5% NaCl solution (5 mL) and the *n*-butanol was removed by evaporation.

4. The solid product was washed with ethyl acetate and dissolved in water (3 mL). The pH was increased to 11 using a small amount of 1 N NaOH. This solution was extracted again with *n*-butanol (2×10 mL) and evaporated to dryness. The solid product was washed again with ethyl acetate and the material was dissolved in water (3 mL). After concentration to 1 mL by N_2 flush and cooling, a precipitate of the sodium salt was obtained, which was collected by centrifugation. Water was decanted off and the solid product was dried under vacuum.

5. The isolated yield of the product (17β-estradiol 3-sulfate as a sodium salt) was 79 mg (57%).

4.9.3 Analytical Data

Analysis by HPLC and NMR showed that the product was > 90% pure.

^{1}H-NMR (400 MHz; DMSO-d_6): δ = 7.12 (d, 1 H), 6.88 (d, 2 H), 6.84 (s, 4 H), 4.47 (s, 17OH), 3.50 (m, 17 H), 2.72 (m, 2 H, 6α, βH), 2.25 (m, 11αH), 2.08 (m, 9 H) 1.92–1.74 (m, 3 H, 7β-, 12β-, 16αH), 1.56 (m, 15αH), 1.40–1.08 (m, 7 H, 16β-, 8-, 7α-, 11β-, 15β-, 14-, 12 H)

FAB-MS: $m/z = 351,1263$ for $C_{18}H_{23}O_5S$-[M-Na]

4.9.4 Conclusion

This recombinant AST is a promising catalyst for the regioselective sulfation of a wide range of phenolic compounds under mild reaction conditions using *p*-nitrophenyl sulfate as a sulfate donor without the need for functional group protection. Steroids, tyrosine-containing peptides, and phenolics are easily sulfated, but surprisingly, and in contrast to all other known bacterial ASTs [95,96], the enzyme from *Desulfitobacterium hafniense* is also able to use some non-phenolic alcohols (1- and 2-phenethylalcohol) as sulfate acceptors.

4.10 Asymmetric Synthesis of Cyclopropanes and Benzosultams via Enzyme-Catalyzed Carbenoid and Nitrenoid Transfer in *E. coli* Whole Cells

Christopher C. Farwell, Hans Renata, and Frances H. Arnold
Division of Chemistry and Chemical Engineering, California Institute of Technology, USA

Enzyme-catalyzed activation of olefins and C−H bonds via direct carbenoid and nitrenoid transfer is a nascent methodology for stereoselective synthesis of cyclopropanes and benzosultams [97,98] (Scheme 4.16). Neither reactivity is observed as part of native biological function, but olefin cyclopropanation and C–H amination are catalyzed by an engineered variant of cytochrome P450 BM3 known as P411$_{BM3}$-CIS that contains mutations to conserved residues T268A and C400S [99]. Expression of P411$_{BM3}$-CIS in *E. coli* BL21 (DE3) results in a highly active and selective catalyst for the generation of cyclopropanes and benzosultams in whole cells, using ethyl diazoacetate (EDA) and arylsulfonyl azides as carbene and nitrene precursors, respectively. Purified P411$_{BM3}$-CIS

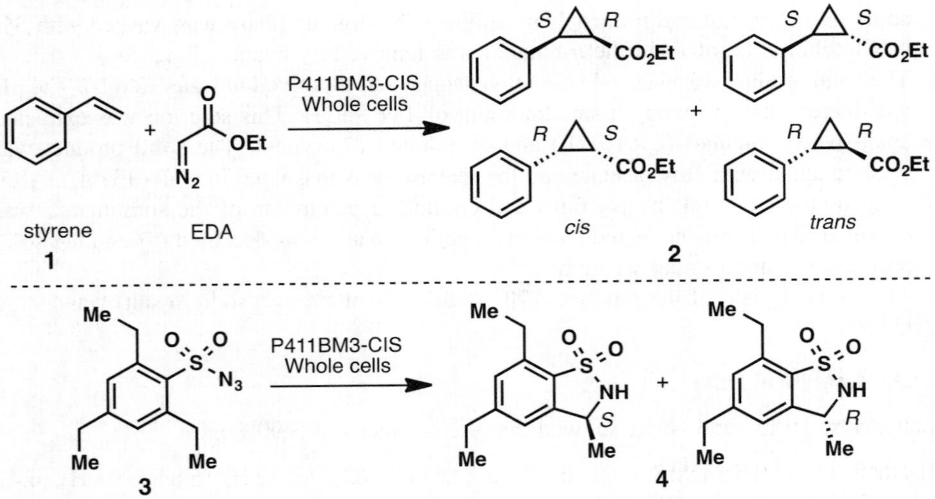

Scheme 4.16 *Biocatalytic olefin cyclopropanation and C–H amination.*

can be used in conversions at small scale, but the whole-cell reaction can be employed at virtually any scale. The intermolecular conversion of styrene **1** to **2** and the intramolecular conversion of 2,4,6-triethylbenzenesulfonyl azide **3** to **4** are shown as representative examples of biocatalytic carbenoid and nitrenoid transfer. This catalyst has also been adopted for carbenoid insertion to N−H bonds [100], nitrenoid insertion to organosulfur heteroatoms [101], and cyclopropanation of variously substituted styrenyl olefins [97], demonstrating the general applicability of these biocatalytic transformations. Since both transformations use the same biocatalyst in similar reaction conditions, we present a combined procedure, with specific details for each reaction listed where appropriate.

4.10.1 Materials and Equipment

4.10.1.1   Cell Culture and Growth

- *E. coli* BL21 (DE3) harboring pET22b-P411$_{BM3}$-CIS
- LB broth powder (20 g.L^{-1})
- LB agar powder (10 g.L^{-1})
- Ampicillin stock solution (100 mg.mL^{-1})
- HyperBroth powder (AthenaES, 45 g.L^{-1})
- HyperBroth glucose additive mix (AthenaES, 200 g.L^{-1})
- δ-Aminolevulinic acid stock solution (δ-ALA, 131 mg.mL^{-1})
- IPTG (118 mg.mL^{-1})
- Two 125 mL and two 1 L glass culture flasks

4.10.1.2   Reaction and Workup

- Nitrogen-free M9 (M9-N) buffer solution (200 mL, M9-N. 1 L contains 31 g Na$_2$HPO$_4$, 15 g KH$_2$PO$_4$, 2.5 g NaCl, 0.24 g MgSO$_4$, 0.01 g CaCl$_2$, and 1 mL micronutrient solution.

Micronutrient solution contains 0.15 mM $(NH_4)_6Mo_7O_{24}$, 20.0 mM H_3BO_3, 1.5 mM $CoCl_2$, 0.5 mM $CuSO_4$, 4.0 mM $MnCl_2$, and 0.5 mM $ZnSO_4$.)
- 250 mM Glucose in M9-N solution (10 mL)
- Glucose oxidase/catalase (10 mL, 14 000/1000 $U.mL^{-1}$)
- Anhydrous sodium sulfate
- HPLC-Grade acetonitrile (1 L)
- HPLC-grade water (1 L)
- HPLC-grade isopropanol (1 L)
- Analytical-grade ethyl acetate (1 L)
- Analytical-grade hexanes (1 L)
- 100% Ethanol
- EDA stock solution (85% in DCM, 1.36 mL, 11.0 mmol)
- Styrene (2.5 mL, 21.7 mmol)
- Arylsulfonyl azide stock solution (80 mM in DMSO, 2.9 mL, 62 mg)
- 98% Argon gas

4.10.1.3 Analytical and General Laboratory Equipment

- HPLC system with UV detection
- SFC system with UV detection
- GC system with flame ionization detection
- Kromasil 100 C18 HPLC column (Peeke Scientific, 4.6×45 mm, 5 µm)
- Chiralpak AD-H HPLC column (Daicel, 4.6×150 mm, 5 µm)
- Cyclosil-B GC column (30 m $\times$ 0.32 mm, 0.25 µm film, J&W Scientific)
- Rotary shakers at 37 °C
- Centrifuge capable of reaching $4000 \times g$
- Silica gel TLC plates (Silica gel 60, F254, Merck)
- Silica gel 60 (AMD silica gel 60, 230–400 mesh)
- Sodium sulfate (anhydrous)
- UV lamp/viewing box

4.10.2 Cell Culture Growth and Expression

1. Cells were streaked on to LB agar plates (100 $µg.mL^{-1}$ ampicillin) from a frozen stock of whole-cell catalyst *E. coli* BL21 (DE3) pET22b-P411$_{BM3}$-CIS and grown at 37 °C overnight to obtain single colonies.
2. Single colonies were picked and inoculated into 25 mL LB medium (containing 100 $µg.mL^{-1}$ ampicillin) in 125 mL sterilized glass flasks capped with aluminum foil and grown overnight at 37 °C with a shake rate of 220 rpm.
3. Expression cultures were inoculated with 25 mL cultures grown in Step 2 into 600 mL HyperBroth containing glucose additive mix (30 mL) and ampicillin (100 $µg.mL^{-1}$). Cultures were grown at 37 °C and 220 rpm shake rate until $OD_{600} \approx 2$. Temperature was then lowered to room temperature, shake rate reduced to 130 rpm, and the cell cultures were allowed to cool for 30 minutes. δ-ALA and IPTG were added to final concentrations of 131 and 118 $mg.L^{-1}$, respectively. Expression was allowed to proceed for 16 hours.

4.10.3 Whole-Cell Cyclopropanation Reaction

1. Expression cultures were centrifuged at $3000 \times g$ for 5 minutes and cell pellets were gently resuspended to $OD_{600} = 70$ in nitrogen-free M9 medium (M9-N).
2. The cell suspension (~54 mL) was made anaerobic by bubbling argon through it in a 500 mL sealed round-bottom flask for 30 minutes.
3. A degassed solution of glucose in M9-N (1.4 mL, 500 mM) was added to the cells to an approximate final concentration of 13 mM, along with glucose oxidase/catalase solution to a final concentration of $700/50$ $U.mL^{-1}$.
4. Neat styrene (2.5 mL, 21.7 mmol) was added dropwise, followed by EDA (85% solution in DCM from Sigma-Aldrich, 1.36 mL, 11.0 mmol).
5. The reaction was stirred at room temperature under argon for 16 hours.

4.10.4 Whole-Cell C–H Amination Reaction

1. Expression cultures were centrifuged at $3000 \times g$ for 5 minutes and cell pellets were gently resuspended to $OD_{600} = 30$ in M9-N medium.
2. The cell suspension (65 mL final volume) was made anaerobic by gently streaming argon through it in a 250 mL round-bottom flask for 30 minutes.
3. A degassed solution of glucose in M9-N ($45 \, g.L^{-1}$) was added to the cells to a final concentration of 13 mM, along with glucose oxidase/catalase solution to a final concentration of $700/50$ $U.mL^{-1}$.
4. 2,4,6-Triethylbenzenesulfonyl azide (2.9 mL, 80 mM, DMSO, 62 mg) was added dropwise over 10 minutes.
5. The reaction was stirred at room temperature under argon for 16 hours.

4.10.5 Whole-Cell Reaction Workup

1. The crude reaction mixture was divided evenly into three 50 mL conical tubes, and the reaction was quenched by the addition of 3 M HCl (1 mL).
2. Each of the aqueous mixtures was extracted with 1 : 1 EtOAc:hexanes (20 mL each) and centrifuged ($4000 \times g$, 5 minutes), and the organics were collected. This step is then repeated two more times.
3. The combined organics were dried over Na_2SO_4, decanted to a round-bottom flask, and concentrated *in vacuo*. For cyclopropanation, excess styrene was removed via azeotrope with H_2O in benzene.
4. For cyclopropanation, silica gel chromatography with 4% EtOAc in hexanes was used to purify cyclopropane products to afford 1.63 g total (78% yield) of the *cis* and *trans* isomers.
5. For C–H amination, silica gel chromatography was used to isolate product **4** by stepwise elution (hexanes, 90/10 hexanes/ethyl acetate, 80/20 hexanes/ethyl acetate, 70/30 hexanes/ethyl acetate) to afford 38.6 mg (69% yield) of both enantiomers.

4.10.6 Analytical Procedure for Whole-Cell Cyclopropanation Reaction

1. Aliquots of the reaction mixture (0.5 mL) were taken in order to assess productivity. Each aliquot was quenched with 3 M HCl (25 μL), mixed with EtOAc (1 mL), and

vortexed for 30 seconds. The mixtures were microcentrifuged at $20\,000 \times g$ for 5 minutes. The organic fraction was passed over a $1:1$ mixture of silica gel 60 and anhydrous sodium sulfate packed in a glass pipet, and $1–2\,\mu L$ were injected for GC-FID analysis.

2. The GC system used a Cyclosil-B (J&W scientific, $30\,m \times 0.25\,mm$, $0.25\,\mu m$ film) column for resolution of cyclopropane diastereomers and enantiomers. The GC injector temperature was set at $300\,°C$. The oven temperature was held at $130\,°C$, $175\,kPa$ for 30 minutes. Retention times for the *cis*-cyclopropane enantiomers are 19.7 (R,S) and 21.0 (S,R) minutes, and retention times for the *trans*-cyclopropane enantiomers are 25.8 (R,R) and 26.4 (S,S) minutes.

3. The diastereoselectivity at the end of the reaction was determined to be $90:10$ *cis/trans* products, with 99% *ee* (S,R) for the *cis* isomer and 47% *ee* (S,S) for the *trans* isomer.

4.10.7 Analytical Procedures for Whole-Cell C–H Amination Reactions

1. Sample aliquots (0.5 mL) were quenched with 3 M HCl (25 µl), mixed with acetonitrile (0.5 mL), and vortexed for 30 seconds. Mixtures were microcentrifuged at $20\,000 \times g$ for 5 minutes and supernatants were filtered using $0.22\,\mu M$ polytetrafluoroethylene syringe filters. 20 µL were injected for HPLC analysis.

2. The HPLC system used a mobile phase of CH_3CN/H_2O with a gradient from 1% CH_3CN to 100% CH_3CN over 25 minutes. Quantification of standards and samples was accomplished using a Kromasil 100 C18 $4.6 \times 50\,mm$, 5 µm HPLC column with a flow rate of $1\,mL.min^{-1}$. UV detection of substrate and product was made at 220 nm and retention times were as follows: triethylbenzenesulfonyl azide **3**, 19.7 minutes; benzosultam **4**, 12.9 minutes.

3. Chiral HPLC analysis of C–H amination products was accomplished using an SFC system, equipped with a Chiralpak AD-H column (Daicel, $4.6 \times 150\,mm$) and eluting with 20% isopropanol in supercritical CO_2. Retention times for the benzosultam **4** enantiomers were 7.6 (S) and 9.3 (R) minutes.

4. Benzosultam **4** was afforded in 89% *ee* (S).

4.10.8 Conclusion

Whole-cell catalysts for intermolecular olefin cyclopropanation and intramolecular C–H amination yield >65% product after purification. These bioconversions have demonstrated scalability, with cyclopropanation in particular efficient in gram-scale reactions. The P411$_{BM3}$-CIS whole-cell catalysts are also highly selective for both olefin cyclopropanation and C–H amination reactions. The *cis/trans* ratio for olefin cyclopropanation of styrene was $90:10$, with *ee* 99% (S,R) for the *cis* diastereomer. *ee* 89% (S) was achieved for C–H amination of **3**. The *cis* selectivity for cyclopropanation is among the highest reported for any cyclopropanation. The P411$_{BM3}$-CIS catalyst and variants thereof have also been employed in carbenoid and nitrenoid chemistries beyond cyclopropanation and C–H amination [100,101]. Further reaction and enzyme engineering will enable a range of carbenoid and nitrenoid reactivities previously inaccessible to biological catalysts.

4.11 Biocatalytic Production of Novel Glycolipids

Karel De Winter, Griet Dewitte, Hai Giang Tran, and Tom Desmet

Centre for Industrial Biotechnology and Biocatalysis, Faculty of Bioscience Engineering, Ghent University, Belgium

Surfactants are amphiphilic molecules that reduce the interfacial tension between liquids, solids, and gases. With a total world production exceeding 13 million tons per year, the bioaccumulation and eco-toxicity of these surfactants have become an issue of major environmental concern [102]. Therefore, over the past decade, glycolipids have gained increasing attention as biosurfactants. Indeed, these compounds can be produced from renewable resources and are known to be readily degraded by nature [103]. However, the structural diversity of available glycolipids is rather limited to date, with only sophorolipids being produced on an industrial scale [104]. This limitation can be overcome by subjecting sophorolipids to a series of chemoenzymatic modifications. In this section, we describe the fermentative production of sophorolipids, followed by alkaline hydrolysis (Scheme 4.17a). Indeed, modification is required, as these biosurfactants are secreted in their acetylated lactonic form [105]. Next, we describe the treatment of sophorolipids with naringinase to obtain the corresponding glucolipids, followed by the modification of the glucolipids using recombinantly produced cellodextrin phosphorylase (CDP) from *Clostridium stercorarium*, allowing the synthesis of cellobiolipids and cellotriolipids (Scheme 4.17c). Sophorolipids can also be directly modified with CDP, yielding glucosophorolipids and cellobiosophorolipids (Scheme 4.17b). Interestingly, these novel glycolipids can be conveniently harvested by simple precipitation.

4.11.1 Procedure 1: Fermentative Production of Sophorolipids

4.11.1.1 Materials and Equipment

- YPD Plate containing *Candida bombicola* ATCC 22214
- Yeast extract (18 g)
- Peptone (4 g)
- Glucose (484 g)
- Sodium citrate trihydrate (20 g)
- NH_4Cl (6 g)
- KH_2PO_4 (4 g)
- K_2HPO_4 (0.64 g)
- $MgSO_4 \, 7\,H_2O$ (2.8 g)
- NaCl (2 g)
- $CaCl_2 \, 2\,H_2O$ (1.08 g)
- Distilled water (5 L)
- NaOH (5 M)
- Rapeseed oil (1 L)
- 4 M Glucose (1 L)
- Ethyl acetate (10 L)
- Hexane (1.5 L)
- Erlenmeyer flask with cotton plug (1 L)
- Fermentor (5 L)

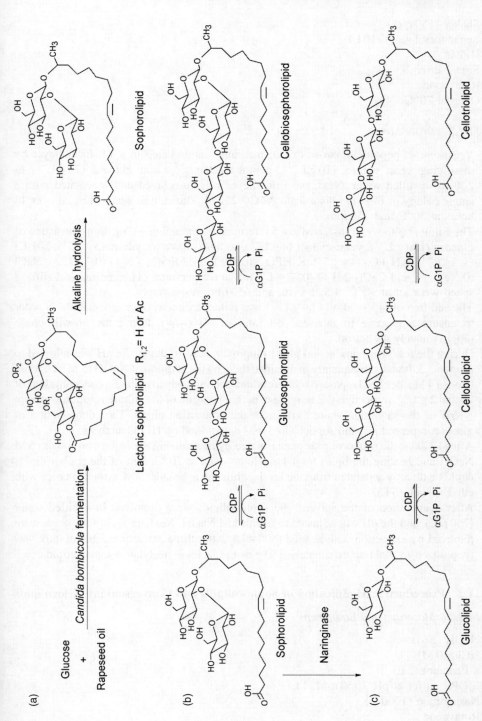

Scheme 4.17 *General scheme for the synthesis of (a) sophorolipid, (b) glucosophorolipid and cellobiosophorolipid, and (c) glucolipid, cellobiolipid, and cellotriolipid.*

- Shaker (150 rpm)
- Separatory funnel (10 L)
- HPLC
- Freeze-drier
- Rotavapor
- Oven at 70 °C

4.11.1.2 Procedure

1. Yeast extract peptone dextrose (YPD) medium was prepared in a 1 L Erlenmeyer by dissolving yeast extract ($10\,g.L^{-1}$), peptone ($20\,g.L^{-1}$), and glucose ($20\,g.L^{-1}$) in 200 mL distilled water. Next, the Erlenmeyer was autoclaved and inoculated with a single colony of *Candida bombicola* ATCC 22214. The culture was incubated over 48 hours at 30 °C and 150 rpm.
2. The culture broth was transferred to a 5 L fermentor containing 4 L medium consisting of glucose ($120\,g.L^{-1}$), yeast extract ($4\,g.L^{-1}$), sodium citrate trihydrate ($5\,g.L^{-1}$), NH_4Cl ($1.5\,g.L^{-1}$), KH_2PO_4 ($1\,g.L^{-1}$), K_2HPO_4 ($0.16\,g.L^{-1}$), $MgSO_4\ 7\,H_2O$ ($0.7\,g.L^{-1}$), NaCl ($0.5\,g.L^{-1}$), and $CaCl_2\ 2\,H_2O$ ($0.27\,g.L^{-1}$). The temperature, pH, aeration, and stirring speed were set at 30 °C, 4.5, 1 vvm, and 650 rpm, respectively.
3. The addition of rapeseed oil ($2.4\,g.L^{-1}$) was initiated directly after inoculation, in order to ensure a glucose to rapeseed oil ratio of 50 (w/w) during the growth phase (approximately 48 hours).
4. During the following production phase (approximately 10 days), the pH was allowed to drop to 3.5. It was subsequently maintained there by the addition of NaOH (5 M). Glucose (from a 4 M stock) and rapeseed oil were added continuously to maintain concentrations of 20 and $2\,g.L^{-1}$, respectively. The oxygen partial pressure of the culture solution was kept $\leq 15\%$ of the saturation value throughout the production phase. The concentration of glucose, rapeseed oil, and sophorolipid was determined by HPLC analysis.
5. After 12 days, the sophorolipids were isolated by neutralizing the culture broth with 5 M NaOH and heating the broth for 1 hour in an oven at 70 °C. Part of the sophorolipids could be directly separated from the broth, while the remainder was extracted twice with ethyl acetate (5 L).
6. After evaporation of the solvent, the sophorolipids were dissolved in distilled water (750 mL), and the pH was adjusted to 8 with 1 M NaOH. Next, traces of fatty acids were removed by extraction with hexane (500 mL). After three extractions, the mixture was lyophilized to yield approximately 350 g of the lactonic acetylated sophorolipids.

4.11.2 Procedure 2: Modification of Sophorolipids and Conversion into Glucolipids

4.11.2.1 Materials and Equipment

- NaOH (5 M)
- HCl (10 M)
- *n*-Pentanol (2 L)
- MOPS buffer at pH 7 (50 mM, 1 L)
- Naringinase (1000 U)
- Rotavapor

- Freeze-drier
- High-speed centrifuge ($5000 \times g$)

4.11.2.2 Procedure

1. The deacetylation and ring opening of the lactonic sophorolipids (Scheme 4.17a) was initiated by dissolving sophorolipid (300 g) in NaOH (5 M, 750 mL). The mixture was refluxed for 10 minute at 100 °C, then cooled to room temperature.
2. Hydrochloric acid (10 M) was added to adjust the pH to 4, and the mixture was extracted twice with *n*-pentanol (1 L). The solvent was evaporated in vacuo and the sophorolipids obtained were dissolved in water and lyophilized. Approximately 250 g of deacetylated sophorolipids was obtained. They were stored at −20 °C.
3. The sophorolipids (150 g) were dissolved in 50 mM MOPS buffer at pH 7 (1 L), containing naringinase (1000 U). The mixture was magnetically stirred and allowed to react for 24 hours at 45 °C, in order to ensure complete conversion.
4. The pH was adjusted to 3 by the addition of hydrochloric acid (10 M), resulting in the precipitation of the formed glucolipids. Finally, approximately 100 g glucolipids was isolated by centrifugation ($5000 \times g$, 10 minutes) and stored at −20 °C.

4.11.3   Procedure 3: Recombinant Production of CDP

4.11.3.1 Materials and Equipment

- Cryovial containing *E. coli* BL21 transformed with the expression plasmid pTrc99aCsCDP [106]
- Tryptone (300 g)
- Yeast extract (150 g)
- NaCl (75 g)
- Ampicillin (500 mg)
- Distilled water (5 L)
- NaOH (5 M)
- IPTG (36 mg)
- Glass test tube (50 mL)
- Erlenmeyer flask with cotton plug (2 L)
- Fermentor (20 L)
- Shaker (200 rpm)
- Spectrophotometer (600 nm)
- High-speed cooled centrifuge ($10\,000 \times g$)

4.11.3.2 Procedure

1. Double Luria–Bertani broth (LB) was prepared by dissolving tryptone ($20\,\text{g.L}^{-1}$), yeast extract ($10\,\text{g.L}^{-1}$), and NaCl ($5\,\text{g.L}^{-1}$) in 15 L distilled water. The double LB was divided into a test tube (5 mL), an Erlenmeyer (500 mL), and a fermentor (14.5 L). All recipients were autoclaved, and ampicillin ($100\,\text{mg.L}^{-1}$ final concentration) was added prior to use.
2. The tube was inoculated with the *E. coli* strain for enzyme production. After 8 hours of incubation at 37 °C and 200 rpm, the culture was transferred to an Erlenmeyer flask and grown overnight under the same conditions.

3. The culture broth was poured into the fermentor for cultivation. The fermentor was operated at 37 °C and 800 rpm, with a constant aeration of 1 vvm. The pH was maintained at 7 by the addition of 5 M NaOH. Antifoam was added manually when required.
4. The optical density of the culture was measured at regular intervals using a spectrophotometer. Enzyme expression was induced by the addition of IPTG (0.01 mM) when the OD_{600} reached 0.6.
5. After 12 hours of cultivation, the culture was centrifuged in a high-speed cooled centrifuge at $10\,000 \times g$ for 15 minutes. The pellets obtained were stored at -20 °C.

4.11.4   Procedure 4: Cell Lysis and Purification of CDP

4.11.4.1   Materials and Equipment

- Lysozyme (750 mg)
- Phenylmethylsulfonyl fluoride (PMSF) (13 mg)
- Ni-NTA resin (200 mL)
- Lysis buffer: 50 mM NaH_2PO_4, 300 mM NaCl, 10 mM imidazole, pH 7.4 (750 mL)
- Equilibration buffer: 50 mM NaH_2PO_4, 300 mM NaCl, 10 mM imidazole, pH 7.4 (1 L)
- Wash buffer: 50 mM NaH_2PO_4, 300 mM NaCl, 25 mM imidazole, pH 7.4 (600 mL)
- Elution buffer: 50 mM NaH_2PO_4, 300 mM NaCl, 400 mM imidazole, pH 7.4 (400 mL)
- 50 mM MES Buffer pH 6.5 (2 L)
- Cell disruptor (sonication)
- High-speed cooled centrifuge ($10\,000 \times g$)
- Fritted chromatography column (1 L)
- Ultrafiltration unit (molecular weight cutoff 30 kDa)

4.11.4.2   Procedure

1. The obtained *E. coli* pellets were resuspended in lysis buffer (750 mL), to which lysozyme (750 mg) and PMSF (0.1 mM) were added. The suspension was incubated on ice for 30 minutes, then sonicated for 3×3 minutes (50% duty cycle). Meanwhile, a fritted chromatography column was loaded with Ni-NTA resin (200 mL), after which the resin was equilibrated with 5 column volumes of equilibration buffer. Note: It's important to ensure that the column doesn't run dry throughout the purification procedure.
2. The mixture was centrifuged for 15 minutes at $10\,000 \times g$ and 4 °C. The cellular debris was discarded and the supernatant was applied to the equilibrated Ni-NTA resin.
3. The resin was washed with 3 column volumes of wash buffer and the His-tagged CDP was collected by applying 2 column volumes of elution buffer.
4. The elution buffer containing the enzyme was poured in a stirred ultrafiltration unit (40 psi) and washed with 2 L MES buffer. The enzyme solution was concentrated to 100 mL and stored at 4 °C.
5. The activity of the enzyme solution was determined discontinuously by measuring the release of phosphate from α-D-glucose 1-phosphate (αG1P) and cellobiose using the method of Gawronski and Benson [107]. 1 U activity was defined as the amount of enzyme that releases 1 µmol phosphate per minutes from 50 mM cellobiose and 50 mM αG1P in a 50 mM MES buffer at pH 6.5 and 37 °C. The purification yielded roughly 25 000 U purified CDP.

4.11.5 Procedure 5: Synthesis and Isolation of Novel Glycolipids

4.11.5.1 Materials and Equipment

- 50 mM MES buffer pH 6.5 (2 L)
- Distilled water (200 mL)
- α-D-Glucose 1-phosphate disodium salt tetrahydrate (214 g)
- High-speed centrifuge (5000 × g)
- Magnetically stirred flask (2 L)
- Freeze-drier

4.11.5.2 Procedure

1. Glucolipid (100 g, 220 mM) and αG1P (124 g, 330 mM) were dissolved in a magnetically stirred flask containing MES buffer at pH 6.5 (1 L). Sophorolipid (100 g, 160 mM) and αG1P (90 g, 240 mM) could also have been used.
2. CDP (10 U.mL^{-1}) was added and the mixtures were incubated at 45 °C.
3. After 24 hours, homogeneous samples were taken and subjected to HPLC analysis. This revealed a conversion of 90 and 83% for the glucolipid and sophorolipid, respectively.
4. The mixture of cellobiolipid and cellotriolipid (approximately 5 : 3 ratio), and the mixture of glucosophorolipid and cellobiosophorolipid (approximately 5 : 3 ratio), from the reaction with glucolipid and sophorolipid, respectively, were harvested by centrifugation (10 minutes at 5000 × g).
5. The supernatant was discarded and both pellets were resuspended in 100 mL distilled water, then centrifuged (10 minutes at 5000 × g). The obtained pellets were lyophilized to yield cellobiolipid and cellotriolipid (100 g) and glucosophorolipid and cellobiosophorolipid (90 g), respectively.

4.11.6 Analytical Data

Samples were first extracted with one volume of MTBE to dissolve all glycolipids. Both phases were analyzed on a Varian ProStar HPLC equipped with a Chromolith Performance RP-18e column (100 × 4.6 mm). Detection was accomplished using an Alltech 2000ES evaporative light-scattering detector (ELSD). The tube temperature, gas flow, and gain were set at 25 °C, 1.5 L.min^{-1}, and 1, respectively. Elution was achieved by means of an acetonitrile/acetic acid (0.5% in MQ water) gradient (5 : 95 to 95 : 5 in 40 minutes). The flow and temperature were set at 1 mL.min^{-1} and 30 °C, respectively. LC/MS analysis was performed on the isolated glycolipids using negative-ion-mode MS on a Quattro LC system (Micromass) (Figure 4.3).

4.11.7 Conclusion

We have desrcibed the production of several new glycolipids on the multi-gram scale. First, sophorolipids were produced through fed-batch fermentation of *Candida bombicola* ATCC 22214. The lactonic acetylated sophorolipids obtained were then subjected to alkaline hydrolysis and partially treated with naringinase to obtain glucolipids. Thanks to the broad acceptor specificity of CDP, the glucolipids and sophorolipids could be efficiently extended with one or two glucose moieties. The resulting novel glycolipids were conveniently harvested by precipitation. These may possess new properties that will broaden the applications of biosurfactants.

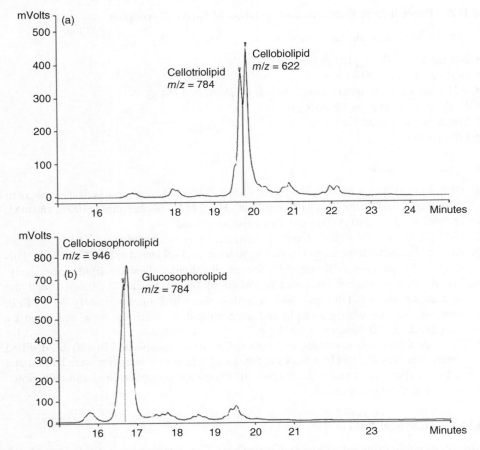

Figure 4.3 Chromatograms of the various glycolipids produced by CDP. Product purification was performed after reacting αG1P with (a) glucolipid or (b) sophorolipid.

4.12 Enzymatic Synthesis of 8-Aza- and 8-Aza-7-Deazapurine 2′-Deoxyribonucleosides

Vladimir A. Stepchenko,[1] Frank Seela,[2] Roman S. Esipov,[3] Anatoly I. Miroshnikov,[3] and Igor A. Mikhailopulo[1]
[1] *Institute of Bioorganic Chemistry, National Academy of Sciences, Belarus*
[2] *Laboratory of Bioorganic Chemistry and Chemical Biology, Center for Nanotechnology, Germany*
[3] *Shemyakin and Ovchinnikov Institute of Bioorganic Chemistry, Russian Academy of Sciences, Russia*

Recently, we have described the versatile enzymatic methods for the synthesis of 8-aza- and 8-aza-7-deazapurine 2′-deoxy-β-D-ribonucleosides **1b-6b** [108]. 8-aza- and 8-aza-7-deazapurine nucleosides (purine numbering throughout) are of great interest as potential drugs and tools in chemistry, chemical biology, and molecular diagnostics [109–112]. Chemical synthesis of such nucleosides suffers from the formation of complex mixtures of *regio-* and *stereo-*isomers (see, e.g., [111,113]), which require tedious chromatographic separation, resulting in low yields and high costs for the desired nucleosides. The use of enzymes, particularly *E. coli* nucleoside phosphorylases, has proven to be an efficient methodology for the synthesis of natural and modified nucleosides (for recent reviews, see, e.g., [114]). We studied two paths of the enzymatic synthesis of 8-aza- and 8-aza-7-deazapurine 2'-deoxy-β-D-ribonucleosides: (i) the transglycosylation reaction, consisting in the transfer of the 2-deoxy-D-ribofuranose moiety of a commercially available natural 2'-deoxynucleoside onto purine analogs **1a–6a**, catalyzed by *E. coli* purine nucleoside phosphorylase (PNP) [115] (path A); and (ii) a cascade one-pot transformation of 2-deoxy-D-ribofuranose into corresponding 2'-deoxy-β-D-ribonucleosides employing the recombinant *E. coli* ribokinase (RK) [116], phosphopentomutase (PPM) [117] and PNP [115] as biocatalysts (path B) (Scheme 4.18).

4.12.1 Materials and Equipment

- 8-Aza-7-deazaadenine (**3a**) (26 mg, 0.193 mmol) (commercially available)
- 2'-Deoxyguanosine (78 mg; 0.292 mmol) (commercially available)
- *K,Na*-Phosphate buffer (10 mM; pH 7.0–7.5)
- Adenosine 5'-triphosphate disodium salt (ATP) (commercially available)
- Manganese (II) chloride tetrahydrate
- Potassium chloride
- Tris·HCl buffer (20 mM; pH 7.5)
- Dimethyl sulfoxide (DMSO)
- Silica gel (35–70 μm)
- Chloroform
- Methanol
- HPLC-grade acetonitrile
- HPLC-grade water
- HPLC-grade trifluoroacetic acid (TFA)
- HPLC system with UV detector
- Kromasil HPLC column (C_{18}, 5 μm, 250 × 4.6 mm)
- Nucleosil HPLC column (C_{18}, 5 μm, 150 × 4.6 mm)

Preparation of the recombinant *E. coli* PNP [115], RK [116], and PPM [117] was described; all the preparations were made many times on a multi-gram scale, either as solutions in phosphate buffer stable for a minimum of 1 year under storage at 4 °C, or as white lyophilized powders with stability of more than 5 years.

4.12.2 Synthesis of 8-Aza-7-Deaza-9-(2-Deoxy-β-D-Ribofuranosyl)-Adenine (3b)

4.12.2.1 Path A (transglysosylation reaction)

1. 8-Aza-7-deazaadenine (26 mg, 0.193 mmol) and 2'-deoxyguanosine (78 mg; 0.292 mmol) were dissolved in *K,Na*-phosphate buffer (10 mL), after which PNP (100 U) was added. The reaction mixture was gently stirred at 50 °C, and the progress

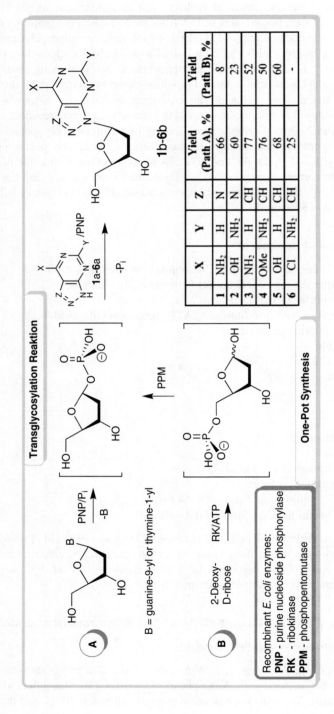

	X	Y	Z	Yield (Path A), %	Yield (Path B), %
1	NH₂	H	N	66	8
2	OH	NH₂	N	60	23
3	NH₂	H	CH	77	52
4	OMe	NH₂	CH	76	50
5	OH	H	CH	68	60
6	Cl	NH₂	CH	25	-

Transglycosylation Reaktion

B = guanine-9-yl or thymine-1-yl

One-Pot Synthesis

Recombinant *E. coli* enzymes:
PNP - purine nucleoside phosphorylase
RK - ribokinase
PPM - phosphopentomutase

Scheme 4.18 *Biocatalytic synthesis of 8-aza- and 8-aza-7-deazapurine 2′-deoxy-β-D-ribonucleosides.*

of the nucleoside formation was monitored by HPLC. After 48 hours, the complete conversion of the starting compound was observed.

2. The reaction mixture was filtered. Silica gel (~2 mL) was added to the filtrate and the mixture was concentrated under reduced pressure. The silica gel containing the products was co-evaporated with ethanol and the powdered residue was purified by chromatography on silica gel using chloroform/methanol 7 : 1 ratio as eluent. 8-aza-7-deaza-2'-deoxyadenosine was obtained as a white amorphous solid (37 mg, 77%). Note: Nucleosides **1b**, **2b**, **4b**, and **5b** were synthesized in a similar way, starting from the corresponding heterocyclic base and thymidine or 2'-deoxyguanosine as glycosyl donor. For yields, see box in Scheme 4.18.

3. For the synthesis of 9-(2-deoxy-β-D-ribofuranosyl)-8-aza-7-deaza-2-amino-6-chloropurine (**6b**), the procedure was modified as follows due to low solubility of heterocycle **6a**: 2-amino-6-chloro-8-aza-7-deazapurine (**6a**; 40 mg, 0.236 mmol) was dissolved in DMSO (2 mL) and added dropwise to a solution of 2'-deoxyguanosine (95 mg; 0.356 mmol) and PNP (270 U) in K,Na-phosphate buffer (18 mL, pH 7.5), previously stirred for 1 hour at 50 °C. The reaction was monitored by HPLC and, after equilibrium was established, was worked up according to the standard protocol.

4.12.2.2 Path B (cascade one-pot synthesis)

1. In order to examine the conversion of 8-aza-7-deazaadenine (**3a**) into corresponding 2'-deoxyriboside **3b** in a cascade of enzymatic transformations, the following stock solutions were prepared:
   a. 10 mM ATP: ATP (139 mg, 0.250 mmol) was dissolved in Tris·HCl buffer (25 mL)
   b. 15 mM MnCl$_2$: Manganese chloride tetrahydrate (74 mg, 0.375 mmol) was dissolved in Tris·HCl buffer (25 mL)
   c. 250 mM KCl: Potassium chloride (465 mg, 6.25 mmol) was dissolved in Tris·HCl buffer (25 mL)
   d. 25 mM 2-Deoxy-D-ribose: 2-deoxy-D-ribose (84 mg; 0.634 mmol) was dissolved in Tris·HCl buffer (25 mL)
   e. 81 mM 8-Aza-7-deazaadenine: 8-aza-7-deazaadenine (11 mg; 0.081 mmol) was dissolved in DMSO (1 mL)

2. Solutions of ATP (400 μL), MnCl$_2$ (400 μL), KCl (400 μL), 2-deoxy-D-ribose (100 μL), and 8-aza-7-deazaadenine (25 μL) were mixed and the volume was adjusted to 2 mL with Tris·HCl buffer. Thus, final concentrations were: 2 mM ATP, 50 mM KCl, 3 mM MnCl$_2$, 20 mM Tris·HCl, 1.3 mM 2-deoxy-D-ribose, and 1 mM 8-aza-7-deazaadenine.

3. RK (9 U), PPM (4 U), and PNP (14 U) were added and the reaction mixture was stirred at 40 °C.

4. After 25 hours, the equilibrium was established at 1 : 1 base/nucleoside ratio, according to HPLC analysis. Note: Other heterocycles than **6a** were tested under the same conditions. For yields, see box in Scheme 4.18.

4.12.3 Analytical Methods

HPLC analysis of the reaction mixtures was performed using Kromasil column (C$_{18}$, 5 μm, 250 × 4.6 mm) and a mixture of CH$_3$CN/H$_2$O/TFA = 10 : 90 : 0.1 (v/v/v) as eluent (isocratic elution, 1 mL · min^{-1}, UV detection at 260 nm). For R_t, see reference [108].

4.12.4 Conclusion

We have described efficient enzymatic methods for the preparation of N^9-2′-deoxy-β-D-ribonucleosides of 8-aza- and 8-aza-7-deazapurines. One consists in the 2-deoxy-D-pentofuranose transfer from 2′-deoxyguanosine or thymidine to heterocyclic bases catalyzed by recombinant *E. coli* PNP. The other makes use of the one-pot cascade transformation of 2-deoxy-D-ribose into the aforementioned nucleosides through intermediate formation of 2-deoxy-D-ribofuranose 5-phosphate (catalyzed by *E. coli* RK; ATP co-factor) and 2-deoxy-α-D-ribofuranose 1-phosphate (catalyzed by *E. coli* PPM without any 1,6-diphosphates of D-hexoses as co-factors), and finally through condensation of the latter with 8-aza- and 8-aza-7-deazapurines catalyzed by the recombinant *E. coli* PNP to afford the desired nucleosides.

The use of *E. coli* PNP as the only biocatalyst of this reaction appears to be advantageous over the cascade method, but the latter may be employed in certain special cases, such as in the synthesis of ^{13}C-enriched [118] and deuterated [119] 2′-deoxyribose moieties of 2′-deoxynucleosides.

It should be stressed that the chemical synthesis of the 2′-deoxy-β-D-ribosides of 8-aza- and 8-aza-7-deazapurines is considerably more laborious, and is accompanied by the formation of *regio*-isomers and α- and β-anomers, which require a tedious chromatographic purification and ultimately give rise to a low yield of the desired isomer [120].

4.13 Phenylalanine Ammonia Lyase-Catalyzed Asymmetric Hydroamination for the Synthesis of L-Amino Acids

Sarah L. Lovelock and Nicholas J. Turner

School of Chemistry, Manchester Institute of Biotechnology, The University of Manchester, UK

Phenylalanine ammonia lyases (PALs) catalyze the enantioselective synthesis of L-amino acids from readily available achiral cinnamic acid derivatives and ammonia (Scheme 4.19). This biocatalytic hydroamination reaction offers a practical and atom-efficient alternative to traditional chemical methods involving metallo- and organocatalysts. Furthermore, PALs do not require the addition of expensive co-factors or recycling systems, making them particularly suitable as industrial biocatalysts. PALs from plant *Petroselinum crispum* (PcPAL), yeast *Rhodotorula glutinis* (RgPAL), and cyanobacteria *Anabaena variabilis* (AvPAL) have broad substrate specificity and accept a range of cinnamic acid analogs [121–125]. While these enzymes have been widely used in small-scale biotransformations, they are also amenable to large-scale production of amino acids. This is highlighted by the recent application of PAL in the synthesis of the non-natural amino acid 2′-chloro-L-phenylalanine on a ton scale by DSM Pharma Chemicals [126]. A decrease in

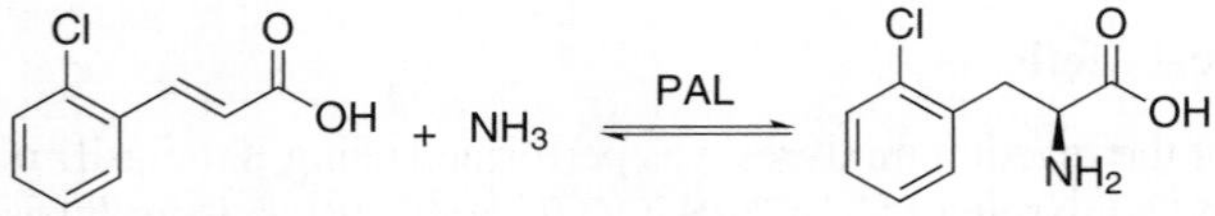

Scheme 4.19 *PAL-catalyzed hydroamination reaction.*

product *ee* has been reported after prolonged reaction times; therefore, in order to maintain good selectivity, the product formation and *ee* should be monitored over time [125].

4.13.1 Materials and Equipment

- Frozen glycerol stock of *E. coli* BL21(DE3) transformed with pET16b-PAL
- LB broth powder (15 g.L^{-1})
- LB agar powder (30 g.L^{-1})
- Auto-induction media powder (35 g.L^{-1})
- Ampicillin (100 μg.mL^{-1})
- 2′-Chlorocinnamic acid (1 g, 5.48 mmol)
- 2.5 M Ammonium carbonate solution (~pH 9.0, not adjusted)
- Aqueous HCl (1 N)
- Diethyl ether
- Water

4.13.2 Initial Culture

1. Colonies were grown on LB agar plates with ampicillin (100 μg.mL^{-1}) by streaking cells from a frozen glycerol stock of *E. coli* BL21(DE3) transformed with pET16b-PAL.
2. Single colonies were inoculated into LB medium (10 mL containing 100 μg.mL^{-1} ampicillin) in 50 mL falcon tubes. Cultures were incubated with shaking at 250 rpm at 37 °C for 16 hours.
3. Auto-induction medium (800 mL with ampicillin 100 μg.mL^{-1}) was inoculated with the initial culture (8 mL) in a 2.5 L baffled flask. The culture was incubated at 18 °C for 3 days with shaking at 250 rpm.
4. PAL-containing *E. coli* BL21(DE3) cells were harvested by centrifugation at 8000 rpm for 20 minutes at 4 °C and then stored at −20 °C until required.

4.13.3 PAL-Catalyzed Hydroamination

1. Wet *E. coli* BL21(DE3) cells expressing PAL (7.5 g) were resuspended in a solution of 2′-chlorocinnamic acid (1 g, 5.48 mmol) in ammonium carbonate solution (2.5 M, 500 mL).
2. The biotransformation was incubated at 30 °C with shaking (250 rpm). Samples (500 μL) were taken at various time intervals for HPLC analysis. After 8 hours, the reaction had reached 80% conversion to give the desired L-amino acid with *ee* > 99% (note that the hydroamination reaction is reversible, and this 80% conversion represents the position of equilibrium under these conditions).
3. The whole-cell biocatalyst was removed by centrifugation at 4000 rpm for 20 minutes at 4 °C. The supernatant was lyophilized to remove the ammonium carbonate solution and the dried sample was dissolved in 1 N HCl (5 mL) to form a hydrochloride salt of the product.
4. The crude product was lyophilized for a second time and the remaining solids were washed repeatedly with diethyl ether to remove the cinnamic acid starting material.
5. For purification (removal of residual salts), the product was dissolved in water and precipitated by the addition of 5 M NH$_4$OH until it reached pH 7.0. The final product was

collected by filtration and washed with ice-cold water to give 2-chloro-L-phenylalanine as a white solid (572 mg, 52% yield, *ee* > 99%).

4.13.4   Analytical Data

^{1}H-NMR (400 MHz D$_2$O+NaOH): δ = 7.32 (1 H, m), 7.20-7.11 (3 H, m), 3.44 (1 H, t, *J* = 7.2 Hz), 2.98 (1 H, dd, *J* = 13.4, 6.4 Hz), 2.81 (1 H, dd, *J* = 13.4, 8.0 Hz)
^{13}C-NMR (400 MHz D$_2$O+NaOH): δ = 182.4, 135.9, 133.8, 131.6, 129.4, 128.2, 127.0, 56.5, 38.6

Product *ee* was determined by isocratic reverse phase chiral HPLC using an Astec Chirobiotic T (4.6 × 250 mm, 5.0 µm) column and 60% methanol/40% water mobile phase at 1.0 mL.min^{-1} and 40 °C. HPLC retention times were as follows: 2′-chlorocinnamic acid, 2.5 minutes; 2′chloro-L-phenylalanine, 6.4 minutes; 2′chloro-D-phenylalanine, 7.7 minutes.

4.13.5   Conclusion

High levels of protein expression in *E. coli* cells and the wide availability of cinnamic acid starting materials make this process synthetically useful and highly scalable. Furthermore, PALs have broad substrate specificity and can be applied to the synthesis of a wide range of L-phenylalanine derivatives with substituents at the 2′, 3′, and 4′ positions. Substrates with alternative aromatic moieties, including pyridyls, furans, thiophenes, benzofurans, and benzothiophenes, are also well tolerated.

References

1. Aresta, M. (ed.) (2010) *Carbon Dioxide as a Chemical Feedstock*, Wiley-VCH, Weinheim.
2. Quadrelli, E.A., Centi, G., Duplan, J.-L., and Perathoner, S. (2011) *ChemSusChem*, **4**, 1194–1215.
3. Sakakura, T., Choi, J.-C., and Yasuda, H. (2007) *Chemical Reviews*, **107**, 2365–2387.
4. Tsuji, Y. and Fujihara, T. (2012) *Chemical Communication*, **48**, 9956–9964.
5. Huang, K., Sun, C.-L., and Shi, Z.-J. (2011) *Chemical Society Reviews*, **40**, 2435–2452.
6. Whiteoak, C.J., Nova, A., Maseras, F., and Kleij, A.W. (2012) *ChemSusChem*, **5**, 2032–2038.
7. Glueck, S.M., Gümüs, S., and Fabian, W.M.F., Faber, K. (2010) *Chemical Society Reviews*, **39**, 313–328.
8. Matsui, T., Yoshida, T., Yoshimura, T., and Nagasawa, T. (2006) *Applied Microbiology and Biotechnology*, **73**, 95–102.
9. Ishii, Y., Narimatsu, Y., Iwasaki, Y. *et al.* (2004) *Biochemical and Biophysical Research Communications*, **324**, 611–620.
10. Wuensch, C., Glueck, S.M., Gross, J. *et al.* (2012) *Organic Letters*, **14**, 1974–1977.
11. Wuensch, C., Schmidt, N., Gross, J. *et al.* (2013) *Journal of Biotechnology*, **168**, 264–270.
12. He, Z. and Wiegel, J. (1995) *European Journal of Biochemistry/FEBS*, **229**, 77–82.
13. He, Z. and Wiegel, J. (1996) *Journal of Bacteriology*, **178**, 3539–3543.
14. Schühle, K. and Fuchs, G. (2004) *Journal of Bacteriology*, **186**, 4556–4567.
15. (a) Wieser, M., Yoshida, T. and Nagasawa, T. (2001) *Journal of Molecular Catalysis B, Enzymatic*, **11**, 179–184; (b) Yoshida, T., Fujita, K. and Nagasawa, T., *Bioscience, Biotechnology, and Biochemistry* (2002) **66**, 2388–2394.
16. Lindsey, A.S. and Feskey, H. (1957) *Chemical Reviews*, **57**, 583–614.

17. Aresta, M., Quaranta, E., Liberio, R. *et al.* (1998) *Tetrahedron*, **54**, 8841–8846.
18. Goto, M., Hayashi, H., Miyahara, I. *et al.* (2006) *The Journal of Biological Chemistry*, **281**, 34365–34373.
19. Kirimura, K., Yanaso, S., Kosaka, S. *et al.* (2011) *Chemistry Letters*, **40**, 206–208.
20. Rodriguez, H., Angulo, I., de las Rivas, B. *et al.* (2010) *Proteins*, **78**, 1662–1676.
21. Gu, W., Yang, J., Lou, Z. *et al.* (2011) *PLoS ONE*, **6**, e16262.
22. Wuensch, C., Gross, J., Steinkellner, G. *et al.* (2013) *Angewandte Chemie (International Edition)*, **52**, 2293–2297.
23. Dresen, C., Richter, M., Pohl, M. *et al.* (2010). *Angewandte Chemie (International Edition)*, **49**, 6600–6603.
24. Kasparyan, E., Richter, M., Dresen, C. *et al.* (2014). *Applied Microbiology and Biotechnology*, **98**. doi: 10.1007/s00253-014-5850-0.
25. Pohl, M., Lingen, B., and Müller, M. (2002). *Chemistry – A European Journal*, **8**, 5288–5295.
26. Müller, M., Gocke, D., and Pohl, M. (2009). *FEBS Journal*, **276**, 2894–2904.
27. Pohl, M., Dresen, C., Beigi, M., and Müller, M. (2012). Enzymatic acyloin and benzoin condensations, in *Enzyme Catalysis in Organic Synthesis* (eds Karlheinz Drauz, Harald Gröger, and Oliver May), 3rd edn, Wiley-VCH, Weinheim, chapter 22 pp. 919–945.
28. Hailes, H., Rother, D., Müller, M. *et al.* (2013). *FEBS Journal*, **280**, 6374–6394.
29. Enders, D., Han, J., and Henseler, A. (2008) *Chemical Communication*, 3989–3991.
30. Maltsev, O.V., Kucherenko, A.S., Beletskaya, I.P. *et al.* (2010) *European Journal of Organic Chemistry*, 2927–2933.
31. García–García, P., Ladépêche, A., Halder, R., and List, B. (2008). *Angewandte Chemie (International Edition)*, **47**, 4719–4721.
32. Hayashi, Y., Itoh, T., Ohkubo, M., and Ishikawa, H. (2008). *Angewandte Chemie (International Edition)*, **47**, 4722–4724.
33. Zandvoort, E., Geertsema, E.M., Baas, B.J. *et al.* (2012) *Angewandte Chemie (International Edition)*, **51**, 1240–1243.
34. Miao, Y., Geertsema, E.M., Tepper, P.G. *et al.* (2013) *ChemBioChem*, **14**, 191–194.
35. Geertsema, E.M., Miao, Y., Tepper, P.G. *et al.* (2013) *Chemistry – A European Journal*, **19**, 14407–14410.
36. Whitman, C.P. (2002) *Archives of Biochemistry and Biophysics*, **402**, 1–13.
37. Poelarends, G.J., Puthan Veetil, V., and Whitman, C.P. (2008) *Cellular and Molecular Life Sciences*, **65**, 3606–3618.
38. Poelarends, G.J. and Whitman, C.P. (2004) *Bioorganic Chemistry*, **32**, 376–392.
39. Baas, B.J., Zandvoort, E., Geertsema, E.M., and Poelarends, G.J. (2013) *ChemBioChem*, **14**, 917–926.
40. Sambrook, J., Fritsch, E.F., and Maniatis, T. (1989) *Molecular Cloning: A Laboratory Manual*, 2nd edn, Cold Spring Harbor Laboratory Press.
41. Waddell, W.J. (1956) *The Journal of Laboratory and Clinical Medicine*, **48**, 311–314.
42. Gotoh, H., Ishikawa, H., and Hayashi, Y. (2007) *Organic Letters*, **9**, 5307–5309.
43. Mager, I. and Zeitler, K. (2010) *Organic Letters*, **12**, 1480–1483.
44. Hynes, P.S., Stupple, P.A., and Dixon, D.J. (2008) *Organic Letters*, **10**, 1389–1391.
45. Palomo, C., Landa, A., Mielgo, A. *et al.* (2007) *Angewandte Chemie (International Edition)*, **46**, 8431–8435.
46. Whitman, C.P., Aird, B.A., Gillespie, W.R., and Stolowich, N.J. (1991) *Journal of the American Chemical Society*, **113**, 3154–3162.
47. Johnson, W.H., Wang, S.C., Stanley, T.M. *et al.* (2004) *Biochemistry*, **43**, 10490–10501.
48. Alza, E. and Pericas, M.A. (2009) *Advanced Synthesis and Catalysis*, **351**, 3051–3056.
49. Riente, P., Mendoza, C., and Pericas, M.A. (2011) *The Journal of Materials Chemistry*, **21**, 7350–7355.

50. Jentzsch, K.I., Min, T., Etcheson, J.I. *et al.* (2011) *The Journal of Organic Chemistry*, **76**, 7065–7075.
51. Qiao, Y., He, J., Ni, B., and Headley, A.D. (2012) *Advanced Synthesis and Catalysis*, **354**, 2849–2853.
52. Harayama, S., Rekik, M., Ngai, K.L., and Ornston, L.N. (1989) *Journal of Bacteriology*, **171**, 6251–6258.
53. Chen, L.H., Kenyon, G.L., Curtin, F. *et al.* (1992) *The Journal of Biological Chemistry*, **267**, 17716–17721.
54. Zandvoort, E., Geertsema, E.M., Baas, B.J. *et al.* (2012) *Angewandte Chemie (International Edition)*, **51**, 1240–1243.
55. Zandvoort, E., Geertsema, E.M., Baas, B.J. *et al.* (2012) *ChemBioChem*, **13**, 1869–1873.
56. Zandvoort, E., Baas, B.J., Quax, W.J., and Poelarends, G.J. (2011) *ChemBioChem*, **12**, 602–609.
57. Narancic, T., Radivojevic, J., Jovanovic, P. *et al.* (2013) *Bioresource Technology*, **142**, 462–468.
58. Sambrook, J. and Russell, W.D. (2001) *Molecular Cloning a Laboratory Manual*, vol. 3, 3rd edn, Cold Spring Harbour Laboratory Press, Cold Spring Harbour, New York, USA.
59. Samanani, N., Liscombe, D.K., and Facchini, P.J. (2005) *The Plant Journal: for Cell and Molecular Biology*, **40**, 302–313.
60. Ruff, B.M., Bräse, S., and O'Connor, S.E. (2012) *Tetrahedron Letters*, **53**, 1071–1074.
61. Pesnot, T., Gershater, M.C., Ward, J.M., and Hailes, H.C. (2012) *Advanced Synthesis and Catalysis*, **354**, 2997–3008.
62. Pesnot, T., Gershater, M.C., Ward, J.M., and Hailes, H.C. (2011) *Chemical Communications*, **47**, 3242–3244.
63. Ilari, A., Franceschini, S., Bonamore, A. *et al.* (2009) *The Journal of Biological Chemistry*, **284**, 897–904.
64. Iwasa, K., Moriyasu, M., Tachibana, Y. *et al.* (2001) *Bioorganic and Medicinal Chemistry*, **9**, 2871–2884.
65. Thomas, C.M. and Ward, T.R. (2005) *Chemical Society Reviews*, **34**, 337.
66. Heinisch, T. and Ward, T.R. (2010) *Current Opinion in Chemical Biology*, **14**, 184.
67. Diamandis, E.P. and Christopoulos, T.K. (1991) *Clinical Chemistry*, **37**, 625.
68. Ward, T.R. (2011) *Accounts of Chemical Research*, **44**, 47.
69. Hyster, T.K., Knörr, L., Ward, T.R., and Rovis, T. (2012). *Science*, **338**, 500.
70. Guimond, N., Gorelsky, S.I., and Fagnou, K. (2011) *Journal of the American Chemical Society*, **133**, 6449.
71. Li, S.-M. (2010) *Natural Product Reports*, **27**, 57–78.
72. Kremer, A., Westrich, L., and Li, S.-M. (2007) *Microbiology*, **153**, 3409–3416.
73. Yu, X., Xie, X., and Li, S.-M. (2011) *Applied Microbiology and Biotechnology*, **92**, 737–748.
74. Unsöld, I.A. and Li, S.-M. (2005) *Microbiology*, **151**, 1499–1505.
75. Liebhold, M., Xie, X., and Li, S.-M. (2012) *Organic Letters*, **14**, 4884–4885.
76. Liebhold, M. and Li, S.-M. (2013) *Organic Letters*, **15**, 5834–5837.
77. Woodside, A.B., Huang, Z., and Poulter, C.D. (1988) *Organic Synthesis*, **66**, 211–215.
78. Jin, J. and Hanefeld, U. (2011) *Chemical Communications*, **47**, 2502–2510.
79. Boersma, A.J., Coquière, D., Geerdink, D. *et al.* (2010) *Nature Chemical*, **11**, 991–995.
80. Sheldon, R.A., Arends, I.W.C.E., and Hanefeld, U. (2007) *Green Chemistry and Catalysis*, Wiley-VCH, Weinheim, pp. 287–290.
81. Resch, V., Seidler, C., Chen, B.-S. *et al.* (2013) *European Journal of Organic Chemistry*, 7697–7705.
82. Feng, X. and Yun, J. (2009) *Chemical Communications*, 6577–6579.
83. Hartmann, E., Vyas, D.J., and Oestreich, M. (2011) *Chemical Communications*, **47**, 7917–7932.
84. Chen, B.-S., Otten, L.G., Resch, V. *et al.* (2013) *Standards in Genomic Sciences*, **9**, 175–184.
85. Holland, H.L. and Gu, J.-X. (1998) *Biotechnology Letters*, **20**, 1125–1126.
86. Cook, C., Liron, F., Guinchard, X., and Roulland, E. (2012) *The Journal of Organic Chemistry*, **77**, 6728–6742.

87. Hickmann, V., Kondoh, A., Gabor, B. *et al.* (2011) *Journal of the American Chemical Society*, **133**, 13471–13480.
88. Eliel, E.L., Bai, X., and Ohwa, M. (2000) *Journal of the Chinese Chemical Society*, **47**, 63–70.
89. Jin, J., Arends, I.W.C.E., and Hanefeld, U. (2012) Addition of water to C=C bonds and its elimination, in *Enzyme Catalysis in Organic Synthesis* (eds K. Drauz and H. Waldmann), third version, Wiley-VCH-Verlag, Weinheim, pp. 467–501.
90. Müller, M. (2013) *Chemie Ingenieur Technik*, **85**, 795–808; Müller, M. (2014) *ChemBioEng Rev.*, **1**, 14.
91. Guilbert, B., Davis, N.J., Pearce, M. *et al.* (1994) *Tetrahedron: Asymmetry*, **5**, 2163–2178.
92. Chapman, E., Best, M.D., Hanson, S.R., and Wong, C.H. (2004) *Angewandte Chemie (International Edition)*, **43**, 3526–3548.
93. Van der Horst, M.A., Van Lieshout, F.T., Bury, A. *et al.* (2012) *Advanced Synthesis and Catalysis*, **354**, 3501–3508.
94. Marhol, P., Hartog, A.F., Van der Horst, M.A. *et al.* (2013) *Journal of Molecular Catalysis B, Enzymatic*, **89**, 24–27.
95. Mozhaev, V.V., Khmelnitsky, Y.L., Sanchez-Riera, F. *et al.* (2002) *Biotechnology and Bioengineering*, **78**, 567–575.
96. Delhom, B., Alvaro, G., Caminal, G. *et al.* (1996) *Biotechnology Letters*, **18**, 609–614.
97. Coelho, P.S., Brustad, E.M., Kannan, A., and Arnold, F.H. (2013) *Science*, **339**, 307–310.
98. McIntosh, J.A., Coelho, P.S., Farwell, C.C. *et al.* (2013) *Angewandte Chemie (International Edition)*, **52**, 9309–9312.
99. Coelho, P.S., Wang, Z.J., Ener, M.E. *et al.* (2013) *Nature Chemical Biology*, **9**, 485–487.
100. Wang, Z.J., Peck, N.E., Renata, H., and Arnold, F.H. (2014) *Chemical Sciences*, **5**, 598–601.
101. Farwell, C.C., McIntosh, J.A., Hyster, T.K. *et al.* (2014) *Journal of the American Chemical Society*, **136**, 8766–8771.
102. Levinson, M.I. (2009) *Handbook of Detergents: Production* (eds U. Zoller and P. Sosis), Taylor and Francis, New York, pp. 1–37.
103. Soberón-Chávez, G. (2010) *Biosurfactants: From Genes to Applications*, Springer, Heidelberg.
104. Develter, W.G.D. and Fleurackers, J.J.S. (2010) *Surfactants from Renewable Resources* (eds M. Kjellin and I. Johansson), John Wiley & Sons, West Susset, pp. 213–238.
105. Lang, S., Brakemeier, A., Heckmann, R. *et al.* (2000) *Chimica Oggi*, **18**, 76–79.
106. Tran, H.G., Desmet, T., De Groeve, M.R.M., and Soetaert, W. (2011) *Biotechnology Progress*, **27**, 326–332.
107. Gawronski, J.D. and Benson, D.R. (2004) *Analytical Biochemistry*, **327**, 114–118.
108. Stepchenko, V.A., Seela, F., Esipov, R.S. *et al.* (2012) *SynLett*, **23**, 1541–1545.
109. Seela, F. and Pujari, S. (2010) *Bioconjugate Chemistry*, **21**, 1629.
110. Jiang, D. and Seela, F. (2010) *Journal of the American Chemical Society*, **132**, 4016.
111. Seela, F., Jiang, D., and Xu, K. (2009) *Organic and Biomolecular Chemistry*, **7**, 3463, and references cited therein.
112. (a) Pokharel, S., Jayalath, P., Maydanovych, O. *et al.* (2009) *Journal of the American Chemical Society*, **131**, 11882; (b) Veliz, E.A., Easterwood, L.M., and Beal, P.A. (2003) *Journal of the American Chemical Society*, **125**, 10867; (c) Wierzchowski, J., Ogiela, M., Iwanska, B., and Shugar, D. (2002) *Analytica Chimica Acta*, **472**, 63.
113. (a) Elliot, R.D. and Montgomery, J.A. (1976) *Journal of Medicinal Chemistry*, **19**, 1186; (b) Kazimierczuk, Z., Cottam, H.B., Revankar, G.R., and Robins, R.K. (1984) *Journal of the American Chemical Society*, **106**, 6379, and references therein.
114. (a) Mikhailopulo, I.A. (2007) *Current Organic Chemistry*, **11**, 317; (b) Mikhailopulo, I.A. and Miroshnikov, A.I. (2010) *Acta Naturae*, **2**, 36; (c) Mikhailopulo, I.A. and Miroshnikov, A.I. (2011) *Mendeleev Communs*, **21**, 57.

115. Esipov, R.S., Gurevich, A.I., Chuvikovsky, D.V. *et al.* (2002) *Protein Expression and Purification*, **24**, 56.
116. Chuvikovsky, D.V., Esipov, R.S., Skoblov, Y.S. *et al.* (2006) *Bioorganic and Medicinal Chemistry*, **14**, 6327.
117. Miroshnikov, A.I., Esipov, R.S., Konstantinova, I.D. *et al.* (2010) *The Open Conference Proceedings Journal*, **1**, 98.
118. van Ouwerkerk, N., Boom, J.H., Lugtenburg, J., and Raap, J. (2000) *The European Journal of Organic Chemistry* **5**, 861.
119. Pathak, T, Bazin, H., and Chattopadhyaya, J. (1986) *Tetrahedron*, **42**, 5427.
120. Seela, F. and Driller, H. (1988) *Helvetica Chimica Acta*, **71**, 757.
121. Gloge, A., Zon, J., Kovari, A. *et al.* (2000) *Chemistry – A European Journal*, **6**, 3386–3390.
122. Paizs, C., Katona, A., and Retey, J. (2006) *The European Journal of Organic Chemistry*, 1113–1116.
123. Paizs, C., Katona, A., and Retey, J. (2006) *Chemistry – A European Journal*, **12**, 2739–2744.
124. Bartsch, S. and Bornscheuer, U.T. (2010) *PEDS*, **23**, 929–933.
125. Lovelock, S.L., Lloyd, R.C., and Turner, N.J. (2014) *Angewandte Chemie (International Edition)*, **53**, 4652–4656.
126. de Lange, B., Hyett, D.J., Maas, P.J.D. *et al.* (2011) *ChemCatChem*, **3**, 289–292.

5

Oxidations

5.1 Semi-Preparative-Scale Drug Metabolite Synthesis with Human Flavin Monooxygenases

Steven P. Hanlon,[1] Matthias Kittelmann,[2] and Margit Winkler[3]

[1]*F. Hoffmann-La Roche Ltd., Switzerland*
[2]*NovartisPharma AG, Switzerland*
[3]*acib GmbH, Austria*

New active pharmaceutical ingredients (APIs) need to be thoroughly tested in the preclinical phase in order to predict their effectiveness, metabolism, and toxicology [1]. In the course of the drug development process, it is therefore necessary to prepare sufficient amounts of authentic metabolites for structure elucidation and as analytical references. Many approved APIs and API candidate molecules contain soft nucleophiles, such as nitrogen or sulfur – typical structural moieties that are metabolized by flavin monooxygenase enzymes (FMOs) in the human body. FMOs are membrane-associated proteins that catalyze substrate oxidation at the expense of NADPH and molecular oxygen [2]. Among the six known FMO isoforms, FMO3 has been ascribed particular importance as the most abundant isoform in the adult human liver [3]. In this section, we describe the use of recombinant human FMO3 as one example of a catalyst for semi-preparative-scale drug metabolite synthesis. Its form as a whole-cell biocatalyst allows economic co-factor recycling and facilitates product isolation. As an example, the N-oxide of Moclobemide (a reversible monoamine oxidase (MAO) inhibitor that is primarily used to treat depression [4]) was prepared on the multi-milligram scale (Scheme 5.1). The experimental details are adapted from reference [5].

Practical Methods for Biocatalysis and Biotransformations 3, First Edition.
Edited by John Whittall, Peter W. Sutton, and Wolfgang Kroutil.
© 2016 John Wiley & Sons, Ltd. Published 2016 by John Wiley & Sons, Ltd.

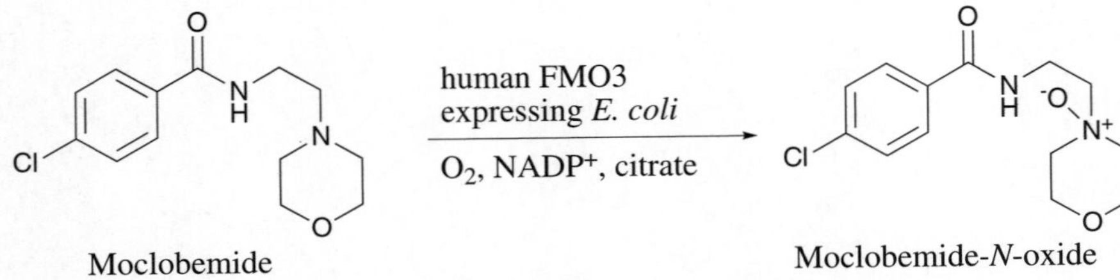

Scheme 5.1 *Whole cell catalyzed N-oxidation of Moclobemide with co-factor recycling.*

5.1.1 Materials and Equipment

- *E. coli* BL21(DE3) harboring plasmid pMS470nFMO3 (frozen glycerol stocks)
- Luria Bertani (LB) medium (bactotryptone $10\,g.L^{-1}$, yeast extract $5\,g.L^{-1}$, NaCl $10\,g.L^{-1}$)
- LB medium containing 15 or $20\,g.L^{-1}$ agar
- Ampicillin ($100\,mg.mL^{-1}$ stock solution in water, filter-sterilized)
- Isopropyl β-D-1-thiogalactopyranoside (IPTG) (1 M stock solution in water, filter-sterilized)
- NADP$^+$ ($787.4\,mg.L^{-1}$)
- Dry ice or liquid nitrogen
- Ice
- Dipotassium hydrogenphosphate
- Potassium dihydogenphosphate
- Magnesium chloride (1 M stock solution in water, filter-sterilized)
- Trisodium citrate (1 M stock solution in water, filter-sterilized)
- Moclobemide (100 mM stock solution in MeOH)
- MeOH
- Spectrophotometer to measure absorption at 600 nm (Eppendorf Biophotometer)
- Syringes (10 and 1 mL)
- 0.22 μm Syringe filters (Rotilabo, CME sterile)
- Sterile loop
- Sterile Petri dish
- 50 mL Falcon tubes
- Baffled 2 L Erlenmeyer flasks
- Eppendorf vessels
- Pipettes and pipette tips
- 24-Well plates (NunclonΔ Surface, Nunc)
- Oxygen-permeable foil (Gas Permeable Adhesive Seals # AB-0718, Thermo Scientific)
- Rotary shaker (5 cm radius) at 27 and 37 °C
- Centrifuge (Eppendorf centrifuge 5810R)
- Tabletop centrifuge (Eppendorf centrifuge 5415)
- HPLC system and UV detection (Agilent)
- HPLC-grade acetonitrile (J.T. Baker)
- Deionized water
- HPLC-grade formic acid (Fluka)

- HPLC-grade ammonium formate (Fluka)
- Zorbax SB-C18 HPLC column (C_{18}, 1.8 μm, 50 × 4.6 mm; Agilent) or Chromolith (C_{18}, 50 × 4.6 mm; Agilent)
- Polypropylene Microtiter plates with sealing foil or glass vials with stopper and septum
- XAD-16 resin (Rohm and Haas, The Dow Chemical Company)
- Aqueous ammonia (25%) (Roth GmbH)
- Diatom granulate (Isolute HM-N, Separtis AG)
- Rotary evaporator (Heidolph)
- Nucleodur 100-10 C18 ec (250 × 20 mm; Machery-Nagel)
- Spot Prep II preparative HPLC system and UV detection (Armen Instrument, Saint-Avé, France)
- Lyophilizer (Martin Christ Gefriertrocknungsanlagen GmbH)

5.1.1.1   Optional

- 150 L Fermentation vessel (Braun Biotech)
- Continuous-flow centrifuge at 4 °C (Heraeus 20 RS)
- Aseol antifoam agent
- HPLC equipped with UV and mass-selective detector

5.1.2   Small-Scale Culture

1. A glycerol stock of *E. coli* BL21(DE3)/pMS470nFMO3 was thawed, and 20 μL were used to inoculate 20 mL of LB medium (containing 100 μg.mL^{-1} ampicillin) in Falcon tubes.
2. Cultures were incubated with shaking at 90 rpm on a rotary shaker at 37 °C overnight.
3. Fresh LB cultures (500 mL) with ampicillin (100 μg.mL^{-1}) in baffled 2 L Erlenmayer flasks were inoculated to OD_{600} 0.05 with overnight culture. These cultures were incubated at 37 °C with shaking at 110 rpm until an optical density of 0.6–0.8 was reached.
4. FMO3 expression was induced by addition of IPTG (1 mM). The cultures were incubated at 30 °C with shaking at 110 rpm for 24 hours.
5. The cells were harvested by centrifugation (10 minutes, 4000 rpm). The cell paste was shock-frozen with liquid nitrogen and stored at −20 °C.

5.1.3   Large-Scale Culture

1. *E. coli* BL21(DE3)/pMS470nFMO3 from a glycerol stock was streaked on LB plates (containing 100 μg.mL^{-1} ampicillin) and grown at 37 °C overnight.
2. A loop of cells was used to inoculate 100 mL of LB medium (containing 100 μg.mL^{-1} ampicillin) in 500 mL baffled flasks.
3. Cultures were incubated with shaking at 90 rpm on a rotary shaker (5 cm radius) at 37 °C for 18 hours.
4. The entire pre-culture was used to inoculate LB medium (100 L containing 100 mg.L^{-1} ampicillin) and Aesol antifoam (0.01% v/v) in a 150 L fermentation vessel. The biomass was cultivated at 37 °C at a stirring speed of 120 rpm and an airflow of 10 L.min^{-1} without pH regulation until an optical density of 0.6–0.8 was reached.
5. FMO3 expression was induced by addition of filter-sterilized IPTG (to a concentration of 1 mM).

6. After 24 hours, the biomass was harvested by continuous-flow centrifugation at 13 000 rpm and 4 °C. The cell paste was shock-frozen in dry ice and stored at −80 °C.

5.1.4 Analytical-Scale Biotransformations

1. The cell paste was thawed and aliquots of 200 µL were transferred to a 24-well multi-dish and dispersed in potassium phosphate buffer (pH 8.5) to give overall volumes of 880 µL per well. The following components were subsequently added:
   a. Trisodium Citrate (50 µL, 1 M), MgCl$_2$ (10 µl, 1 M) and NADP$^+$ (50 µL, 1 mM)
   b. Substrate (10 µL, 100 mM in methanol or water)
2. The plate was covered with oxygen-permeable foil and agitated at 37 °C and 900 rpm on a Titramax for 16–20 hours.

5.1.5 Preparative-Scale Biotransformations

1. The cell paste was thawed (25 g) and dispersed in potassium phosphate buffer (pH 8.5) in a 2 L baffled flask to give an overall volume of 445 mL. The following components were subsequently added:
   a. Trisodium Citrate (25 mL, 1 M), MgCl$_2$ (5 ml, 1 M) and NADP$^+$ (25 mL, 1 mM)
   b. Moclobemide (50 mg in 2.5 mL MeOH, 0.186 mmol)
2. The flask(s) were agitated at 27 °C and 120 rpm for 24 hours.
3. The cells were separated by centrifugation at 4000 rpm and 4 °C for 45 minutes. The supernatant was subsequently shock-frozen and stored at −20 °C.
4. For product isolation, the supernatant was thawed in a water bath at 50 °C and subsequently mixed with aqueous ammonia (5 mL, 25%) and XAD-16 resin (100 g) for 30 minutes at room temperature.
5. After filtration over gauze, the resin was treated with acetonitrile (MeCN) (500 mL) for 15 minutes and filtered again. This procedure was repeated.
6. Diatom granulate (10 g) was added to the combined extracts and the solvent was removed under reduced pressure.
7. The dry material was filled into a 50 × 25 mm pre-column and subjected to preparative reverse phase chromatography at room temperature. The mobile phase consisted of the following gradient of ammonium formate (10 mM in H$_2$O) and MeCN: 0–5 minutes, 5% MeCN isocratic; 22.5 minutes, 20% MeCN; 25 minutes, 38% MeCN; 28 minutes, 100% MeCN; followed by 2 minutes' re-equilibration at 5% MeCN isocratic. Product elution was monitored at 237 nm from 8 to 17 minutes, and the combined fractions were evaporated under reduced pressure to about 100 mL and then lyophilized.
8. Residual ammonium formate was removed by another lyophilization step after dissolution of the residue in MeCN (2 mL).
9. Moclobemide-N-oxide (65 mg) was obtained in 55% yield. Product identity and purity (>90%) were confirmed by HPLC and NMR, respectively.

5.1.6 Analytical Methods

Aliquots of 0.5 mL of biotransformation samples were thoroughly mixed with 0.5 mL of MeOH. The mixtures were vortexed and centrifuged at 16.200 × g for 5 minutes, then 200 µL was transferred into 96-well Polypropylene Microtiter plates or glass vials and 10 µL was injected for HPLC analysis.

The HPLC system used a mobile phase consisting of MeCN and formic acid (0.1% v/v in water) at a flow rate of $1.2 \, mL.min^{-1}$ at room temperature. Starting from 30% MeCN, the analytes were separated on a Chromolith column by a gradient: 0–3.50 minutes, 80% MeCN; 3.51–4.00 minutes, 100% MeCN; 4.01–4.50 minutes, re-equilibration at 30% MeCN isocratic. Substrate and product were detected at 260 nm and, after splitting the flow to approximately $0.8 \, mL.min^{-1}$, on a mass-selective detector. The retention times of Moclobemide and the N-oxide were 2.0 and 2.4 minutes, respectively.

Alternatively, a flow rate of $1.5 \, mL.min^{-1}$ at 50 °C was used with the same mobile phase. Starting from 0% MeCN, the analytes were separated on a Zorbax SB-C18 column by a gradient: 0–2 minutes, 50% MeCN; 0.7 minutes, re-equilibration at 0% MeCN. Substrate and product were detected at 240 nm. The retention times of Moclobemide and the N-oxide were 1.35 and 1.40 minutes, respectively.

5.1.7 Conclusion

Recombinant human FMO3 is the first example of a membrane-associated flavoprotein to be used as a metabolic mimic for generation of an API metabolite on preparative scale. *E. coli* strains harboring other human FMO isoforms, such as FMO5, are also accessible using the same [5] or a slightly modified expression strategy [6]. Hence, a panel of non-CYP drug-metabolizing flavin monooxygenase enzymes is now available, which offers the possibility of studying the action of a single FMO species on drugs and drug candidates. Whereas eukaryotic expression strategies are also very useful in generating enzymes for screening or reaction phenotyping applications (e.g., FMO1, FMO3, or FMO5 BD supersomes), our strains have the advantage of being cheap, easy to handle, and usable at scale. We used Moclobemide as a model compound, but hFMO-expressing strains are routinely screened for activity toward drugs and drug candidates in development, and these reactions may be upscaled whenever the respective metabolites are required.

5.2 Biobased Synthesis of Industrially Relevant Nitriles by Selective Oxidative Decarboxylation of Amino Acids by Vanadium Chloroperoxidase

Andrada But[1], Jerôme Le Nôtre[1], Elinor L. Scott[1], Ron Wever[2], and Johan P. M. Sanders[1]

[1]*Biobased Chemistry and Technology, Wageningen University, The Netherlands*
[2]*Van 't Hoff Institute for Molecular Sciences, University of Amsterdam, The Netherlands*

Nitriles are important chemicals for a variety of applications. They are widely used as intermediates in a number of chemical syntheses, as building blocks for polymer production (acrylonitrile), and as industrial solvents (acetonitrile). One example of nitrile industrial production is based on the introduction of the nitrile functionality to a hydrocarbon backbone via ammoxidation [7] – an energy-intensive process that involves the use of ammonia as nitrogen source and the formation of toxic compounds as side products. Other methods are available, but toxic chemicals (HCN) and/or controlled forcing reaction conditions (high temperature and pressure) are required.

Consequently, there is a need to develop rapid and facile methods of synthesizing nitriles in one pot under energy-efficient reaction conditions and without toxic reagents or byproducts. In addition, the demand for biobased materials – including nitrogen-containing chemicals like nitriles – is continuously increasing, as is the need to reduce our dependency on fossil resources. In the context of a biobased economy, where the use of renewable feedstocks is required, amino acids represent an attractive source of nitrogen-containing chemicals, and they are present in widely available protein-rich materials, including waste streams from the biofuel industry, such as dried distillers' grains with solubles and slaughter and poultry waste. For example, glutamic acid is one of the most abundant amino acids in a number of waste streams [8].

As shown previously, amino acids can be easily converted into nitriles and aldehydes by oxidative decarboxylation, using either an enzymatic approach (heme bromoperoxidase/NaBr/H_2O_2) or a chemical approach (NaOCl/cat.NaBr). But due to the instability of the heme haloperoxidases toward hydrogen peroxide in the enzymatic approach and the unfavorable results of the lifecycle assessment of the chemical approach, it has been necessary to find a more suitable system [9]. In this section, we demonstrate that by using the very stable vanadium chloroperoxidase (VCPO) [10] in the presence of hydrogen peroxide and a catalytic amount of sodium bromide, it is possible to convert glutamic acid into 3-cyanopropanoic acid with high conversion and selectivity (Scheme 5.2).

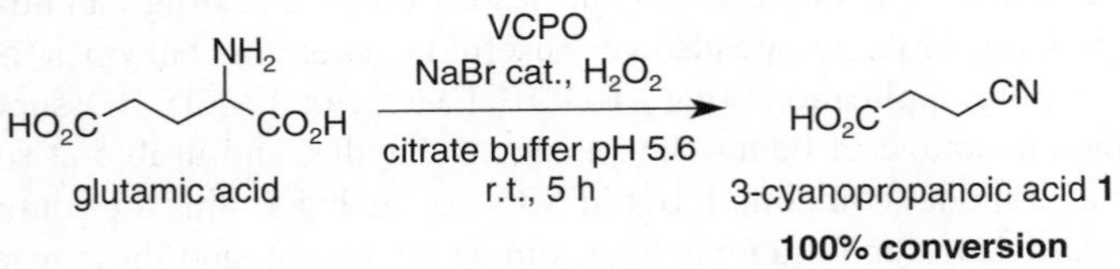

Scheme 5.2 *Enzymatic oxidative decarboxylation of glutamic acid into 3-cyanopropanoic acid 1 by VCPO.*

This compound can be further converted into succinonitrile, which is the precursor of diaminobutane (DAB) – a compound used in the production of Nylon 4/6. 3-cyanopropanonic acid can also be converted into acrylonitrile – an important building block in the production of various polymers.

This enzymatic system, VCPO/NaBr/H_2O_2, can also be applied to convert other amino acids into nitriles. For example, we showed that it is possible to convert phenylalanine into phenylacetonitrile [11]. In this conversion, a side product, the corresponding aldehyde, is also formed together with phenylacetonitrile (Scheme 5.3).

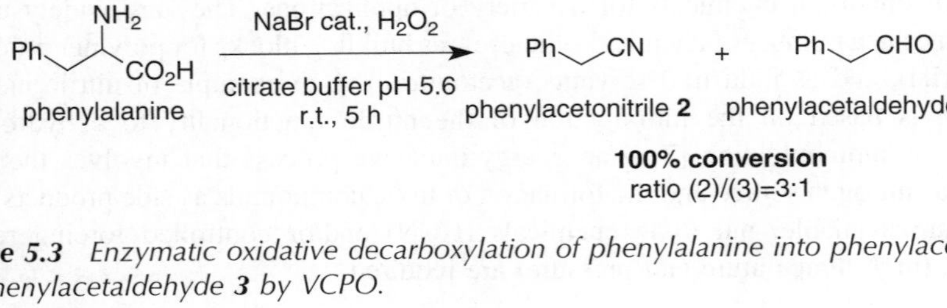

Scheme 5.3 *Enzymatic oxidative decarboxylation of phenylalanine into phenylacetonitrile 2 and phenylacetaldehyde 3 by VCPO.*

5.2.1 Materials and Equipment

- Recombinant VCPO (0.6 mg.mL^{-1}, 9 µM) [12]
- Citrate buffer (20mM, pH 5.6)
- L-Glutamic acid (1.47 mg, 5 mM)
- L-Phenylalanine (1.65 mg, 5 mM)
- NaBr (2 mM)
- Hydrogen peroxide solution (34.5–36.5%)
- Na$_2$S$_2$O$_3$
- Acetonitrile
- Methanol
- HPLC with RI detection (Alltech IOA-1000 Column for Organic acids (9 µm particle size, 7.8 × 300 mm))
- UHPLC with UV detection (Acquity HPLC BEH C18 column (1.7 µm particle size, 2.1 × 150 mm) + VanGuard Acquity HPLC BEH C18 precolumn (1.7 µm particle size, 2.1 × 5.0 mm)) [10,13]

5.2.2 Oxidative Decarboxylation of Amino Acids into Nitriles by VCPO

1. A solution was prepared (2 mL) containing 5 mM amino acid, 2 mM NaBr, and 18 nM VCPO in 20 mM citrate buffer at pH 5.6. The solution was stirred at 300 rpm at room temperature (21 °C).
2. Small aliquots (2 µL (1.2 M) × 15) of H$_2$O$_2$ (18 mM) were added to this reaction mixture at intervals of 20 minutes. To prevent evaporation during the reaction, the vials were closed with a cap. The gas pressure was equilibrated when the covers were opened for the addition of hydrogen peroxide.
3. The reaction was stopped after 5 hours by adding Na$_2$S$_2$O$_3$ to the mixture.

5.2.3 Analytical Method

Amino acids were analyzed by UHPLC-UV after derivatization [13]. 3-cyanopropanoic acid was analyzed directly by HPLC-RI. In the case of phenylacetonitrile, 6 mL of acetonitrile/methanol (1/1, v/v) was added to the crude mixture to obtain a homogeneous solution, which was analyzed without derivatization by UHPLC-UV. In addition to the phenylacetonitrile, about 25% phenylacetaldehyde was formed, which was analyzed in the same procedure.

5.2.4 Conclusion

Nitriles of industrial relevance like 3-cyanopropanoic acid can be obtained from amino acids by an oxidative decarboxylation reaction catalyzed by the enzyme VCPO. The enzyme is a selective tool in the conversion of glutamic acid into 3-cyanopropanoic acid, with 100% conversion and selectivity. Bromide is regenerated during the process, so only a low concentration is needed. Initial experiments suggest that it is possible to convert a mixture of amino acids using this enzymatic system. Thus, amino acids present in the hydrolysate of protein-rich waste streams may be converted into a mixture of industrially interesting nitriles that can be easily removed from the water phase by extraction and further separated by distillation.

5.3 Terminal Oxygenation of Fatty Acids by a CYP153A Fusion Construct Heterologously Expressed in *E. coli*

Sumire Honda Malca, Daniel Scheps, Bettina M. Nestl, and Bernhard Hauer

Institute of Technical Biochemistry, University of Stuttgart, Germany

ω-Hydroxy fatty acids (ω-OHFAs) are valuable chemicals for the fragrance industry and for several commodity and advanced plastic applications. Saturated ω-OHFAs serve as building blocks for the synthesis of polymers that exhibit high durability, water resistance, and chemical compatibility with polyamides and polyesters [14]. Unsaturated ω-OHFAs can be cross-linked with different types of monomers or decorated with flame-retardant or bioactive functionalities at their double-bond sites, making them attractive for industrial and medical applications [15]. One way to synthesize ω-OHFAs is to hydroxylate the terminal C$-$H bond of fatty acids or fatty acid methyl esters; however, this is a difficult reaction to achieve using chemical catalysts, and, if achieved, it is limited in terms of regio- or chemoselectivity [16]. Cytochrome P450 monooxygenases from the bacterial CYP153A subfamily are alkane or primary alcohol hydroxylases with high ω-regioselectivities [17–20]. CYP153A from *Marinobacter aquaeolei* (CYP153A$_{M.aq}$) additionally catalyzes the ω-hydroxylation of C_8-C_{18} saturated and 9-*cis/trans*-C_{14}-C_{18} monounsaturated fatty acids. A variant, G307A, possesses 2- to 20-fold higher catalytic efficiency toward different saturated fatty acids under *in vitro* conditions [21]. To achieve higher electron coupling efficiency and higher conversion, a fusion chimera consisting of the improved CYP153A$_{M.aq(G307A)}$ variant and the reductase domain (CPR$_{BM3}$) of P450 BM3 from *Bacillus megaterium* was created [22]. The uptake of C_9-C_{12} fatty acid methyl esters by *E. coli* cells can be facilitated by the co-expression of the heterologous *alkL* gene, which encodes an outer-membrane transport protein [23]. We aimed to provide a protocol that requires shake flasks instead of a bioreactor. Experimental details are adapted from reference [22] for the ω-hydroxylation of C12 dodecanoic acid on a 0.2 L scale (Scheme 5.4).

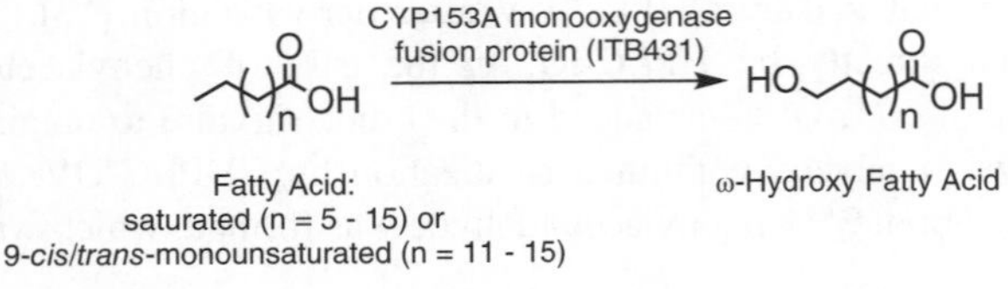

Scheme 5.4 *Biocatalytic synthesis of ω-OHFAs.*

5.3.1 Materials and Equipment

- Chemically competent *E. coli* HMS174(DE3) cells
- Plasmid pITB431 (pET28a(+)_CYP153A$_{M.aq(G307A)}$-CPR$_{BM3}$) or plasmid pITB407 (pColaDuet1-CYP153A$_{M.aq(G307A)}$-CPR$_{BM3}$ and AlkL)
- LB broth (autoclaved)
- LB agar (autoclaved)
- TB medium: $90 \times$ TB broth and $10 \times$ TB buffer ($125.4 \, g.L^{-1}$ dipotassium hydrogen phosphate and $23.1 \, g.L^{-1}$ potassium dihydrogen phosphate), autoclaved
- Kanamycin sulfate ($30 \, mg.mL^{-1}$ in water, filter-sterilized)

- 0.1 M IPTG (in water, filter-sterilized)
- 1 M δ-Aminolevulinic acid (in water, filter-sterilized)
- 0.2 M Potassium phosphate buffer pH 7.4 (autoclaved)
- D-(+)-Glucose monohydrate (200 g.L^{-1} in water, filter-sterilized)
- Glycerol, 99% (500 g.L^{-1} in water, autoclaved)
- Diethyl ether
- Dimethyl sulfoxide (DMSO)
- Methyl *tert*-butyl ether (MTBE)
- Decanoic acid (8.61 mg.mL^{-1} in DMSO)
- Dodecanoic acid (50 mg.mL^{-1} in DMSO)
- 12-Hydroxy dodecanoic acid (21.6 mg.mL^{-1} in DMSO)
- 1,12-Dodecanedioic acid (23 mg.mL^{-1} in DMSO)
- N,O-Bis(trimethylsilyl)trifluoroacetamide containing 1% trimethylchlorosilane
- Orbitary shakers at 37 and 30 °C
- Centrifuge
- Photometer
- Gas chromatograph coupled to FID, equipped with a fused silica capillary GC column (5% diphenyl, 95% dimethyl polysiloxane)

5.3.2 Cultivation, Protein Expression, and Cell Harvesting

1. Chemically competent *E. coli* HMS174(DE3) cells were transformed with plasmid pITB407 or pITB407 and streaked on to a LB agar plate with 30 µg.mL^{-1} kanamycin to obtain single colonies.
2. One colony of the freshly plated transformants was inoculated in LB medium (5 mL) containing 30 µg.mL^{-1} kanamycin.
3. Two 2 L Erlenmeyer flasks, each containing TB medium (400 mL) with 30 µg.mL^{-1} kanamycin, were inoculated with the pre-culture (2 mL). Cultures were incubated at 37 °C and 180 rpm. At an OD$_{600\ nm}$ of 1.2–1.5, cells were induced for recombinant protein expression with 0.1 mM IPTG and 0.5 mM δ-aminolevulinic acid.
4. At an OD$_{600\ nm}$ of 9–10, induced resting cells were harvested by centrifugation at 4 °C and 4225 × g for 20 minutes. After washing twice with 0.2 M potassium phosphate buffer pH 7.4 containing 4 g.L^{-1} glycerol, they were resuspended in the same solution to a final wet biomass concentration of 100 g.L^{-1}. The cell suspension was used immediately for the biotransformation step, or alternatively was stored at 4 °C and 140 rpm for 2–3 days (fed with 4 g.L^{-1} glycerol per day). Note: Functional CYP expression can be verified in whole cells [24] using the carbon monoxide-differential spectral method by Omura and Sato.

5.3.3 Whole-Cell Biotransformation with Resting Cells

1. A 0.2 L cell suspension containing 50 g.L^{-1} wet biomass in 0.2 M potassium phosphate buffer with 4 g.L^{-1} glycerol and 4 g.L^{-1} glucose was prepared in a 1 L Erlenmeyer flask.
2. Dodecanoic acid was added to the cell suspension in a final concentration of 1 g.L^{-1} from the 50 g.L^{-1} stock solution in DMSO.
3. The reaction was incubated at 30 °C and 180 rpm for 8 hours. Glucose and glycerol (4 g.L^{-1} each) were added after 4 hours. Aliquots of 0.5 mL were taken for analyses at time points 0, 4, and 8 hours (see Analytical Methods).

5.3.4 Analytical Methods

Conversions were stopped with 30 μL concentrated HCl (37% HCl). 1 mM C10 decanoic acid added as internal standard from the 8.61 mg.mL^{-1} (50 mM) stock solution in DMSO. The reaction mixtures were extracted twice, with 0.5 mL diethyl ether each time. The organic phases were collected, dried over Na_2SO_4 (anhydrous), and evaporated. Samples were resuspended in MTBE (60 μL). 1% trimethylchlorosilane in *N,O*-bis(trimethylsilyl)trifluoroacetamide (60 μL) was added and the samples were incubated at 70 °C for 30 minutes for derivatization.

Samples were analyzed on a GC-FID instrument equipped with the indicated fused silica capillary column and hydrogen as carrier gas (flow rate 0.8 mL.min^{-1}; linear velocity 30 cm.s^{-1}). The injector and detector temperatures were set at 250 and 310 °C, respectively. The column oven was set at 130 °C for 2 minutes, raised to 250 °C at a rate of 10 °C. min^{-1}, held isotherm for 3 minutes, and then raised to 300 °C at 40 °C.min^{-1}. Substrate conversion and product formation were quantified from the peak areas using calibration curves estimated from a series of standard solutions prepared from the dodecanoic acid, 12-hydroxy dodecanoic acid, and 1,12-dodecanedioic stock solutions. The standards were treated in the same manner as the samples.

5.3.5 Conclusion

The monooxygenase CYP153A$_{M.aq}$ catalyzes the oxidation of 0.2 mM C_{12}-C_{14} saturated and 9-*cis/trans*-C_{16}-C_{18} monounsaturated fatty acids with 64–93% conversion and >95% ω-regioselecitivity under *in vitro* conditions [21]. The conversion level of 1 mM tetradecanoic acid with the enzyme variant G307A is higher by 40% than that of the wild-type enzyme [21]. Whole-cell experiments with the engineered fusion construct toward 5 mM dodecanoic acid result mainly in the formation of 12-hydroxy dodecanoic, with 49% conversion and ≥98% ω-regioselectivity. The dicarboxylic acid product accounts for 5–8% of the total product, which might be formed by the presence of constitutive dehydrogenases in *E. coli* [22]. In another experiment with whole cells expressing the fusion construct and the *alkL* gene, 4 g.L^{-1} ω-hydroxy dodecanoic acid was obtained from methyl dodecanoate after 28 hours using a two-phase system (1 L bioreactor, 5: 1 aqueous/organic) [22]. To conclude, the broad substrate range of this biocatalyst enables its application in the selective terminal hydroxylation of fatty acids of different chain lengths and saturation levels using *E. coli* as production host.

5.4 Enantioselective Oxidative C−C Bond Formation in Isoquinoline Alkaloids Employing the Berberine Bridge Enzyme

Verena Resch, Joerg H. Schrittwieser, and Wolfgang Kroutil
Department of Chemistry, Organic and Bioorganic Chemistry, University of Graz, Austria

Constructing complex core structures found in natural products is still a major challenge in asymmetric synthesis. Nature uses and offers enzymes which are able to perform reactions that are – to date – not possible employing classical chemistry. One example from alkaloid metabolism is the berberine bridge enzyme (BBE) (EC 1.21.3.3), a redox enzyme that converts (*S*)-reticuline as the natural substrate possessing a 1-benzyl-1,2,3,

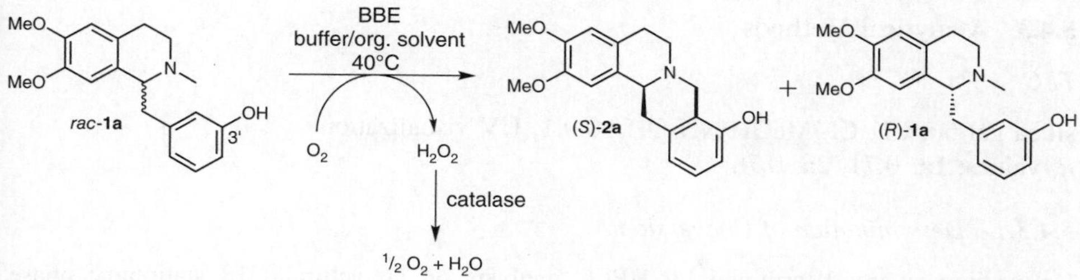

Scheme 5.5 *Biocatalytic transformation of rac-**1a** as a non-natural substrate resulted in the formation of (S)-**2a** via C−C-bond formation. The substrate enantiomer (R)-**1a** stayed untouched in this kinetic resolution.*

4-tetrahydroisoquinoline backbone to (*S*)-scoulerine, a berbine derivative [25,26]. This transformation represents an asymmetric oxidative intramolecular C−C coupling involving C−H activation at the expense of molecular oxygen (Scheme 5.5). The reaction conditions have been optimized [27] for the transformation of non-natural substrates in organic synthesis [28–30].

5.4.1 Materials and Equipment

- BBE (purified, 1.5 mL enzyme solution, final concentration $= 1\,g.L^{-1} = 0.017$ mM; expressed as previously described) [29,31]
- Substrate *rac*-**1a** (500 mg, 1.6 mmol; synthesized as previously described) [28,29]
- Toluene (17.5 mL, grade 99.9%)
- Tris-HCl buffer (7.5 mL, 10 mM Tris-HCl, pH 9.0, 10 mM MgCl$_2$)
- Crude catalase (125 mg, final concentration 5 g.L^{-1}, from bovine liver, Sigma-Aldrich (C-10, Lot.: 81H7146, 1600–2000 U.mg^{-1} solid)
- Round-bottom flask (50 mL)
- Rotary shakers at 40 °C
- Ethyl acetate (3 × 10 mL)
- Na$_2$SO$_4$ anhydrous
- Silica gel (pore size 60 Å, 0.040–0.063 mm particle size)
- Silica gel TLC plates (silica gel 60 F$_{254}$, Merck)

5.4.2 Procedure

1. Substrate **1a** (500 mg, 1.6 mmol) was dissolved in toluene (17.5 mL) and buffer (7.5 mL, 10 mM Tris-HCl, pH 9.0, 10 mM MgCl$_2$) containing BBE (1.5 mL enzyme solution, final concentration $= 1\,g.L^{-1} = 0.017$ mM) and crude catalase (125 mg, final concentration 5 g.L^{-1}). The mixture was shaken in a light-shielded round-bottom flask (50 mL) at 200 rpm and 40 °C for 24 hours.
2. The reaction was stopped by phase separation, followed by extraction of the aqueous phase with ethyl acetate (3 × 10 mL). The combined organic layers were dried (Na$_2$SO$_4$) and the solvent was evaporated under reduced pressure to give the crude product.
3. Chromatography over silica using CH$_2$Cl$_2$/MeOH/NH$_4$OH (97 : 2 : 1) as eluent afforded (*S*)-**2a** (207 mg, 42%, *ee* > 97%) and (*R*)-**1a** (249 mg, 49%, *ee* > 97%), both as off-white solid foams.

5.4.3 Analytical Methods

TLC

silica gel 60; CH_2Cl_2/MeOH/NH_4OH 90:9:1; UV visualization
R_f-values: **1a**: 0.71, **2a**: 0.78

5.4.3.1 Determination of Conversion

Conversions were determined by HPLC analysis on an achiral C18 stationary phase (Phenomenex, LUNA C18, 0.46 × 25 cm, 5 µm). Eluent: buffer (30 mM $HCOONH_4$, pH 2.8)/methanol/acetonitrile = 67/18/15 (isocratic); flow rate: 0.5 mL.min^{-1}; column temperature: 20 °C; detection wavelength: 280 nm. Retention times: **1a**, 8.4 minutes; **2a**, 11.6 minutes.

5.4.3.2 Determination of Enantiomeric Excess

Optical purity of **1a** was determined by HPLC analysis on a chiral column (Chiralcel OJ from Daicel Chemical Industries, 0.46 × 25 cm). Eluent: *n*-heptane/2-propanol = 70/30 (+0.1% formic acid); flow rate: 0.5 mL.min^{-1} (isocratic); column temperature: 20 °C; detection wavelength: 280 nm. Retention times: (*S*)-**1a**, 14.2 minutes; (*R*)-**1a**, 18.8 minutes.

Optical purity of **2a** was determined by HPLC analysis on a chiral column (Chiralpak AD from Daicel Chemical Industries, 0.46 × 25 cm). Eluent: *n*-heptane/2-propanol = 70/30; flow rate: 0.5.mL min^{-1}; column temperature: 18 °C; detection wavelength: 280 nm. Retention times: (*R*)-**2a**, 11.0 minutes; (*S*)-**2a**, 18.4 minutes.

5.4.3.3 Melting Points

(*R*)-**1a**: 151–153 °C
(*S*)-**2a**: 90–95 °C

5.4.3.4 Optical Rotation

(*R*)-**1a**: $[\alpha]_D^{20}$: −109.4; $c = 1$ (g.100 mL^{-1}), $CHCl_3$
(*S*)-**2a**: $[\alpha]_D^{20}$: −273.4; $c = 1$ (g.100 mL^{-1}), $CHCl_3$

5.4.3.5 NMR

(*R*)-**1a**: ^{1}H-NMR (CDCl$_3$; 300 MHz): δ 2.52 (3H, s, NCH$_3$), 2.64–2.97 (4H, m, CH$_2$), 3.20–3.30 (2H, m, CH$_2$), 3.50 (3H, s, OCH$_3$), 3.77–3.86 (4H, s + m overlap, OCH$_3$ + CH), 5.96 (1H, s, Ar), 6.53–6.57 (2H, s + m overlap, Ar), 6.65–6.67 (2H, m, Ar), 7.05–7.10 (1H, m, Ar). ^{13}C-NMR (CDCl$_3$; 75 MHz): δ 24.3, 41.4, 41.8, 45.9, 55.4, 55.7, 64.7, 111.1, 111.2, 113.8, 116.7, 121.4, 124.7, 128.3, 129.4, 141.0, 146.3, 147.5, 157.1
(*S*)-**2a**: ^{1}H-NMR (CDCl$_3$; 300 MHz): δ 2.57–2.63 (2H, m, CH$_2$), 2.86 (1H, dd, $J_1 = 16.2$ Hz, $J_2 = 11.5$ Hz, CH$_2$), 3.08–3.15 (2H, m, CH$_2$), 3.22 (1H, dd, $J_1 = 16.5$ Hz, $J_2 = 3.5$ Hz, CH$_2$), 3.40 (1H, d, $J = 15.4$ Hz, CH$_2$), 3.54–3.61 (1H, m, CH), 3.77 (3H, s, OCH$_3$), 3.80 (3H, s, OCH$_3$), 4.13 (1H, d, $J = 15.7$ Hz, CH$_2$), 6.34 (1H, d, $J = 7.9$ Hz, Ar), 6.53 (1H, s, Ar), 6.59 (1H, d, $J = 7.6$ Hz, Ar), 6.65 (1H, s, Ar), 6.82 (1H, t, $J = 7.7$ Hz, Ar). ^{13}C-NMR (CDCl$_3$; 75 MHz): δ 28.5, 36.2, 51.4, 53.5, 55.9, 56.1, 59.1, 108.5, 111.3, 112.5, 120.2, 121.4, 126.5, 126.9, 127.1, 129.1, 135.5, 147.5, 147.6, 152.7

5.4.4 Conclusion

The BBE was successfully employed for the transformation of a variety of different benzylisoquinoline alkaloids. Beside the non-natural substrate **1a**, BBE accepts numerous structurally related isoquinoline alkaloids (Scheme 5.6). In general, substrates with variations on the isoquinoline ring (**1a–1j**) are readily accepted. No C−C bond coupling product was detected if the OH functionality at the 3′ position (R^4) was altered. For the natural substrate **1b**, the reaction is not only enantioselective, but also highly regioselective; thus, no formation of a regioisomeric product (**3**) is detected. In cases of non-natural substrates, the regioisomeric product is usually found as a minor side product. A complete switch in the regioselectivity can be achieved by substrate engineering, using substrates in which the usual reaction position on the aromatic moiety for the C−C bond formation is blocked (**1l–1m**) by a fluorosubstituent.

Employing the BBE as an asymmetric catalyst in organic synthesis allows the preparation of numerous enantiomerically pure berbine and isoquinoline alkaloid derivatives that represent pharmaceutically interesting compounds. In addition, BBE has not only been used in the transformation of non-natural substrates, but also in the first asymmetric total synthesis of naturally occurring (*S*)-scoulerine (**2b**) [29].

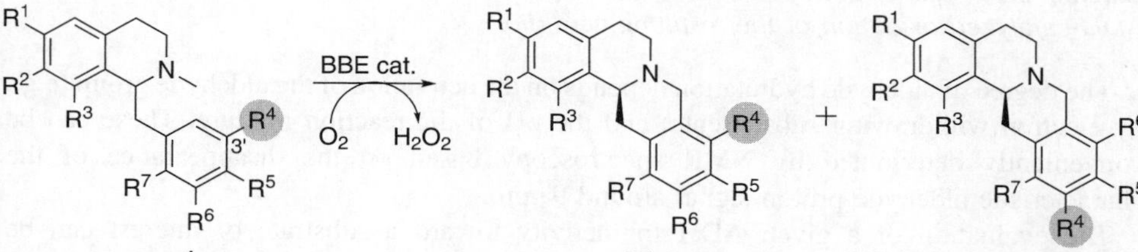

	R^1	R^2	R^3	R^4	R^5	R^6	R^7
a	OMe	OMe	H	OH	H	H	H
b	OMe	OH	H	OH	OMe	H	H
c	H	H	H	OH	H	H	H
d	OMe	H	H	OH	H	H	H
e	OCH$_2$O		H	OH	H	H	H
f	OMe	OH	H	OH	H	H	H
g	OH	OMe	H	OH	H	H	H
h	OMe	OMe	OMe	OH	H	H	H
i	OH	OMe	H	OH	OMe	H	H
j	OMe	OMe	H	OH	H	OH	H
k	OMe	OMe	H	OH	H	H	F
l	OMe	OMe	H	H	H	OH	F
m	OMe	OH	H	H	H	OH	F
n	OH	OMe	H	H	H	OH	F

2a = manibacanine
1b = reticuline
2b = scoulerine
3b = coreximine
2f = anibacanine
3f = pseudoanibacanine
3i = isocoreximine

Scheme 5.6 *Substrate scope of the BBE using natural and non-natural substrates. The position of the catalytically essential OH-group is highlighted in gray. Substrates and products that are described as natural products are listed with their trivial names.*

5.5 Oxidation of Aldehydes Using Alcohol Dehydrogenases

Frank Hollmann
*Department of Biotechnology, Delft University of Technology,
The Netherlands*

Alcohol dehydrogenases (ADHs) (EC 1.1.1) are well known as catalysts for the reduction of carbonyl groups into their corresponding alcohols and for the reverse reaction, the oxidation of alcohols into aldehydes and ketones [32–34].

ADH-catalyzed oxidations proceed via hydride abstraction from the starting material [34]. Aldehyde protons generally cannot be abstracted as hydrides, which is why aldehydes are usually inert to ADH-catalyzed oxidation. However, provided the aldehyde substrate is sufficiently hydrated to the *gem*-diol, ADHs can be very efficient catalysts for the oxidation of aldehydes (Scheme 5.7).

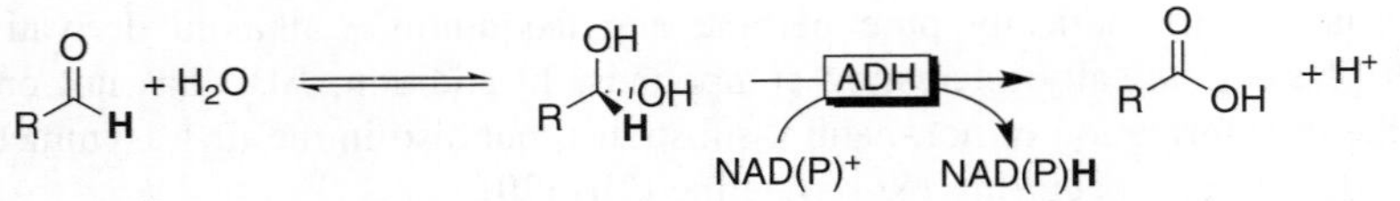

Scheme 5.7 Hydratation equilibrium of aldehydes and schematic representation of ADH-catalyzed oxidation of the resulting gem-diol.

The degree of aldehyde hydratation depends on the activation of the aldehyde group (e.g., by electron withdrawing substituents) and the pH of the reaction mixture. These can be conveniently determined by NMR spectroscopy based on the disappearance of the characteristic aldehyde proton signal around 9 ppm.

The evaluation of a given ADH for activity toward a substrate of interest can be conveniently carried out by UV-spectroscopy, by following the appearance of the characteristic absorption of reduced nicotinamides at 340 nm.

For preparative oxidation, use of an efficient NAD(P)$^+$ regeneration system is strongly recommended, as, first, the nicotinamide co-factors are expensive, preventing their use in stoichiometric amounts, and, second, reduced nicotinamide accumulates in the course of the reaction. The latter fact is easily utilized by the (same) ADH to reduce the aldehyde starting material into the corresponding alcohol, resulting in a Cannizarro-type disproportionation of the aldehyde starting material [35].

Suitable NAD(P)$^+$ regeneration systems comprise (i) co-application of a reducible cosubstrate (such as acetone) [36] and (ii) (chemo)enzymatic systems for the aerobic re-oxidation of NAD(P)H [37–40].

Also, α-substituted aldehydes such as 2-aryl propionaldehydes undergo rapid racemization (especially in slightly alkaline media), thereby enabling dynamic kinetic resolutions. Provided the ADH-catalyst used is enantioselective, enantiopure 2-aryl-propionic acids (profens) can be obtained [38].

5.5.1 Materials and Equipment

- ADH: The ADH used in this study was obtained from c-LEcta (Leipzig, Germany). It was typically added as lyophilized powder (1–10 g.L^{-1} total volume) to the reaction mixtures. It is worth mentioning here that various ADHs are commercially available from different suppliers and that the exact reaction conditions (temperature, co-factor, and buffer requirements) will depend on the enzyme used (see respective product data sheets)

- Nicotinamide co-factor: Depending on the requirements of the ADH used, either NAD^+ or $NADP^+$ were added to the reaction mixture from 100 mM stock solutions in buffer to a final concentration of 0.1–5.0 mM
- Buffer: We used Tris-HCl buffer (50 mM, adjusted to pH 8 with 1 N NaOH). Note: The buffer should be chosen according to the ADH requirements (in any case, we recommend slightly alkaline conditions (pH 8–9) for high activity of the ADH in oxidative direction)
- Substrate: Racemic Flurbiprofenaldehyde (**1**) was obtained from Chiracon GmbH (Luckenwalde, Germany) and added as solid to the reaction mixture. Note: Some aldehydes may be too reactive to be added at once (with some aldehydes, we observed significant reduction of the ADH stability). If low productivities are observed, addition of the aldehyde is recommended; in case of liquid aldehydes, distillation is recommended prior to use.
- Acetone (1–1.5 eq with respect to the aldehyde starting material)
- Na_2SO_4
- HCl (3 M)
- Ethyl acetate
- TLC plates (silica gel 60 F254, Merck)
- Round-bottom flask (250 mL)
- Rotary evaporator

5.5.2 Semi-Preparative Oxidation of Flurbiprofenaldehyde (1) (Scheme 5.8)

1. Tris buffer (75 ml, 50 mM, pH 8) containing 5 mM of NADH was supplemented with Flurbiprofenaldehyde (1.01 g) and ADH-9 (0.5 g) and gently stirred for 122 hours. The enzyme and starting material were added portion-wise at intervals of approximately 12 hours (100 and 50 mg, respectively). For co-factor regeneration, 0.6 mL of acetone was added per day.
2. At intervals, TLC reaction control was performed, indicating almost complete conversion of the starting material (until 70 hours), after which it gradually increased. Throughout the reaction, a slight TLC signal for the corresponding alcohol was observed.
3. After 122 hours, the reaction mixture was worked up by addition of 250 mL of saturated Na_2SO_4 and 250 mL HCl (3 M) and then by extraction with ethyl acetate (two times, 250 mL each). The combined organic phases were dried over Na_2SO_4 and the solvent

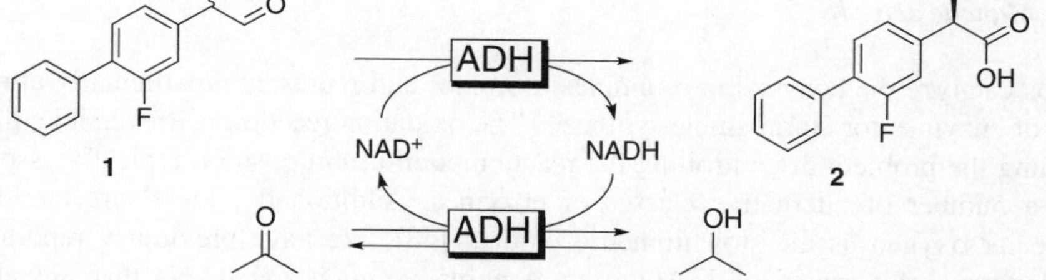

Scheme 5.8 *Oxidative dynamic kinetic resolution of racemic Flurbiprofenealdehyde.*

was removed *in vacuo*, resulting in 790 ± 23 mg of crude product. This crude product was composed of 59.2 ± 3.2% Flurbiprofen (46% overall yield), 8.1 ± 0.4% flurbiprofene alcohol, and 19.6 ± 0.1% starting material (as determined by HPLC analysis; the values shown represent the average of two independent reactions).

5.5.3 Analytical Methods

5.5.3.1 Achiral HPLC Analysis

TFA (50 µL 10%) and DMSO (100 µL) were added to 100 µL of the reaction solution, which was then short-mixed. After 5 minutes of centrifugation at 16 000× g, 140 µL of the supernatant was transferred to an HPLC vial with micro-insert and analyzed.

Conditions: 9 minutes' isocratic 85% (5% acetonitrile with 0.1% TFA) and 15% (95% acetonitrile with 0.1% TFA) at 1.0 mL.min^{-1} on an XTerra RP18 column (3.5 µm, 4.6 × 150 mm) at 35 °C, followed by detection at 210 nm

5.5.3.2 Chiral HPLC Analysis of Flurbiprofen

TFA (50 µL 10%) and *n*-heptane (200 µL) were added to 100 µL of reaction solution, which was then mixed vigorously for 45 seconds. After 30 seconds of centrifugation at 16 000× g, 350 µL of the supernatant was transferred to a fresh tube and dried by addition of MgSO$_4$(s). After 30 seconds' further centrifugation at 16 000× g, the supernatant was transferred to an HPLC vial and analyzed.

Conditions: 15 minutes' isocratic 97% *n*-heptane, 3% iso-propanol, and 0.1% TFA at 1.0 mL.min^{-1} on a Chiralpak AD-H column (4.6 × 250 mm, 5 µm) at 40 °C, followed by detection at 210 nm

5.5.4 Conclusion

Sometimes old dogs can be taught some new tricks. Such is the case with ADH-catalyzed oxidations of aldehydes: for a long time considered a lab curiosity [41–47], it has now been revived as a synthetically useful reaction [35,38].

5.6 MAO-Catalyzed Deracemization of Racemic Amines for the Synthesis of Pharmaceutical Building Blocks

Diego Ghislieri and Nicholas J. Turner
School of Chemistry, Manchester Institute of Biotechnology, The University of Manchester, UK

MAOs catalyze the conversion of amines to imines and represent a particularly attractive class of enzymes for chiral amine synthesis. The oxidation reaction is irreversible, thereby avoiding the problem of controlling the reaction equilibrium position typically associated with a number of alternative classes of enzymes. Additionally, these enzymes utilize molecular oxygen as the stoichiometric oxidant [48]. We have previously reported the development of variants of MAO from *Aspergillus niger* (MAO-N) that are able to selectively oxidize a range of chiral amines [49–51]. When coupled with a non-selective

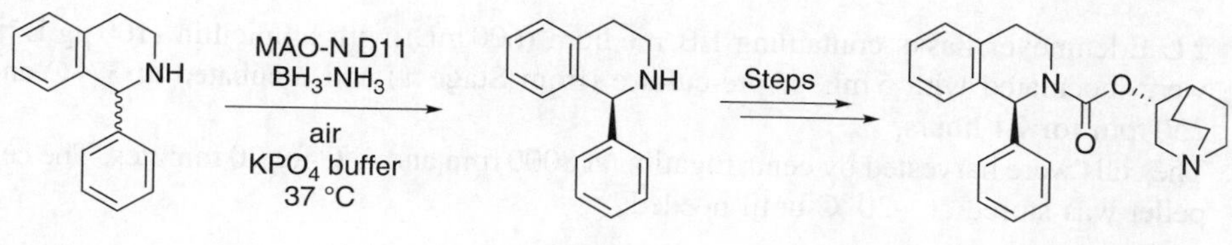

Scheme 5.9 *MAO-catalyzed deracemization of racemic amines for the synthesis of pharmaceutical building blocks.*

chemical reducing agent, the MAO-N variants have been shown to mediate the deracemization of chiral primary, secondary, and tertiary amines. Recently, by combining rational structure-guided engineering with high-throughput screening, it has been possible to expand the substrate scope of MAO-N to accommodate amine substrates containing bulky aryl substituents [52]. In this section, we show in detail the deracemization of 1-phenyl-tetrahydroisoquinoline (**1**), a key intermediate for the enantioselective synthesis of Solifenacin (**2**) (Scheme 5.9).

5.6.1 Materials and Equipment

- *E. coli* BL21(DE3) containing MAO-N D11 variant
- LB broth medium ($10\,g.L^{-1}$ tryptone, $5\,g.L^{-1}$ yeast extract, $10\,g.L^{-1}$ NaCl)
- Ampicillin ($100\,\mu g.mL^{-1}$ stock solution in water, filter-sterilized)
- Potassium dihydrogen phosphate
- Potassium hydrogen phosphate
- *rac*-1-Phenyltetrahydroisoquinoline (1 g, 4.82 mmol)
- BH_3NH_3 (656 mg, 19.32 mmol)
- Aqueous NaOH (4 mL, 10 M)
- *tert*-Butyl methyl ether (600 mL)
- HPLC-grade hexane
- HPLC-grade iPrOH
- HPLC-grade diethylamine
- 1 mL Syringes
- Sterile loop
- 25 ml Falcon tube
- 2 L Erlenmeyer flasks
- Rotary shakers at 37 °C
- High-speed centrifuge
- HPLC system with UV detection
- CHIRALCEL OD-H column

5.6.2 MAO-N Culture

1. A single colony of *E. coli* BL21(DE3) containing MAO-N D11 variant was inoculated into a 15 mL Falcon tube with LB medium (5 mL, containing $100\,\mu g.mL^{-1}$ ampicillin). The tube was incubated at 37 °C for 16 hours, with shaking at 250 rpm.

2. 2 L Erlenmeyer flasks containing LB medium (600 mL) with ampicillin (100 µg.L^{-1}) were inoculated with 5 mL of pre-culture (from Stage 1) and incubated at 37 °C and 250 rpm for 24 hours.
3. The cells were harvested by centrifugation at 8000 rpm and 4 °C for 20 minutes. The cell pellet was stored at −20 °C until needed.

5.6.3 Whole-Cell MAO-N Catalyzed Deracemization

1. In a 500 mL screw-cap bottle, *rac*-1-phenyltetrahydroisoquinoline (1 g, 4.82 mmol) and BH$_3$NH$_3$ (656 mg, 19.32 mmol, 4 eq.) were dissolved in KPO$_4$-buffer (320 mL, 1 M, pH 7.8).
2. Cell pellets from *E. coli* cultures (25 g) containing MAO-N D11C variant were added to the solution.
3. The bottle was placed in a shaking incubator and shaken at 37 °C and 250 rpm, and 0.5 mL samples were withdrawn for analysis at various time intervals. Conversion was complete after 48 hours.
4. Aqueous NaOH (4 mL, 10 M) and *tert*-butyl methyl ether (400 mL) were added to the reaction mixture and, after vigorous mixing, the enzyme was removed by filtration through a Celite pad. The two layers were separated and the aqueous phase was extracted with *tert*-butyl methyl ether (2 × 100 mL). The combined organic phases were dried over MgSO$_4$ and concentrated under vacuum. (*S*)-1-phenyltetrahydroisoquinoline (900 mg, 90% yield, 98% *ee*) was obtained as a yellowish solid.

[α]^{20}$_D$: +33.9° (c = 1.0, CH$_2$Cl$_2$)
^{1}H-NMR (400 MHz; CDCl3): δ 7.45 − 7.30 (m, 5H), 7.24 − 7.19 (m, 2H), 7.15 − 7.06 (m, 1H), 6.84 (d, *J* = 7.8 Hz, 1H), 5.17 (s, 1H), 3.34 (dt, *J* = 15.8, 7.1 Hz, 1H), 3.21 − 3.05 (m, 2H), 2.96 − 2.83 (m, 1H), 2.03 (s, 1H)
^{13}C-NMR (100 MHz; CDCl3): δ 145.0, 138.4, 135.5, 129.1, 129.1, 128.5, 128.2, 127.4, 126.3, 125.7, 62.2, 42.4, 29.9

5.6.4 Analytical Methods

Standard solutions were prepared by dissolving the required compound in 100 mL of *tert*-butyl methyl ether. Biotransformation samples were prepared as follows: aqueous NaOH-solution (20 µL, 10 M) was added to a sample (500 µL) of the reaction mixture in an Eppendorf tube, followed by *tert*-butyl methyl ether (1 mL). After vigorous mixing by means of a vortex mixer, the sample was centrifuged at 13 200 rpm for 1 minute, the organic phase was separated and dried with MgSO$_4$, and 20 µL was injected for HPLC analysis.

Quantitation of standards and samples was achieved by isocratic elution (hexane/iPrOH/DEA, 97 : 3 : 0.1 v/v/v) over a CHIRALCEL OD-H column at a flow rate of 1 mL.min^{-1}. HPLC retention times were detected at 220 nm and standards were as follows: (*S*)-1-phenyltetrahydroisoquinoline, 10.4 minutes; (*R*)-1-phenyltetrahydroisoquinoline, 15.6 minutes.

5.6.5 Conclusion

Using the MAO-N D11 variant, deracemization of *rac*-1-phenyltetrahydroisoquinoline (*rac*-**1**) was carried out on a preparative scale, leading to the formation of (*S*)-1-phenyl-tetrahydroisoquinoline (*S*)-**1**) in 90% isolated yield and with *ee* = 98% after 48 hours.

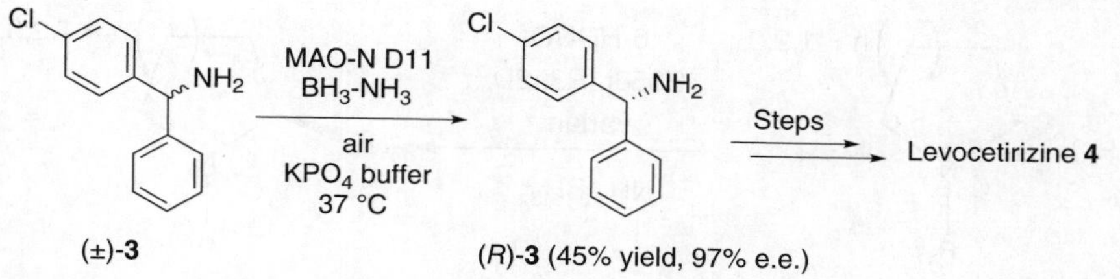

Scheme 5.10 *Asymmetric synthesis of the generic APIs Solifenacin and Levocetirizine.*

R = -H (±)-Eleagnine **5**
R = -OMe (±)-Leptaflorin **6**

R = -H (*R*)-**5** (93% yield, 99% e.e.)
R = -OMe (*R*)-**6** (99% e.e.)

(±)-Coniine **7**

(*R*)-**7** (85% yield, 90% e.e.)

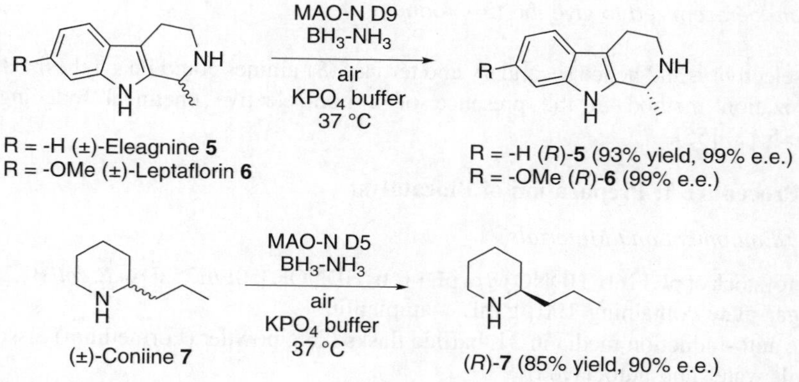

Scheme 5.11 *Asymmetric synthesis of the natural products (R)-coniine, (R)-eleagnine, and (R)-leptaflorine.*

The substrate specificity of the MAO-N is not limited to the tetrahydroisoquinoline motif: a full range of racemic primary, secondary, and tertiary amines can be deracemized using a toolbox of different MAO-N variants. These engineered MAO-N biocatalysts have already been applied for gram-scale deracemization reactions for the efficient asymmetric synthesis of the generic APIs Solifenacin and Levocetirizine (Scheme 5.10), as well as the natural products (*R*)-coniine, (*R*)-eleagnine, and (*R*)-leptaflorine (Scheme 5.11).

5.7 Synthesis of (*S*)-Amines by Chemo-Enzymatic Deracemization Using an (*R*)-Selective Amine Oxidase

Rachel S. Heath, Marta Pontini, Beatrice Bechi, and Nicholas J. Turner
School of Chemistry, Manchester Institute of Biotechnology, The University of Manchester, UK

Previous examples of amine oxidases used for the deracemization of chiral amines have shown (*S*)-selectivity. These include variants of MAO-N [53] and cyclohexylamine oxidase (CHAO) [54]. Recently, an amine oxidase from *Arthrobacter nicotinovorans* (6-hydroxy-D-nicotine oxidase (6-HDNO)) was shown to be (*R*)-selective toward nicotine and other closely related tertiary amines. The substrate scope of the enzyme was widened to include secondary amines such as pyrrolidines, piperidines, and tetrhydroisoquinolines by the introduction of a double mutation into the active site. The resulting variant displayed high

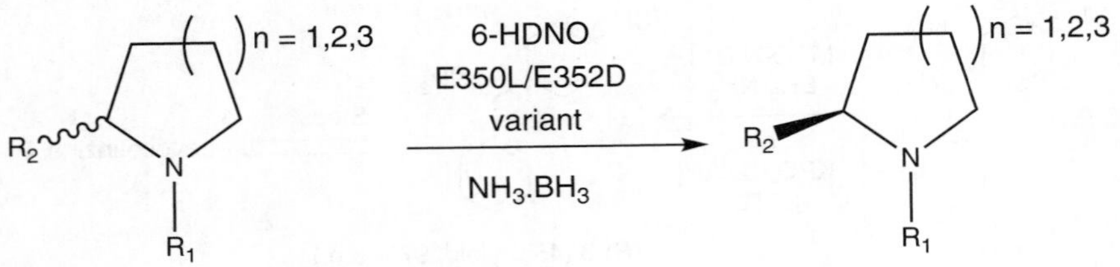

Scheme 5.12 *5-, 6-, and 7-membered rings containing secondary and tertiary amines are substrates for the 6-HDNO variant. In the presence of a non-selective reducing agent racemic amines are deracemized to give the (S)-product.*

enantioselectivity, and hence secondary and tertiary (*S*)-amines could be synthesized by this deracemization method in the presence of a non-selective chemical reducing agent (Scheme 5.12) [55].

5.7.1 Procedure 1: Preparation of Biocatalyst

5.7.1.1 Equipment and Materials

- Glycerol stock of pET16b_HDNO *wt* or pET16b_HDNOE350L/E352D in *E. coli* BL21 (DE3)
- LB agar plate containing $100\,\mu g.mL^{-1}$ ampicillin
- Sterile auto-induction media in 2 L baffled flasks (28 g powder (Formedium) dissolved in 800 mL water and autoclaved)
- 1 M Potassium phosphate buffer (pH 8.0)
- Ampicillin solution ($100\,mg.mL^{-1}$)
- Shaking incubator and static incubator
- Sterile loop

5.7.1.2 Procedure

1. An ampicillin plate was streaked with cells from the glycerol stock and grown overnight at 37 °C.
2. A single colony was picked from the plate and used to inoculate LB medium (10 mL) containing ampicillin (10 μL), and was grown overnight at 37 °C, with shaking at 250 rpm.
3. Sterile auto-induction media (800 mL) containing 800 μL ampicillin was inoculated with 8 mL of overnight culture. The culture was left to grow for 50 hours at 26 °C, with shaking at 200 rpm.
4. The cells were harvested by centrifugation at 4000 rpm for 20 minutes, washed with phosphate buffer, and stored at −20 °C until use. (Cells stored this way remained active for at least 1 year.)

5.7.2 Procedure 2: Biotransformation

5.7.2.1 Equipment and Materials

- Frozen *E. coli* cells containing expressed HDNO *wt* or HDNO E350L/E352D (from Procedure 1)
- *rac*-6-hydroxy nicotine
- ammonia borane (NH_3BH_3)

- 1 M Potassium phosphate buffer (pH 8.0)
- 10 M NaOH
- Dichloromethane
- MgSO$_4$
- Shaking incubator
- HPLC or GC
- Falcon tubes
- 500 mL Duran bottles

5.7.2.2 Biotransformation with 6-Hydroxy Nicotine

1. Cells containing *wt* HDNO (10 g) were suspended in 1 M potassium phosphate buffer (pH 8.0, 100 mL), in a 500 mL Duran bottle.
2. *rac*-6-hydroxy nicotine (100 mg) was added along with NH$_3$BH$_3$ (400 mg)
3. The bottle was place in an incubator at 37 °C, with shaking at 250 rpm, and 300 μL samples were taken at intervals and basified to pH 14 with 10 M NaOH and then extracted into dichloromethane (1 mL). Samples were analyzed by HPLC.
4. After 6 hours, HPLC analysis (CHIRALCEL DAICEL IA column; flow rate 1.5 mL. min^{-1}; UV 280 nM; eluent: hexane/iPrOH 95:5 + 0.1% DEA; *Rt* [(R)] = 19.8 minutes, *Rt* [(S)]) showed full conversion to the (S)-enantiomer.
5. The reaction was basified with 10 M NaOH (2 mL) and extracted three times with dichloromethane (3 × 100 mL). The combined organic phases were dried with MgSO$_4$, concentrated under reduced pressure, and purified by chromatography over silica (CH$_2$Cl$_2$/MeOH 90 : 10) to give the (S)-6-hydroxy nicotine product (76% yield, *ee* 99%).

Biotransformations can also be carried out on a smaller scale. The following is a general method for use with the wild type or the variant and applicable to a variety of substrates (examples shown in Figure 5.1, with *ee* and yields in Table 5.1).

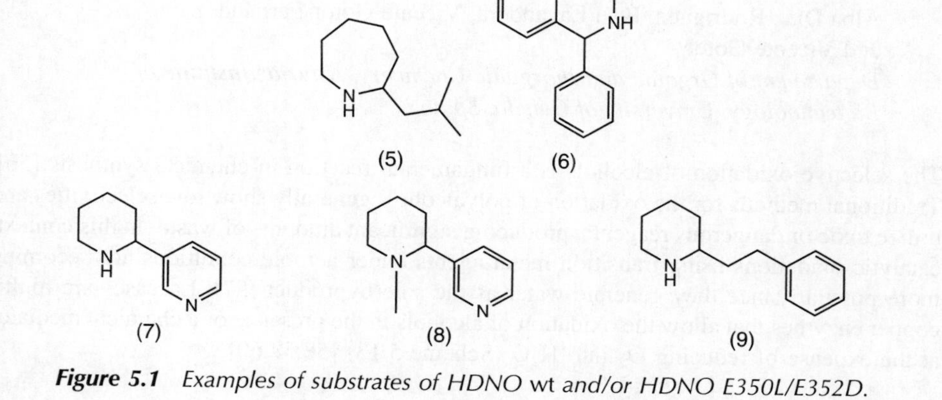

Figure 5.1 *Examples of substrates of HDNO wt and/or HDNO E350L/E352D.*

Table 5.1 *Preparative-scale deracemization reactions using either wild-type HDNO or the E350L/E352D variant.*

Substrate	ee (%)	Product[a]	Isolated yield/%
(1)	>99	(S)	55[b]
(2)	>99.	(S)	93
(3)	95	(S)	60[b]
(4)	>99	(S)	71
(5)	>99	(S)	n.d.[d]
(6)	84	(R)[c]	86
(7)	>99	(S)	55
(8)	97	(S)	60[b]
			75
(9)	90	(S)	83

[a] Absolute configuration of product determined by comparison with authentic standards.
[b] Values in italics indicate data from WT enzyme.
[c] Apparent change in selectivity due to Cahn-Ingold-Prelog convention.
[d] Reaction scaled but product not isolated in pure form.

5.7.2.3 Small-Scale Biotransformation

1. Cells (500 mg) were resuspended in 1 M potassium phosphate buffer, pH 8.0, to a final volume of 5 mL in a 50 mL Falcon tube. It is important to carry out the biotransformation in the appropriate volume for the vessel, in order to provide enough headspace for oxygen. The FAD co-factor in HDNO is regenerated from its reduced form using molecular oxygen.
2. Substrate and NH_3BH_3 were added to final concentrations of 10 and 40 mM, respectively.
3. 300 μL samples were taken at intervals, basified to pH 14 with 10 M NaOH, and extracted into dichloromethane (1 mL).
4. Samples were analyzed by HPLC or GC to determine the *ee* and the time for completion.

More details on analysis methods and substrate scope can be found in reference [55].

5.8 Selective Oxidation of Diols into Lactones under Aerobic Conditions Using a Laccase-TEMPO Catalytic System in Aqueous Medium

Alba Díaz-Rodríguez, Iván Lavandera, Vicente Gotor-Fernández, and Vicente Gotor
Department of Organic and Inorganic Chemistry, Asturias Institute of Biotechnology, University of Oviedo, Spain

The selective oxidation of alcohols is a fundamental reaction in chemical synthesis [56]. Traditional methods for the oxidation of polyalcohols generally show low selectivities and utilize toxic or dangerous reagents, producing significant amounts of waste. In this context, catalytic oxidations using transition metal agents under aerobic conditions are becoming more popular, since they generate water as the sole byproduct [57]. Laccases are multi-copper enzymes that allow the oxidation of alcohols in the presence of a chemical mediator at the expense of reducing O_2 into H_2O (Scheme 5.13) [58,59,60].

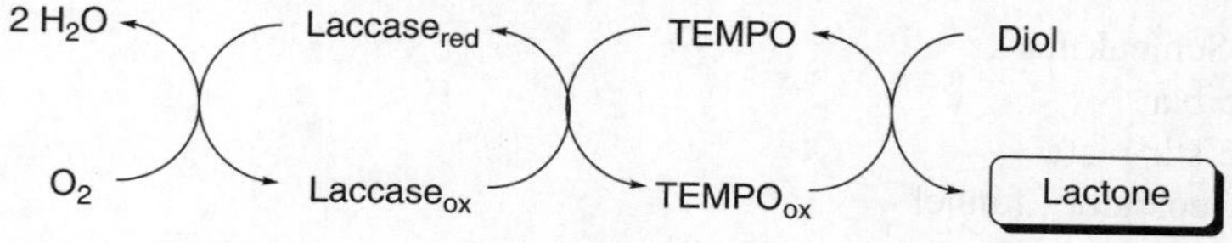

Scheme 5.13 *Representation of the oxidation of diols using a laccase/TEMPO-mediated protocol.*

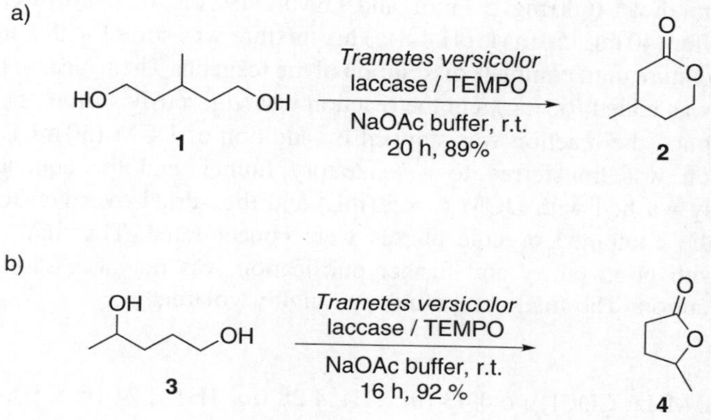

Scheme 5.14 *Catalytic oxidation of 3-methyl-1,5-pentanediol and 1,4-pentanediol using the laccase/TEMPO system.*

Recently, we have developed an efficient and robust catalytic system to quantitatively oxidize aliphatic diols using the *Trametes versicolor* laccase/TEMPO system and ambient air in aqueous media (Scheme 5.14). Oxidations have been performed in a non-stereoselective fashion, but with complete regio/monoselectivity, leading to the corresponding lactones (through intermediate hemiacetal formation) with excellent purity after a simple extraction protocol [61]. This catalytic chemoenzymatic system is also suitable for the selective oxidation of primary hydroxyl and amino groups in both aqueous and biphasic media [62].

In this section, we report on the preparation of two lactones with interesting applications in industry (a substituted δ-valerolactone **2** and γ-valerolactone **4**) through the laccase/TEMPO-mediated oxidation of the corresponding diols using a commercially available enzyme [61].

5.8.1 Procedure 1: Preparation of 3-Methyltetrahydro-2-Pyranone 2

5.8.1.1 Materials and Equipment

- 3-Methyl-1,5-pentanediol **1** (600 mg, 5.1 mmol)
- 2,2,6,6-Tetramethylpiperidine-1-oxyl (TEMPO, 92 mg, 0.59 mmol)
- Laccase from *Trametes versicolor* from Sigma-Aldrich (13.6 U.mg^{-1}, 60 mg)
- Acetate buffer (50 mM, pH 4.8, 40 mL)
- Dichloromethane (DCM, 150 mL)
- Na$_2$SO$_4$ anhydrous (1 g)

- 150-mL Schlenk flask
- Magnetic bar
- Magnetic stir plate
- 250 mL Separatory funnel
- Rotatory evaporator
- 250 mL round-bottom flask

5.8.1.2 Procedure

1. The prochiral diol **1** (600 mg, 5.1 mol) and TEMPO (92 mg, 0.59 mol) were dissolved in NaOAc buffer (40 mL, 50 mM, pH 4.8). This mixture was stirred with a magnetic bar at room temperature until complete dissolution of the reagents. Then, laccase from *Trametes versicolor* was added (60 mg) and the reaction was vigorously stirred open to air.
2. After 20 hours, the reaction was stopped by addition of DCM (50 mL).
3. The solution was transferred to a separatory funnel, and the aqueous phase was successively washed with DCM (2×50 mL) and then dried over Na_2SO_4 (1 g). After filtration, the combined organic phases were concentrated. The final lactone **2** was obtained with good purity and further purification was not necessary (typical yield = 89%). Caution: The final compound was highly volatile.

^{1}H-NMR (300 MHz; CDCl$_3$): δ 4.43 (m, 1H), 4.28 (m, 1H), 2.71 (m, 1H), 2.11 (m, 2H), 1.93 (m, 1H), 1.52 (m, 1H), 1.06 (d, 3H, $J = 6.1$ Hz)
^{13}C-NMR (75 MHz; CDCl$_3$): δ 171.7, 69.0, 38.6, 31.1, 27.0, 21.9
IR (neat): 1729, 1465, 1170 cm^{-1}
MS (ESI$^+$, m/z): 115 ((M + H)$^+$, 100%), 137 ((M + Na)$^+$, 20%)

5.8.2 Procedure 2: Preparation of γ-valerolactone (GVL, 4)

5.8.2.1 Materials and Equipment

- 1,4-Pentanediol **3** (500 mg, 4.8 mmol)
- 2,2,6,6-Tetramethylpiperidine-1-oxyl (TEMPO, 60 mg, 0.38 mmol)
- Laccase from *Trametes versicolor* from Sigma-Aldrich (13.6 U.mg^{-1}, 60 mg)
- Acetate buffer (50 mM, pH 4.8, 30 mL)
- Dichloromethane (DCM, 150 mL)
- Na_2SO_4 anhydrous (1 g)
- 150 mL Schlenk flask
- Magnetic bar
- Magnetic stir plate
- 250 mL Separatory funnel
- Rotatory evaporator
- 250 mL Round-bottom flask

5.8.2.2 Procedure

1. The racemic diol **3** (500 mg, 4.8 mol) and TEMPO (60 mg, 0.38 mol) were dissolved in NaOAc buffer (30 mL, 50 mM, pH 4.8). The mixture was stirred with a magnetic bar at

Table 5.2 *GC analytical retention times for the separation of diols and lactones.*[a]

Diol	Lactone	$t_{R\ diol}$ (minutes)	$t_{R\ lactone}$ (minutes)
		6.3	7.3
		4.3	3.6

[a] Hewlett Packard HP-1 column (30 m × 0.32 mm × 0.25 μm, 12 psi). GC program: 50/3/5/220/0 (initial temp (°C)/time (minutes)/slope (°C.min⁻¹)/final temperature (°C)/time (minutes)).

room temperature until complete dissolution of the reagents. Then, laccase from *Trametes versicolor* (60 mg) was added and the reaction was vigorously stirred open to air.

2. After 16 hours, the reaction was stopped by addition of DCM (50 mL).
3. The solution was transferred to a separatory funnel, and the aqueous phase was successively washed with DCM (2 × 50 mL) and then dried over Na_2SO_4 (1 g). After filtration, the combined organic phases were concentrated in a rotatory evaporator. The final lactone **4** was obtained with good purity, and further purification was not necessary (typical yield = 92%). Caution: The final compound was highly volatile.

^{1}H-NMR (300 MHz; CDCl$_3$): δ 4.74 (m, 1H), 2.64 (m, 2H), 2.46 (m, 1H), 1.93 (m, 1H), 1.50 (d, J = 6.9 Hz, 3H)
^{13}C-NMR (75 MHz; CDCl$_3$): δ 174.3, 74.4, 26.9, 26.3, 18.2
IR (neat): 1768, 1268, 1170, 738 cm^{-1}
MS (ESI$^+$, m/z) 101 [(M + H)$^+$, 100%], 123 [(M + Na)$^+$, 25%]

5.8.3 Analytics

The reaction progress was monitored using a GC chromatograph equipped with an FID on an achiral HP-1 column, with N_2 as carrier gas. The conversion and purity of the final lactones were analyzed by NMR spectroscopy (Table 5.2).

5.8.4 Conclusion

We have described the employment of the *Trametes versicolor* laccase/TEMPO pair to efficiently oxidize interesting aliphatic diols in a regio- and/or monoselective fashion, yielding valuable lactones with excellent purity and yield after a simple extraction. This is a practical methodology for the aerobic oxidation of 1,4- and 1,5-diols in aqueous medium, using laccases and in the absence of a base. This catalytic system is compatible with the presence of unprotected secondary alcohols, observing the selective oxidation of the primary alcohol as shown for the preparation of γ-valerolactone.

References

1. Cashman, J.R. (2008) *Expert Opinion on Drug Metabolism and Toxicology*, **4**, 1507–1521.
2. Tynes, R.E. and Hodgson, E. (1985) *Archives of Biochemistry and Biophysics*, **240**, 77–93.
3. Brunelle, A., Bi, Y.A., Lin, J. *et al.* (1997) *Drug Metabolism and Disposition: The Biological Fate of Chemicals*, **25**, 1001–1007.
4. Moreau, J.L., Jenck, F., Martin, J.R. *et al.* (1993) *Pharmacopsychiatry*, **26**, 30–33.
5. Hanlon, S.P., Camattari, A., Abad, S. *et al.* (2012) *Chemical Communications*, **48**, 6001–6003.
6. Geier, M., Bachler, T., Hanlon, S.P. *et al.* (2015) *Microbial Cell Factories*, **14**, 82.
7. (a) Pollak, P., Romeder, G., Hagedorn, F., and Gelbke, H.-P. (2000) Nitriles, in *Ullmann's Encyclopedia of Industrial Chemistry* (eds M. Bohnet, C.G. Brinker, and B. Cornils), Wiley-VCH, Weinheim; (b) Weissermell, K. and Arpe, H.-J. (1993) *Industrial Organic Chemistry*, VCH.
8. (a) Kim, Y., Mosier, N.S., Hendrickson, R. *et al.* (2008) *Bioresource Technology*, **99**, 5165–5176; (b) Lammens, T.M., Franssen, M.C.R., Scott, E.L., and Sanders, J.P.M. (2012) *Biomass and Bioenergy*, **44**, 186–181.
9. (a) Nieder, M. and Hager, L.P. (1985) *Archives of Biochemistry and Biophysics*, **240**, 121–127; (b) Le Nôtre, J., Scott, E.L., Franssen, M.C.R., and Sanders, J.P.M. (2011) *Green Chemistry*, **13**, 807; (c) Lammens, T.M., Gangarapu, S., Franssen, M.C.R., Scott, E.L., and Sanders, J.P.M. (2011) *Biofuels Bioproducts and Biorefinery*, **6**, 177–187; (d) Lammens, T.M., Potting, J., Sanders, J.P.M., and de Boer, I.J.M. (2001) *Environmental Science & Technology*, **45** 8521–8528.
10. Renirie, R., Pierlot, C., Aubry, J.-M. *et al.* (2003) *Advanced Synthesis and Catalysis*, **45**, 849–858.
11. But, A., Le Nôtre, J., Scott, E.L. *et al.* (2012) *ChemSusChem*, **5**, 1199–1202.
12. Hemrika, W., Renirie, R., Macedo-Ribeiro, S. *et al.* (1999) *The Journal of Biological Chemistry*, **274**, 23820–23827.
13. Hanczko, R., Jambor, A., Perl, A., and Molnar-Perl, I. (2007) *Journal of Chromatography A*, **1163**, 25–42.
14. Liu, C., Liu, F., Cai, J. *et al.* (2011) *Biomacromolecules*, **12**, 3291.
15. Yang, Y.X., Lu, W.H., Zhang, X.Y. *et al.* (2010) *Biomacromolecules*, **11**, 259.
16. Labinger, J.A. and Bercaw, J.E. (2002) *Nature*, **417**, 507.
17. Bordeaux, M., Galarneau, A., Fajula, F., and Drone, J. (2011) *Angewandte Chemie-International Edition in English*, **50**, 2075.
18. Fujii, T., Narikawa, T., Sumisa, F. *et al.* (2006) *Bioscience, Biotechnology, and Biochemistry*, **70**, 1379.
19. Scheps, D., Honda Malca, S., Hoffmann, H. *et al.* (2011) *Organic and Biomolecular Chemistry*, **9**, 6727.
20. van Beilen, J.B., Funhoff, E.G., van Loon, A. *et al.* (2006) *Applied and Environmental Microbiology*, **72**, 59.
21. Honda Malca, S., Scheps, D., Kühnel, L. *et al.* (2012) *Chemical Communications*, **48**, 5115.
22. Scheps, D., Malca, S.H., Richter, S. *et al.* (2013) *Microbial Biotechnology*, **6**, 694.
23. Julsing, M.K., Schrewe, M., Cornelissen, S. *et al.* (2012) *Applied and Environmental Microbiology*, **78**, 5724.
24. Johnston, W.A. and Gilliam, E.M.J. (2013) in *Cytochrome P450 Protocols*, 3rd edn (eds I.R. Phillips, E.A. Shephard, and P.R.O. d. Montellano), Springer, New York, p. 189.
25. Rink, E. and Boehn, H. (1975) *FEBS Letters*, **49**, 396–399.
26. Winkler, A., Lyskowski, A., Riedl, S. *et al.* (2008) *Nature Chemical Biology*, **4**, 739–741.
27. Resch, V., Schrittwieser, J.H., Wallner, S. *et al.* (2011) *Advanced Synthesis and Catalysis*, **353**, 2377–2383.

28. Schrittwieser, J.H., Resch, V., Sattler, J.H. *et al.* (2011) *Angewandte Chemie-International Edition in English*, **50**, 1068–1071.
29. Schrittwieser, J.H., Resch, V., Wallner, S. *et al.* (2011) *The Journal of Organic Chemistry*, **76**, 6703–6714.
30. Resch, V., Lechner, H., Schrittwieser, J.H. *et al.* (2012) *Chemistry – A European Journal*, **18**, 13173–13179.
31. Winkler, A., Lyskowski, A., Riedl, S. *et al.* (2008) *Nature Chemical Biology*, **4**, 739–741.
32. Kroutil, W., Mang, H., Edegger, K., and Faber, K. (2004) *Current Opinion in Chemical Biology*, **8**, 120–126.
33. Kroutil, W., Mang, H., Edegger, K., and Faber, K. (2004) *Advanced Synthesis and Catalysis*, **346**, 125–142.
34. Faber, K. (2011) *Biotransformations in Organic Chemistry*, 6th edn, Springer, Berlin.
35. Wuensch, C., Lechner, H., Glueck, S.M. *et al.* (2013) *ChemCatChem*, **5**, 1744–1748.
36. Lavandera, I., Kern, A., Resch, V. *et al.* (2008) *Organic Letters*, **10**, 2155–2158.
37. Könst, P., Kara, S., Kochius, S. *et al.* (2013) *ChemCatChem*, **5**, 3027–3032.
38. Könst, P., Merkens, H., Kara, S. *et al.* (2012) *Angewandte Chemie-International Edition in English*, **51**, 9914–9917.
39. Gargiulo, S., Arends, I.W.C.E., and Hollmann, F. (2011) *ChemCatChem*, **3**, 338–342.
40. Aksu, S., Arends, I.W.C.E., and Hollmann, F. (2009) *Advanced Synthesis and Catalysis*, **351**, 1211–1216.
41. Abeles, R.H. and Lee, H.A. (1960) *The Journal of Biological Chemistry*, **235**, 1499–1503.
42. Henehan, G.T.M., Chang, S.H., and Oppenheimer, N.J. (1995) *Biochemistry*, **34**, 12294–12301.
43. Henehan, G.T.M. and Oppenheimer, N.J. (1993) *Biochemistry*, **32**, 735–738.
44. Trivić, S., Leskova, V., and Winston, G.W. (1999) *Biotechnology Letters*, **21**, 231–234.
45. Mee, B., Kelleher, D., Frias, J. *et al.* (2005) *FEBS Journal*, **272**, 1255–1264.
46. Höllrigl, V., Hollmann, F., Kleeb, A. *et al.* (2008) *Applied Microbiology and Biotechnology*, **81**, 263–273.
47. Velonia, K. and Smonou, I. (2000) *Journal of the Chemical Society-Perkin Transactions 1*, 2283–2287.
48. Ghislieri, D. and Turner, N.J. (2014) *Topics in Catalysis*, **57**, 284.
49. Alexeeva, M., Enright, A., Dawson, M.J. *et al.* (2002) *Angewandte Chemie-International Edition in English*, **41**, 3177.
50. Carr, R., Alexeeva, M., Enright, A. *et al.* (2003) *Angewandte Chemie-International Edition in English*, **42**, 4807.
51. Dunsmore, C.J., Carr, R., Fleming, T., and Turner, N.J. (2006) *Journal of the American Chemical Society*, **128**, 2224.
52. Ghislieri, D., Green, A.P., Pontini, M. *et al.* (2013) *Journal of the American Chemical Society*, **135**, 10863.
53. (a) Ghislieri, D., Green, A.P., Pontini, M. *et al.* (2013) *Journal of the American Chemical Society*, **135**, 10863–10869; (b) Alexeeva, M., Enright, A., Dawson, M.J. *et al.* (2002) *Angewandte Chemie-International Edition in English*, **41**, 3177–3180; (c) Rowles, I., Malone, K.J., Etchells, L.L. *et al.* (2012) *ChemCatChem*, **4**, 1259–1261; (d) Carr, R., Alexeeva, M., Dawson, M.J. *et al.* (2005) *ChemBioChem*, **6**, 637–639.
54. (a) Leisch, H., Grosse, S., Iwaki, H. *et al.* (2012) *Canadian Journal of Chemistry*, **90**, 39–45; (b) Li, G., Ren, J., Iwaki, H. *et al.* (2014) *Applied Microbiology and Biotechnology*, **98**, 1681–1689; (c) Li, G., Ren, J., Yao, P. *et al.* (2014) *ACS Catalysis*, **4**, 903–908.
55. Heath, R.S., Pontini, M., Bechi, B., and Turner, N.J. (2014) *ChemCatChem*, **6**, 897–1117.
56. Sheldon, R.A. (2015) *Catalysis Today*, **247**, 4–13.
57. Cardona, F. and Parmeggiani, C. (2012) *Green Chemistry*, **14**, 547–564.
58. Fabbrini, M., Galli, C., Gentili, P., and Macchitella, D. (2001) *Tetrahedron Letters*, **42**, 7551–7553.

59. Arends, I.W.C.E., Li, Y.-X., Ausan, R., and Sheldon, R.A. (2006) *Tetrahedron*, **62**, 6659–6665.
60. Mogharabi, M. and Faramarzi, M.A. (2014) *Advanced Synthesis & Catalysis*, **356**, 897–927.
61. Díaz-Rodríguez, A., Lavandera, I., Kanbak-Aksu, S. *et al.* (2012) *Advanced Synthesis and Catalysis*, **354**, 3405–3408.
62. Díaz-Rodríguez, A., Martínez-Montero, L., Lavandera, I., *et al.* (2014) *Advanced Synthesis & Catalysis*, **356**, 2321–2329.

6

Reductions

6.1 Tetrahydroxynaphthalene Reductase: Broad Substrate Range of an NADPH-Dependent Oxidoreductase Involved in Reductive Asymmetric Naphthol Dearomatization

Michael A. Schätzle, Syed Masood Husain, and Michael Müller

Institute of Pharmaceutical Sciences, Albert-Ludwigs-Universität Freiburg, Germany

The biosynthesis of 1,8-dihydroxynaphthalene melanin, a virulence factor of many filamentous fungi, is a complex, matrix-like [1] network [2,3]. Many metabolites have been shown to be derived from intermediate polyhydroxynaphthalenes, or have been proposed to be the products of, for example, oxidation of the respective polyhydroxynaphthalenes to the naphthoquinones, followed by a double reduction via the hydronaphthoquinones [2–5]. One of the enzymes from the rice blast fungus *Magnaporthe grisea* is the NADPH-dependent tetrahydroxynaphthalene reductase (T_4HNR). We recently reported the T_4HNR-catalyzed reduction of tetralones (e.g., **3**, **5** in Scheme 6.1), cyclohexanones, and 2-hydroxynaphthoquinones (such as lawsone, **7**), as well as the known reduction of polyhydroxynaphthalenes (such as **1**) [6,7]. The use of 2-tetralone derivatives as model substrates instead of the physiological tetrahydroxynaphthalene, in combination with a site-selective point mutation, revealed major enzyme–substrate interactions [6]. Experimental details are taken primarily from references [6] and [7].

6.1.1 Procedure 1: T₄HNR-Catalyzed Reduction of 1,3,6-Trihydroxynaphthalene 1

6.1.1.1 *Materials and Equipment*

- NADP-Na (54 mg, 71 µmol)
- 1,3,6-Trihydroxynaphthalene (**1**, 125 mg, 710 µmol)
- 2-Propanol (3.5 mL)

Practical Methods for Biocatalysis and Biotransformations 3, First Edition.
Edited by John Whittall, Peter W. Sutton, and Wolfgang Kroutil.
© 2016 John Wiley & Sons, Ltd. Published 2016 by John Wiley & Sons, Ltd.

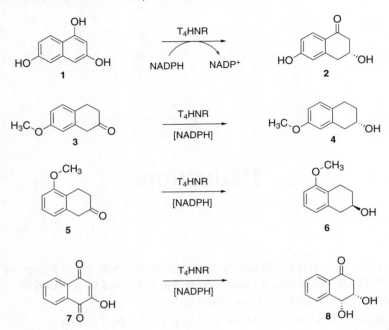

Scheme 6.1 T_4HNR-catalyzed reduction of naphthol **1**, 2-tetralones **3** and **5**, and 2-hydroxynaphthoquinone **7**.

- D-Glucose (639 mg, 3.55 mmol)
- K-Phosphate (KPi) buffer (67.5 mL, 50 mM, pH 7.0, 1 mM ethylenediaminetetraacetic acid (EDTA), 1 mM dithiothreitol)
- T_4HNR (8 U)
- Glucose dehydrogenase (GDH) (215 U)

6.1.1.2 Procedure

1. Nitrogen was bubbled through the buffer solution and 2-propanol for 30 minutes, followed by degassing under reduced pressure before use.
2. Trihydroxynaphthalene **1** (125 mg, 710 μmol) was dissolved in 2-propanol (3.5 mL) and added to a solution of D-glucose (639 mg, 3.55 mmol) and NADP-Na (54 mg, 71 μmol) in KPi buffer (67.5 mL, 50 mM, pH 7.0, 1 mM EDTA, 1 mM dithiothreitol).
3. T_4HNR (8 U) and GDH (215 U) were added slowly and the reaction mixture was stirred under nitrogen for 24 hours at room temperature.
4. The solution was acidified to pH 6 (HCl aq.), extracted with ethyl acetate (3×70 mL), and dried over $MgSO_4$, and the solvent was removed under reduced pressure. ^{1}H-NMR spectroscopy of the crude product showed a conversion of 49%.
5. The crude product was purified by chromatography over silica using cyclohexane/ethyl acetate (3 : 7) as eluent to obtain 47 mg (27 μmol, 39%) of (*R*)-8-deoxyscytalone (**2**, *ee* > 99%).

6.1.1.3   Analytical Data

^{1}H-NMR (400 MHz; acetone-d_6): δ 8.01 (s, 1H, OH), 7.81 (d, J = 8.4 Hz, 1H, H-8), 6.78
(dd, J = 8.4, 2.4 Hz, 1H, H-7), 6.74 (d, J = 2.4 Hz, 1H, H-5), 4.29–4.37 (m, 1H, H-3), 4.25
(s, 1H, OH), 3.18 (dd, J = 16.0, 3.8 Hz, 1H, H-4), 2.91 (dd, J = 16.0, 7.6 Hz, 1H, H-4),
2.78 (ddd, J = 16.4, 3.8, 1.2 Hz, 1H, H-2), 2.55 (ddd, J = 16.4, 8.5, 0.8 Hz, 1H, H-2)

^{13}C-NMR (100.6 MHz; acetone-d_6): δ 195.5 (C-1), 163.2 (C-6), 145.1 (C-4a), 129.8 (C-8),
126.5 (C-8a), 116.1 (C-5), 115.2 (C-7), 67.2 (C-3), 48.3 (C-2), 39.5 (C-4)

Circular dichroism (c = 126 μmol × L^{-1}, l = 0.2 cm, acetonitrile): λ (Mol. CD) (nm) = 224
(−5.81), 258 (0.08), 292 (−1.39), 328 (−0.17), 338 (0.32)

Product *ee* was determined by chiral phase HPLC-DAD (Chiralcel AS-H, 15 °C, 0.5 mL ×
min^{-1}, *n*-hexane/2-propanol = 90:10 (v/v)): t_R = 49.80 minutes (*R*), t_R = 58.92 minutes (*S*)

6.1.2   Procedure 2: T$_4$HNR-Catalyzed Reduction of 2-Tetralones

6.1.2.1   Materials and Equipment

- NADP-Na (119 mg, 156 μmol)
- 7-Methoxy-2-tetralone (**3**, 915 mg, 5.2 mmol)
- 2-Propanol (8.5 mL)
- D-Glucose (2.81 g, 15.6 mmol)
- KPi buffer (165 mL, 50 mM, pH 7.0, 1 mM EDTA, 1 mM dithiothreitol)
- T$_4$HNR (40 U)
- GDH (1.5 kU)

6.1.2.2   Reduction of Tetralone 3

1. Nitrogen was bubbled through the buffer solution and 2-propanol for 30 minutes,
   followed by degassing under reduced pressure before use.
2. Tetralone **3** (915 mg, 5.2 mmol) was dissolved in 2-propanol (8.5 mL) and added to a
   solution of D-glucose (2.81 g, 15.6 mmol) and NADP-Na (119 mg, 156 μmol) in KPi
   buffer (165 mL, 50 mM, pH 7.0, 1 mM EDTA, 1 mM DTT).
3. T$_4$HNR (40 U) and GDH (1.5 kU) were added slowly and the reaction was stirred under
   nitrogen for 24 hours at room temperature.
4. The solution was acidified to pH 6 (HCl aq.), extracted with ethyl acetate (3 × 70mL),
   and dried over MgSO$_4$, and the solvent was removed under reduced pressure. ^{1}H-NMR
   analysis of the crude product showed a conversion of >95%.
5. The crude product was purified by chromatography over silica using cyclohexane/ethyl
   acetate (2 : 1) as eluent to obtain 851 mg (4.8 mmol, 92%) of (*S*)-7-methoxy-2-tetralol
   (**4**, *ee* > 99%).

6.1.2.3   Analytical Data

^{1}H-NMR (400 MHz; CDCl$_3$): δ 7.00 (d, J = 8.5 Hz, 1H, H-5), 6.70 (dd, J = 8.5, 2.7 Hz, 1H,
H-6), 6.61 (d, J = 2.7 Hz, 1H, H-8), 4.10–4.17 (m, 1H, H-2), 3.77 (s, 3H, OCH$_3$), 3.05
(dd, J = 16.5, 5.0 Hz, 1H, H-1), 2.88 (d″t″, J = 16.7, 5.8 Hz, 1H, H-4), 2.70–2.81 (m, 2H,
H-4 and H-1), 1.99–2.08 (m, 1H, H-3), 1.71–1.86 (m, 2 H, H-3 and OH).

^{13}C-NMR (100.6 MHz; CDCl$_3$): δ 157.7 (C-7), 135.3 (C-8a), 129.5 (C-5), 127.7 (C-4a), 113.9 (C-8), 112.4 (C-6), 67.2 (C-2), 55.2 (OCH$_3$), 38.6 (C-1), 31.7 (C-3), 26.1 (C-4). GC-MS (column FS-Supreme-5 (CS Chromatographie-Service), 30 m × 0.250 mm × 0.250 μm; T_{GC} (injector) = 250 °C, T_{MS}(ion source) = 200 °C, time program (oven): $T_{0\,min}$ = 60 °C, $T_{3\,min}$ = 60 °C, $T_{14\,min}$ = 280 °C (heating rate 20 °C × min^{-1}), $T_{19\,min}$ = 280 °C): t_R = 10.28 minutes; EI-MS (70 eV, EI); m/z (%): 178 (61) [M$^+$], 160 (100), 145 (55), 134 (70), 129 (37), 115 (35), 103 (21), 91 (62), 77 (26), 65 (21), 51 (12).

Product *ee* was determined by chiral phase HPLC-DAD of the Mosher acid derivative synthesized from **4** (Chiralcel OD-H, 20 °C, 0.5 mL × min^{-1}, *n*-hexane/2-propanol = 98 : 2 (v/v)): t_R = 15.42 minutes (*S,S*), t_R = 18.06 minutes (*S,R*); *de* > 99% (*S,S*)-derivative.

6.1.2.4   Reduction of Tetralone 5

Enzymatic reduction of 5-methoxy-2-tetralone **5** (92 mg, 0.52 mmol) following the same procedure as for tetralone **3**, and purification by chromatography over silica using cyclohexane/ethyl acetate (3 : 1) as eluent yielded (*R*)-5-methoxy-2-tetralol **6** (70 mg, 75%) with an *ee* of 99%.

6.1.3   Procedure 3: T$_4$HNR-Catalyzed Reduction of 2-Hydroxynathoquinones

6.1.3.1   Materials and Equipment

- NADP-Na (48 mg, 60 μmol)
- Lawsone (2-hydroxy-*p*-naphthoquinone, **7**, 100 mg, 570 μmol)
- 2-Propanol (3.9 mL)
- D-Glucose (517 mg, 2.8 mmol)
- KPi buffer (78 mL, 50 mM, pH 7.0, 1 mM EDTA, 1 mM dithiothreitol)
- T$_4$HNR (16 U)
- GDH (300 U)

6.1.3.2   Procedure

1. Nitrogen was bubbled through KPi buffer (78 mL, 50 mM, pH 7.0, 1 mM EDTA, 1 mM DTT) for 30 minutes.
2. Glucose (517 mg, 2.8 mmol), NADP-Na (48 mg, 60 μmol), and GDH (300 U) were added and the mixture was stirred slowly under nitrogen at room temperature.
3. Lawsone **7** (100 mg, 570 μmol) in 2-propanol (3.9 mL) was added slowly while stirring, followed by T$_4$HNR (16 U), and the mixture was stirred slowly for 24 hours.
4. The solution was acidified to pH 6 with 10% of H$_2$SO$_4$ (0.5 mL), and ethyl acetate (10 mL) was added. The solution was filtered through a plug of silica and washed with ethyl acetate (20 mL). The aqueous layer was extracted with ethyl acetate (2 × 50 mL) and the combined organic layer was washed with brine (30 mL). The organic layer was dried over MgSO$_4$ and filtered, and the solvent was removed under reduced pressure. ^{1}H-NMR spectroscopy of the crude product showed a conversion of 99%.
5. The crude product was purified by chromatography over silica using chloroform/methanol (9 : 1) as eluent to obtain (3*S*,4*R*)-3,4-dihydroxy-3,4-dihydronaphthalen-1 (2*H*)-one **8** (92 mg, 524 μmol, 92%).

6.1.3.3 Analytical Data

^{1}H-NMR (400 MHz; acetone-d_6 + D$_2$O): δ 7.86 (dd, J = 7.9, 1.2 Hz, 1H, CH$_{arom}$), 7.60–7.68 (m, 2H, CH$_{arom}$), 7.40 (dt, J = 7.5, 1.5 Hz, 1H, CH$_{arom}$), 4.96 (d, J = 2.8 Hz, 1H, H-4), 4.37 (td, J = 6.5, 3.2 Hz, 1H, H-3), 2.90 (dd, J = 16.9, 6.5 Hz, 1H, H-2), 2.80 (dd, J = 16.9, 3.5 Hz, 1H, H-2)

^{13}C-NMR (100.6 MHz; acetone-d_6 + D$_2$O): δ 198.2 (C-1), 145.5, 133.1 (C$_{arom}$), 135.6, 130.2, 129.3, 127.3 (CH$_{arom}$), 71.5, 71.4 (C-3 and C-4), 45.0 (C-2)

Circular dichroism (c = 2.97 mmol × L^{-1}, l = 0.2 cm, acetonitrile): λ (Mol. CD) (nm) = 220 (−7.41), 240 (−0.15), 256 (−2.40), 266 (−1.53), 288 (−4.97), 318 (1.62)

6.1.4 Conclusion

An extraordinarily broad substrate range was shown for T$_4$HNR, which allowed for detailed studies on the catalytic features of this and related enzymes [6,7]. Compounds such as polyhydroxynaphthalenes, hydroxynaphthoquinones, anthrones, oxanthrones, and different cyclic ketones were identified as suitable substrates for polyhydroxynaphthalene redcutases and related enzymes [8]. On the basis of our findings, one can state that "classical" oxidoreductases have been treated too restrictively concerning alternate activities.

6.2 Chemoenzymatic Synthesis of Diastereo- and Enantiomerically Pure 2,6-Disubstituted Piperidines via Regioselective Monoamination of 1,5-Diketones

Robert C. Simon,[1] Ferdinand Zepeck,[2] and Wolfgang Kroutil[1]

[1] *Department of Chemistry, Organic and Bioorganic Chemistry, University of Graz, Austria*

[2] *Sandoz GmbH, Austria*

Functionalized chiral piperidines are popular key targets in a vast number of synthetic protocols and are among the most prevelant skeletal fragments in natural products. In particular, 2,6-disubstituted piperidines constitute an important class, due to their range of biological activity and their abundant occurrence in nature [9]. Various sophisticated chemical synthetic routes by which to access this scaffold have been described in literature [10], as have enzymatic approaches utilizing, for example, ω-transaminases (ω-TAs) [11]; these enzymes are pyridoxal-5′-phosphate (PLP)-dependent biocatalysts that have proven to be remarkably efficient in the synthesis of chiral amines via asymmetric reductive amination, kinetic resolution, or deracemization [12].

Starting from designated 1,5-diketones (**1** in Scheme 6.2), an asymmetric reductive monoamination furnished the corresponding Δ1-piperideine (**3**) after a spontaneous ring closure at high conversion under excellent regio- and stereocontrol. A subsequent reduction yielded the analoguous diastereo- and enantiomerically pure 2,6-disubstituted piperidine (**4**). The concept was demonstrated for the synthesis of both enantiomers of the alkaloid dihydropinidine [13] (R = n-C$_3$H$_7$) using complementary ω-TAs (*Chromobacterium violaceum* [14] for the (*S*)- and *Arthrobacter* sp. [15] for the (*R*)-enantiomer),

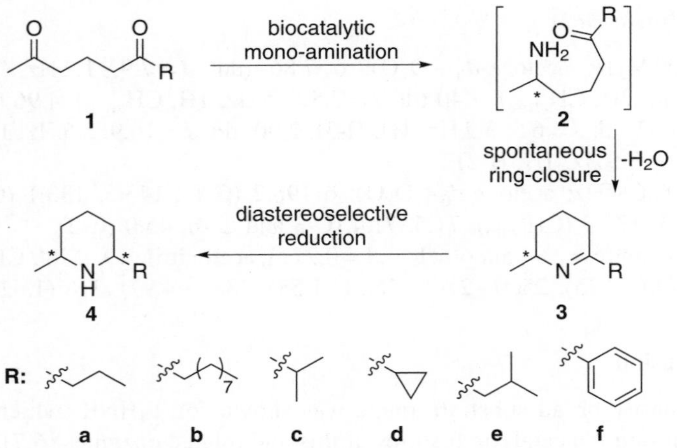

Scheme 6.2 *Chemoenzymatic synthesis of 2,6-disubstituted piperidines with a regioselective monoamination as the key step.*

which can be extended to access the epimers [11b] and related natural products as well [11c].

6.2.1 Procedure 1: Cultivation and Expression of ω-TAs

6.2.1.1 Materials and Equipment

- *E. coli* BL21(DE3) harboring plasmid of designated ω-TA (frozen glycerol stocks)
- LB media: yeast extract (5 g.L^{-1}), trypton (10 g.L^{-1}), NaCl (5 g.L^{-1})
- Ampicillin (100 mg.L^{-1})
- Isopropyl thiogalacto pyranoside (120 mg.L^{-1})
- Phosphate buffer (KH$_2$PO$_4$ (4.72 g.L^{-1}), K$_2$HPO$_4$ (11.36 g.L^{-1}))
- Pyridoxal-5′-phosphate (247 mg.L^{-1})
- Heatable shaker
- Centrifuge

6.2.1.2 Expression of ω-TAs in E. coli

6.2.1.2.1 BL21(DE3)

1. LB-ampicillin medium (330 mL, 100 mg.L^{-1} ampicillin) was inoculated with a cell suspension from an overnight culture (5 mL) of *E. coli/*ω-TA and incubated at 37 °C (120 rpm). As soon as the OD$_{600}$ reached a value between 0.5 and 0.7 (approx. 3 hours), overexpression of the ω-TA was induced by the addition of isopropyl β-D-1-thiogalactopyranoside (IPTG) (120 mg.L^{-1}).
2. Flasks were shaken for an additional 3 hours at 120 rpm and 20 °C and were harvested by centrifugation at 12·10^3 rpm and 4 °C for 20 minutes.
3. The pellets were washed twice with phosphate buffer (pH 7, 100 mM, 1.0 mM PLP), frozen with liquid nitrogen, and lyophilized. The crude cells obtained were stored at 4 °C until use.

6.2.2 Procedure 2: Biocatalytical Synthesis of 2,6-Disubstituted Piperideines

6.2.2.1 Materials and Equipment

- Lyophilized cell preparation containing the overexpressed ω-TA
- Nonane-2,6-dione (78 mg, 0.5 mmol)
- Pyridoxal-5′-phosphate (2.47 mg)
- Nicotinamide adenine dinucleotide (6.63 mg)
- Ammonium formate (94.6 mg)
- Formate dehydrogenase (110 U)
- Alanine dehydrogenase (120 U)
- Ethyl acetate (20 mL)
- 15 mL Sarstedt tube
- Shaker
- Rotary evaporator

6.2.2.2 Biotransformation of Nonane-2,6-dione (1a) (Scheme 6.3)

1. Lyophilized cells of *E. coli* containing the overexpressed ω-TA from either *Chromobacterium violaceum* or *Arthrobacter* sp. (225 mg) were rehydrated in a potassium phosphate buffer (10 mL, pH 7.0, 100 mM) containing PLP (1.0 mM), NAD$^+$ (1.0 mM), ammonium formate (150 mM), FDH (110 U), Ala-DH (120 U), and D-/L-alanine (500 mM) at 22 °C for 30 minutes. Nonane-2,6-dione **1a** (78 mg, 0.5 mmol) was added and the reaction was shaken for 26 hours at 30 °C.
2. Saturated Na$_2$CO$_3$ solution was added (1.00 mL) and the reaction mixture was extracted with EtOAc (4 × 5 mL). Combined organic layers were dried over Na$_2$SO$_4$, filtrated, and converted in the next step, without further purification.

6.2.2.3 Analytical Data for 2-Methyl-6-Propyl-2,3,4,5-Tetrahydropyridine 3a

The product **3a** was obtained in >99% conversion with an optical purity *ee* > 99%, as judged by GC (*vide infra*). It can be characterized using Et$_2$O for extraction, followed by careful concentration on a rotary evaporator (650 mbar, 35 °C).

^{1}H-NMR (300 MHz; CDCl$_3$): δ$_H$ (ppm) 0.80 (t, $^3J_{9,8}$ = 7.4 Hz, 3 H, 9-H), 1.08 (d, $^3J_{Me,2}$ = 6.4 Hz, 3 H, Me at C-2), 1.45 (m$_c$, 4 H, 4-H and 8-H), 1.60 (m$_c$, 2 H, 3-H), 1.98 (m$_c$, 2 H, 5-H), 2.04 (t, $^3J_{7,8}$ = 7.4 Hz, 2 H, 7-H), 3.34 (m$_c$, 1 H, 2-H)
^{13}C-NMR (75 MHz; CDCl$_3$): δ$_C$ (ppm) 14.0 (C-9), 18.5 (C-4), 20.5 (C-8), 23.7 (Me at C-2), 28.6 (C-5), 29.3 (C-3), 42.9 (C-7), 52.8 (C-2), 171.6 (C-6)

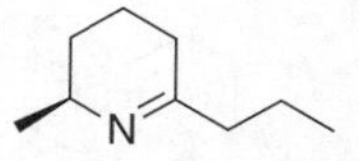

(S)-**3a**
conv. >99%,
ee >99% (S)
(*Chromobacterium violaceum*)

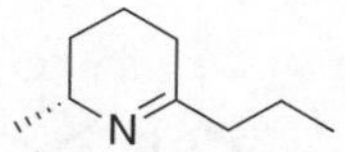

(R)-**3a**
conv. >99%,
ee >99% (R)
(*Arthrobacter* sp.)

Scheme 6.3 *Biotransformation of nonane-2,6-dione (1a).*

GC-MS (EI+, 70 eV): m/z (%) = 139 [M$^+$] (22), 124 [C$_8$H$_{14}$N$^+$] (46), 111 [C$_7$H$_{13}$N$^{\bullet+}$] (58), 96 [C$_6$H$_{10}$N$^+$] (100)

Determination of conversion by means of achiral GC analysis: column HP-5 (Agilent); carrier gas = nitrogen; temperature program = from 60 to 150 °C, with slope 5 °C.min^{-1}; R_t (**3a**) = 10.08 minutes, R_t (**1a**) = 18.85 minutes

Determination of *ee* by means of chiral GC analysis: column DEX-CB (Agilent; CP-Chirasil); carrier gas = nitrogen, temperature program = from 80 °C (5 minutes isotherm) to 90 °C, with slope 0.5 °C.min^{-1}; R_t [(2*R*)-**3a**] = 11.05 minutes, R_t [(2*S*)-**3a**] = 11.45 minutes

6.2.3   Synthesis of Optically Pure 2-Methyl-6-propyldihydropiperidine 4a (Scheme 6.4)

6.2.3.1   *Materials and Equippment*

- Palladium on activated charcoal (~8 mg)
- Hydrogen (balloon)
- Celite 545 (500 mg)
- Ether-HCl-solution (2 N)
- 25 mL Round-bottom flask
- Funnel
- Magnetic stirrer
- Rotary evaporator

6.2.3.2   *Diastereoselective Reduction*

1. The crude solution of the biotransformation containing the corresponding cyclic imine **3a** was treated with palladium on activated charcoal (10% wt). The mixture was stirred vigorously and was hydrogenated for 4 hours.
2. After completion of the reaction, detected by GC or GC-MS, the solution was filtered through a small plug of Celite 545 and cooled to 0 °C. Ether-HCl solution was added dropwise (~5 eq.), the solvent was removed, and the precipitate was collected. If necessary, all products can be recrystallized from CHCl$_3$.

6.2.3.3   *Analytical Data for (2S,6R)-2-Methyl-6-Propyldihydropiperidine*

The product was obtained as colorless solid in the form of its HCl salt in 94% yield (82 mg, 0.47 mmol, *de* > 99%). Melting point: 243–245 °C.

^{1}H-NMR (300 MHz; CDCl$_3$): δ_H (ppm) 0.85 (t, $^3J_{9,8}$ = 7.2 Hz, 3 H, 9-H), 1.28–1.48 (m, 3 H, 8-H and 4-H$_a$), 1.50 (d, $^3J_{Me,2}$ = 6.2 Hz, 3 H, Me at C-2), 1.52–1.65 (m, 2 H, 5-H$_a$),

(2*S*,6*R*)-**4a**
ee >99%, de > 99%

(2*R*,6*S*)-**4a**
ee >99%, de > 99%

Scheme 6.4   *Synthesis of pptically pure 2-methyl-6-propyldihydropiperidine (4a).*

1.68–1.78 (m, 3 H, 3-H and 7-H$_a$), 1.80–1.94 (m, 2 H, 4-H$_b$ and 5-H$_b$), 1.96–2.10 (m, 1 H, 7-H$_b$), 2.88 (m$_c$, 1 H, 6-H), 3.05 (m$_c$, 1 H, 2-H) 9.02 (brd, 2 H, NH$_2$$^+$)

^{13}C-NMR (75 MHz; CDCl$_3$): δ_C (ppm) = 13.8 (C-9), 18.8 (C-4), 19.5 (Me at C-2), 22.9 (C-4), 27.5 (C-5), 30.7 (C-3), 35.2 (C-7), 54.5 (C-6), 58.3 (C-2). GC-MS (EI+, 70 eV): m/z (%) = 98 [C$_6$H$_{12}$N$^+$] (100). Optical rotation: [α]$_D$^{20} = −12.2 (c 0.5, EtOH, de > 99%)

6.2.4 Conclusion

Various enantiocomplementary ω-TAs were employed for the regioselective monoamination of selected 1,5-diketones **1a–1f**. The products were generally obtained at high conversion with excellent regio- and stereocontrol. Notably, the Δ1-piperideines can be converted to the *syn*- or *anti*-piperidine diastereomers, depending on the reduction conditions employed [11]. This approach represents a novel route by which to access the piperidine scaffold that can be found in multiple natural products and related bioactive compounds.

6.3 Asymmetric Amination of Ketones Employing ω-TAs in Organic Solvents

Francesco G. Mutti,[1] Christine Fuchs,[2] and Wolfgang Kroutil[2]
[1] *School of Chemistry, Manchester Institute of Biotechnology, The University of Manchester, United Kingdom*
[2] *Department of Chemistry, Organic and Bioorganic Chemistry, University of Graz, Austria*

ω-TAs are PLP-dependent enzymes that have been extensively employed for the kinetic resolution of racemic amines and for the amination of carbonylic compounds in aqueous buffers (or in the presence of an organic co-solvent) or in biphasic systems (aqueous buffer–organic solvent). Protocols for the kinetic resolution of racemic amines and for the asymmetric amination of prochiral ketones in an aqueous buffer have been published in the previous volume of this series [16].

In contrast, biocatalysis in organic solvents offers several advantages over traditional biotransformations in aqueous media, such as increased solubility of hydrophobic substrates, shorter and simplified work-up procedures, easy catalyst recycling, and enhanced enzyme stability at elevate temperature [17].

Lyophilized crude cell-free extracts of ω-TAs have been recently employed for the asymmetric amination of ketones in organic solvents [18]. Wild-type ω-TAs and variants were equally good at catalyzing the amination of ketones in an organic solvent following optimized reaction conditions without the need for immobilization, as described elsewhere [19]. The highest reaction rate was reached with methyl *tert*-butyl ether (MTBE) as a solvent, at a water activity (a$_w$) of 0.6. The amination of a panel of prochiral ketones was carried out in MTBE using 2-propylamine as an achiral and inexpensive amine donor (Scheme 6.5).

6.3.1 Part 1: Preparation of the Catalyst

6.3.1.1 Materials and Equipment

- LB medium (330 mL: 10 g.L^{-1} tryptone (Oxoid LP0042), 5 g.L^{-1} yeast (Oxoid LP0021), 5 g.L^{-1} NaCl)

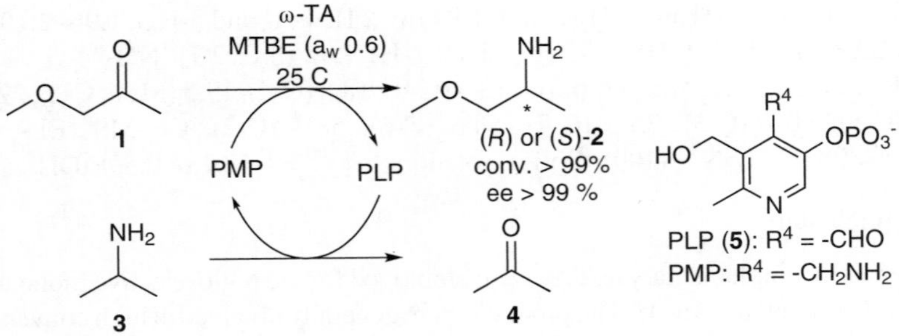

Scheme 6.5 *Biocatalytic amination using ω-TAs in MTBE with defined water activity (a_w 0.6).*

- Ampicillin
- *E. coli* cells harboring the plasmid coding for the ω-TA from *Arthrobacter* sp. (*R*-selective, ArR-ωTA) [20] or *Arthrobacter citreus* (*S*-selective, ArS-ωTA) [21]
- IPTG
- Phosphate buffer (pH 7, 100 mM) containing pyridoxal-5′-phosphate (PLP 0.5 mM and 1 mM)
- EDTA (1 mM in phosphate lysis buffer)
- Lysozyme from chicken egg white (>40 000 U.mg^{-1}; 1.06 mg.mL^{-1} in phosphate lysis buffer)
- Phenylmethanesulfonyl fluoride (1 mM in phosphate lysis buffer)
- Incubator for *E. coli* cell cultivation
- Orbital shaker
- Lyophilizer

6.3.1.1.1 Optional

- (*R*)-Phenyl-ethylamine (100 mM in 1 mL phosphate buffer–PLP)
- Sodium pyruvate (100 mM in 1 mL phosphate buffer–PLP)
- GC-FID

6.3.1.2 Procedure

1. LB-ampicillin medium (330 mL, 100 mg.L^{-1} ampicillin) was inoculated with a cell suspension from an overnight culture (5 mL) and cells were grown at 37 °C and 120 rpm until the OD$_{600}$ reached 0.7 (~3 hours). Overexpression of the ω-TA was induced by IPTG (1 mM). Flasks were shaken overnight at 120 rpm and 20 °C.
2. Cells were harvested by centrifugation at 12 000 rpm and 4 °C for 20 minutes. The pellets were washed with phosphate buffer (pH 7, 100 mM) containing PLP (0.5 mM), shock-frozen with liquid nitrogen, and lyophilized. The cells obtained were stored at 4 °C.

It is recommended that the level of enzyme expression be estimated by SDS-gel page. A low level of overexpression results in a catalyst preparation of mediocre quality (i.e., reduced ω-TA content). Ideally, the activity of the whole-cell catalyst obtained might be determined as follows: The activity of ArR-ωTA (U.mg^{-1}$_{lyophilized\ whole\ cells}$) was assayed using enantiopure (*R*)-phenyl-ethylamine (100 mM) as amine donor and sodium pyruvate

(100 mM) as carbonyl acceptor. A typical sample was prepared using 10 mg of lyophilized cells in phosphate buffer (1 mL, pH 7, 100 mM, PLP 1 mM). Samples were prepared and quenched after different time periods (30, 60, 90, 120, 180, 240, and 300 seconds) and analyzed by GC-FID on an achiral phase. The apparent units were calculated by linear regression within the linear range. The enzymatic activity after this step is typically around $1.90\,\text{U.mg}^{-1}_{\text{lyophilized cells}}$. One unit is defined as the amount of ω-TA that will deaminate 1 μmol of (*R*)-phenyl-ethylamine in 1 minute at 30 °C and pH 7.

3. Lyophilized cells of *E. coli* (500 mg) containing the overexpressed ArR-ωTA were suspended in lysis buffer (8 mL, phosphate buffer pH 7, 100 mM, 1 mM EDTA, 1 mM PLP, lysozyme from chicken egg white $1.06\,\text{mg.mL}^{-1}$ >40 000 U.mg^{-1}, 1 mM phenyl-methanesulfonyl fluoride) and homogenized by stirring with a glass rod to give a smooth suspension. Lysis was performed while shaking on an orbital shaker at 150 rpm and 22 °C for 3 hours.

4. The suspension was centrifuged (18 000 rpm, 38 700 g, 15 minutes, 4 °C). The supernatant was shock-frozen in liquid nitrogen in a rotating round-bottom flask (250 mL) and freeze-dried for 2 hours, yielding a crude powder extract, which was used for amination in organic solvent without any further purification. As an alternative, lyophilized cells of *E. coli* can be disrupted with a French press, using the same lysis buffer without lysozyme. The lyophilized crude preparation was stored at −20 °C.

It is recommended that this protocol be followed exactly, since slight changes could significantly lower the performance of the catalyst during the biocatalytic amination. The lyophilization process is crucial to optimal catalyst preparation. Ideally, the activity of the lyophilized crude preparation might be determined again after Step 4, as previously described, and then compared with the activity after Step 1. The enzymatic activity after Step 2 is typically around $1.60\,\text{U.mg}^{-1}_{\text{lyophilized crude preparation}}$.

6.3.2 Part 2: Amination in MTBE on Preparative Scale

6.3.2.1 Materials and Equipment

- Lyophilized crude preparation of ArS-ωTA from part 1 (20 mg)
- MTBE, water-saturated (22.6 mL)
- Distilled water (~0.36 mL)
- Methoxy-acetone **1** (commercial purity 95%, 100 mg, 1.13 mmol)
- 2-Propylamine **2** (commercial purity 99%, 0.289 mL, 3.36 mmol)
- HCl in diethyl ether solution (2 N, 2 mL)
- Activated 3 Å molecular sieves
- 4-(*N,N*-Dimethylamino)pyridine (5 mg)
- Acetic anhydride (100 μL)
- Sodium sulfate anhydrous
- Round-bottom flask (50 mL)
- Orbital shaker or Eppendorf thermomixer
- Optional: Karl–Fisher titration apparatus
- Claisen apparatus equipped with a Vigreux column for fractional distillation
- GC-FID

6.3.2.2 Procedure

1. The lyophilized crude enzyme preparation of the (*S*)-selective ArS-ωTA (452 mg) was suspended in water-saturated MTBE (22.6 mL) in a 50 mL glass flask. The suspension containing the catalyst was vigorously shaken for 30 minutes at 25 °C.Note: During this time, the solid catalyst absorbed part of the water from the organic solvent until phase equilibrium was reached. It is recommended that the suspension be shaken on the orbital shaker at the highest speed available (up to 900 rpm). Elevate shaking speed is even more important during the reaction (as is generally true for heterogeneous catalyzed reactions).
2. An additional amount of water (about 0.36 mL to reach a_w 0.6) was added and the sample was shaken for another 30 minutes to equilibrate the system.Note: The amount of water added at this step may vary slightly depending on the dryness of the catalyst. Ideally, the final water content may be measured by Karl–Fischer titration, by taking a small aliquot from the organic phase. The residual water concentration in MTBE should be 0.65 (v/v). This value corresponds to an a_w of 0.6.
3. Ketone substrate **1** (100 mg, 1.13 mmol) and 2-propylamine **3** (3 eq., 0.289 mL, 3.36 mmol) were added. The reaction mixture was vigorously shaken at 900 rpm and 25 °C. After 4 hours, quantitative conversion (>99%) was detected by GC-FID.
4. The enzyme preparation was filtered and the organic phase was dried with anhydrous Na_2SO_4 or, better, $MgSO_4$.
5. The ethereal solution was distilled using a Claisen apparatus equipped with a Vigreux column to remove the unreacted 2-propylamine (b.p. 32.4 °C). Note that the other components of the mixture have the following boiling points (P 1 atm): MTBE (b.p. 55.2 °C), acetone (b.p. 56.2 °C) and product (*S*)-methoxy-2-propylamine (*S*)-**2** (b.p. 92.5–93.5 °C).
6. The organic mixture was cooled to 4 °C and an ethereal solution of HCl (2 N, 2 mL) was added, resulting in a cloudy solution, due to the formation of (*S*)-**2**·HCl. The remaining solvent was then evaporated, yielding a pale yellow oil of (*S*)-**2**·HCl, which was washed twice with *n*-pentane (2 × 10 mL) (yield 84%). The product formed a pale yellow solid upon standing overnight at 4 °C. (*S*)-**2**·HCl was identified by H^1- and ^{13}C-NMR.

6.3.3 Analytical Data

^{1}H-NMR (300 MHz; CDCl$_3$): δ 1.37 (d, broad, 3H, CH$_3$-CH), 3.34 (s, 3H, CH$_3$-O), 3.51 (m, broad, 3H, O-CH$_2$ and CH-N), 8.23 (s, broad, 3H, NH$_3^+$)
^{13}C-NMR (300 MHz; CDCl$_3$): δ 15.2, 47.7, 59.1, 73.1

6.3.3.1 Determination of Conversion

Column: Agilent J&W DB-1701 (30 m, 250 μm, 0.25 μm); carrier gas He
GC program parameters: injector 250 °C; constant pressure 1 bar; temperature program 60 °C/hold 6.5 minutes; 100 °C/rate 20 °C.min^{-1}/hold 1 minute
Ketone **1**, 4.0 minutes; amine **2**, 3.4 minutes

6.3.3.2 Determination of Optical Purity

Column: Varian Chrompack Chirasel Dex-CB (25 m, 320 μm, 0.25 μm); carrier gas H$_2$
GC program parameters: injector 200 °C; constant flow 1.7 mL.min^{-1}; temperature program 60 °C/hold 2 minutes; 100 °C/rate 5 °C.min^{-1}/hold 2 minutes; 160 °C/rate 10 °C.min^{-1}/hold 0 minutes
Acetylated derivatives of: (*S*)-**2**, 11.8 minutes; (*R*)-**2**, 12.9 minutes

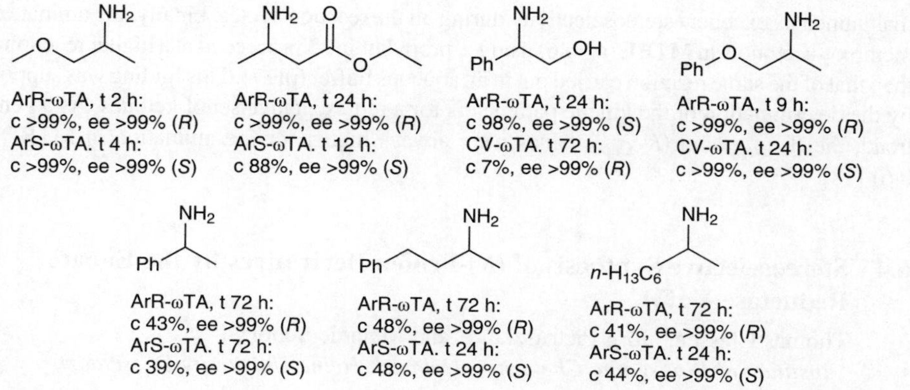

NH2
NH2 O
NH2
Ph OH
NH2
Ph O
ArR-ωTA, t 2 h:
c >99%, ee >99% (R)
ArS-ωTA. t 4 h:
c >99%, ee >99% (S)
ArR-ωTA, t 24 h:
c >99%, ee >99% (R)
ArS-ωTA. t 12 h:
c 88%, ee >99% (S)
ArR-ωTA, t 24 h:
c 98%, ee >99% (S)
CV-ωTA. t 72 h:
c 7%, ee >99% (R)
ArR-ωTA, t 9 h:
c >99%, ee >99% (R)
CV-ωTA. t 24 h:
c >99%, ee >99% (S)
NH2
Ph
NH2
Ph
NH2
n-H13C6
ArR-ωTA, t 72 h:
c 43%, ee >99% (R)
ArS-ωTA. t 72 h:
c 39%, ee >99% (S)
ArR-ωTA, t 72 h:
c 48%, ee >99% (R)
ArS-ωTA. t 24 h:
c 48%, ee >99% (S)
ArR-ωTA, t 72 h:
c 41%, ee >99% (R)
ArS-ωTA. t 24 h:
c 44%, ee >99% (S)

Scheme 6.6 *Amines prepared from their corresponding ketones (50 mM) in MTBE (a_w 0.6) using 3 eq. of 2-propylamine as amino donor. Conversions can be increased in all cases applying a larger excess of 2-propylamine (10–20 eq.).*

6.3.4 Conclusion

ω-TAs are capable of operating efficiently in organic solvents at a defined water activity (e.g., a_w 0.6 for ArR-ωTA) without the need for enzyme immobilization. Furthermore, the inexpensive 2-propylamine proved to be an efficient amine donor for each ω-TA tested, which is not the case in aqueous buffer. Scheme 6.6 summarizes the results obtained for the amination of a panel of prochiral ketones using various ω-TAs in MTBE (a_w 0.6). Both enantiopure (*R*)- and (*S*)-configured amines were obtained employing stereo-complementary ω-TAs. Additionally, employing an organic solvent as a reaction medium simplified the work-up procedure: the enzyme was decanted or filtered off from the reaction medium and the amine product was recovered by distillation and/or precipitation as its hydrochloride salt. Moreover, the same aliquot of enzyme could be recycled several times, affording quantitative conversion for the amination of the target substrate within the same reaction time (Figure 6.1); the catalyst

100
80
60
40
20
0
c [%]
1
2
3
4
5
6
7
8
9
10
Cycle

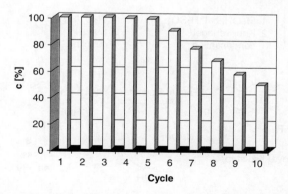

Figure 6.1 *Recycling of ω-TA and amination of methoxy-acetone (**1**, 50 mM) in MTBE (1 mL, a_w 0.6) using 2-propylamine (**3**, 3 eq.) as amine donor and ArR-ωTA (crude preparation, 20 mg, 0.82 U.mg^{-1} $_{\text{crude preparation}}$). The enzyme was recycled 10 times and the reaction time was 4 hours per cycle.*

maintained its excellent stereoselectivity during all the reaction cycles. Finally, the amination of methoxy-acetone **1** in MTBE (a_w 0.6) using 2-propylamine **3** proceeded at a higher reaction rate than that of the same reaction carried out in an aqueous buffer (pH 7). This finding was supported by the determination of the kinetic parameters for a representative set of ketones. As a general trend, the $(k_{cat})_{app}$ and $(K_M)_{app}$ values were always higher for the amination in MTBE (a_w 0.6) [18].

6.4 Stereoselective Synthesis of (R)-Profen Derivatives by the Enoate Reductase YqjM

Thomas Classen,[2] Jörg Pietruszka,[1,2] and Melanie Schölzel[1]
[1] *Institute of Bioorganic Chemistry, Heinrich Heine University Düsseldorf, Germany*
[2] *IBG-1: Biotechnology Research Center Jülich, Germany*

Chiral 2-arylpropionic acid derivatives such as compound **3** in Scheme 6.7 can be easily obtained from the prochiral compound **2** by reducing the activated double bond with the enoate reductase YqjM from *Bacillus subtilis*. The unsaturated ester **2** is prepared from its corresponding aryl compound **1** by esterification and a consecutive Knoevenagel condensation. The route offers convenient access to profen derivatives, and especially to (R)-flurbiprofen methyl ester (**4**); (R)-flurbiprofen is a potential agent for the treatment of

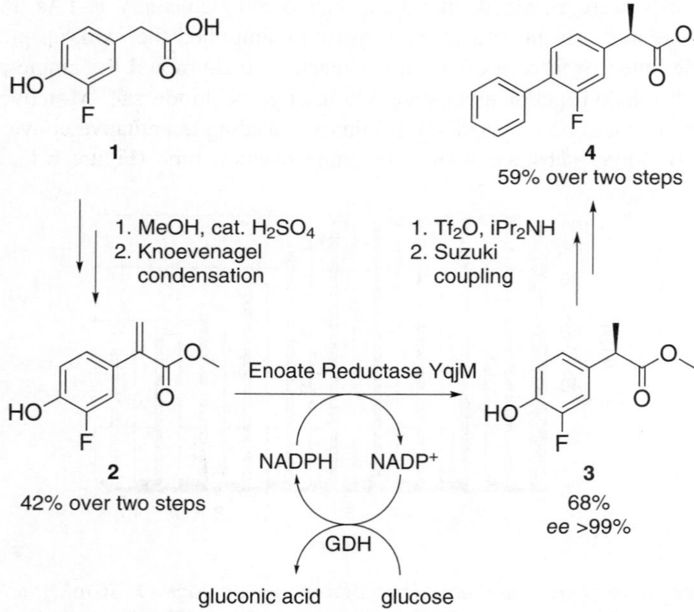

Scheme 6.7 Chemoenzymatic synthesis of the (R)-flurbiprofen methyl ester (**4**).

Alzheimer's disease [22,23]. The biocatalyst has been produced by heterologous expression in *Escherichia coli* and purified using the *N*-terminal-fused hexa-histidine tag. Optionally, addition of extra flavinmononucleotide (FMN) co-factor to the protein increases the specific activity by a factor of eight.

The present protocol for the conversion of compound **3** is primarily based on the work of Pietruszka and Schölzel [24], while the expression and purification of the enoate reductase YqjM is described in detail by Classen *et al.* [25].

6.4.1 Chemoenzymatic Synthesis of the (*R*)-Flurbiprofen Methyl Ester (4)

6.4.1.1 *Materials and Equipment*

- *E. coli* BL21(DE3) harboring the pHT::yqjM-vector (pET22b amp^R p$_{T7}$(6xhis:tev: yqjM)) [25]
- LB$_{amp}$ agar plate (1% w/v tryptone, 0.5% w/v yeast extract, 1% w/v NaCl, 100 µg.mL^{-1} ampicillin)
- 1 L TB$_{amp}$ medium (1.2% w/v tryptone, 2.4% w/v yeast extract, 0.4% w/v glycerol, 0.231% w/v KH$_2$PO$_4$, 1.254% w/v K$_2$HPO$_4$)
- IPTG 1 M in water (sterile-filtered)
- Buffer P$_0$: 20 mM potassium phosphate buffer pH 6.5
- Buffer P$_{30}$: P$_0$ with 30 mM imidazole pH 6.5 from 2 M stock solution
- Buffer P$_{250}$: P$_0$ with 250 mM imidazole pH 6.5 from 2 M stock solution
- GDH (e.g., from Evocatal)
- Glucose
- Disodium NADP$^+$
- 2-(3-Fluoro-4-hydroxyphenyl)acetic acid (**1**)
- Methanol
- Concentrated sulfuric acid
- Paraformaldehyde
- Anhydrous dimethylformamide
- Potassium carbonate
- 2-Methyl-tetrahydrofuran
- Ethyl acetate (EtOAc)
- MTBE
- MgSO$_4$
- 3.5 L Fernbach flask (other flasks may work as well)
- Orbital shaker 25 °C, 120 rpm
- Cooled centrifuge for 50 mL reaction tubes (18 000 rcf)
- French press cell disruptor
- Peristaltic pump (optional)
- MPLC with UV/Vis detector (280 nm and optional 400 nm)
- 5 mL Ni-NTA column (e.g., NiNTA superflow cartridge from Qiagen)
- Ultrafiltration device (e.g., Vivaspin 20 MWCO 10 kDa from Satorius)
- PD10-desalting column (e.g., from GE Healthcare)
- Optional: FMN
- 50 mL Conical reaction tubes

- Lyophilizer (optional)
- 1.5 mL Reaction tubes
- Equipment for chemical synthesis under inert conditions: Schlenk line equipped with vacuum and dry nitrogen, Schlenk flasks, and heat gun to flame-dry the flasks at 250 °C
- Silica gel column
- Eluent petroleum ether/EtOAc = 90 : 10, 80 : 20, and 50 : 50

6.4.1.2   Expression

1. TB$_{amp}$-medium (5 mL) was inoculated with several colonies of *E. coli* BL21(DE3) pHT:: yqjM and incubated overnight at 37 °C under mild shaking.
2. For the main culture, TB$_{amp}$-medium (1 L) in a baffled Fernbach flask was inoculated with the entire overnight culture and incubated for 8–10 hours at 25 °C and 120 rpm.
3. The main culture was then induced with 1M IPTG to a final concentration of 100 μM and incubated for a further 14 hours at 25 °C and 120 rpm.
4. The whole cells were harvested by centrifuging the suspension at 8000 rcf and 4 °C for 20 minutes, yielding around 16 g wet cell mass per liter.
5. The pellets were either stored long-term frozen at −20 °C (up to 3 months) or used directly for purification.

6.4.1.3   Purification of YqjM

1. Wet cells (10 g) were resuspended in buffer P$_0$ (50 mL)
2. The resulting 20% w/v suspension was subjected to cell disruption by passing the entire volume through a French press twice (other methods of cell disruption such as ultra-sonication work as well).
3. The disrupted cells were pelleted by centrifugation (18 000 rcf/4 °C/20 minutes).
4. The clarified supernatant was applied to a 5 mL NiNTA column equilibrated with buffer P$_0$ using a peristaltic pump, circulating the cell-free crude extract through the column five times. The loaded column was subjected to an elution program using a protein MPLC system: 6 column volumes (CV) with P$_0$ (this step is optional and can be skipped without loss of purity or yield), 6 CV with P$_{30}$, and, finally, 6 CV of P$_{250}$ to elute the protein. Note: The elution chromatogram can be monitored at wavelengths of either 280 nm (all proteins) or 400 nm (only YqjM). If no automatic chromatographic system is available, the protocol is also applicable to gravity flow- or syringe-driven chromatography.
5. The yellow eluate fractions, containing the desired protein, were combined and concentrated using ultrafiltration (10 kDa MWCO) up to 10 mg · mL^{-1}. Note: The eluate can optionally be supplemented with 5 mM FMN during the concentration step to complement the fraction of apo-YqjM with FMN. This procedure is able to increase the specific activity by a factor of eight.
6. Using a (microvolume)-photometer (blanked on ultrafiltrator flow-through), the degree of FMN saturation (*S*) and the protein concentration (β) were estimated using the empirical formulae:

$$S = 3.39 \frac{A_{458\,nm}}{A_{280\,nm}} = 4.53 \frac{A_{432\,nm}}{A_{280\,nm}} \quad \text{and} \quad \beta = \frac{A_{280\,nm}}{(1.015 \cdot S + 0.8)}\, mg/mL$$

7. To remove the remaining imidazole (and the unbound FMN, if used), the buffer was exchanged to P_0 using a PD10 desalting column. Note: Imidazole can decrease the stability of the protein in solution and cause impurities during the following work-up. Remaining FMN will interfere with photometric protein concentration, but not the reaction. The removal can be omitted. However, to remove the small compounds totally, two passes are necessary. This procedure yields up to 400 mg pure protein with $2 \, U \cdot mg^{-1}$ (one unit is defined as the amount of enzyme necessary to convert 1 µmol of cyclohex-2- enone in 1 minute under the given conditions (30 °C, 20 mM KPi pH 6.5, 0.15 mM NADPH, 1 mM cyclohex-2-enone, photometrically monitored via the NADPH consumption at 340 nm).

8. The protein was stored as lyophilisate or used directly. The solution was stored in the fridge for up to four weeks.

6.4.1.4 Synthesis of Enoate 2

1. Acid **1** (993 mg, 5.84 mmol) was dissolved in methanol (20 mL) and treated with a few drops of concentrated sulfuric acid.
2. After stirring at room temperature overnight, the solvent was removed under reduced pressure.
3. The residue was dissolved in EtOAc (50 mL) and washed with water (20 mL). The organic layer was dried over $MgSO_4$ and filtered, and the solvent was removed under reduced pressure. Chromatography over silica using EtOAc/petroleum ether (50 : 50) as eluent afforded the methyl 2-(3-fluoro-4-hydroxyphenyl)acetate (1.05 g, 5.71 mmol, 98%) as a colorless solid. $R_f = 0.62$ (EtOAc : petroleum ether, 50 : 50); ^{1}H-NMR (600 MHz; CDCl$_3$): δ 3.55 (s, 2 H, 2-H), 3.70 (s, 3 H, OMe), 5.21 (s, 1 H, OH), 6.90–6.95 (m, 2 H, 5'-H and 6'-H), 7.02 (d, $^3J_{2,F}$ = 11.3 Hz, 1 H, 2'-H); ^{13}C-NMR (151 MHz; CDCl$_3$): δ 40.1 (C-2), 52.2 (OMe), 116.5 (d, $^2J_{2,F}$ = 18.7 Hz, C-2'), 117.3 (d, $^4J_{6,F'}$ = 2.1 Hz, C-6'), 125.7 (d, $^3J_{5,F}$ = 3.4 Hz, C-5'), 126.6 (d, $^3J_{1,F}$ = 6.4 Hz, C-1'), 142.7 (d, $^2J_{4,F}$ = 14.2 Hz, C-4'), 150.8 (d, $^1J_{3,F}$ = 237.8 Hz, C-3'), 172.0 (C-1); ^{19}F-NMR (CDCl$_3$; 565 MHz): δ −(8.65–8.69) (m, F); IR (ATR, film): $\tilde{\nu}$ (cm^{-1}) = 3455, 3016, 2970, 2949, 1740, 1519, 1440, 1366, 1276, 1229, 1217, 1144, 1112; MS (EI, 70 eV): m/z = 185 (35) [M$^+$], 125 (100) [M-C$_2$H$_3$O$_2$$^+$].
4. Under an atmosphere of dry nitrogen, paraformaldehyde (163 mg, 5.44 mmol, 2.0 equiv.) was added to a solution of methyl 2-(3-fluoro-4-hydroxyphenyl)acetate (500 mg, 2.72 mmol, 1.0 equiv.) and potassium carbonate (940 mg, 6.80 mmol, 2.5 equiv.) in anhydrous dimethylformamide (15 mL). The reaction mixture was heated to 70 °C.
5. After complete conversion was detected (by GC-MS), water was added. The mixture was extracted with EtOAc (3 × 50 mL) and the organic portions were combined and washed with water (20 mL) and brine (20 mL), respectively, and dried over $MgSO_4$.
6. The solvent was removed under reduced pressure and the residue was purified by chromatography over silica using EtOAc/petroleum ether (10 : 90) as eluent to afford product **2** (226 mg, 1.15 mmol, 42%) as a light-yellow solid. $R_f = 0.38$ (petroleum ether/ EtOAc, 70 : 30); ^{1}H-NMR (600 MHz; DMSO-d6): δ 3.76 (s, 3 H, Me), 5.98 (s, 1 H,

3-H_a), 6.14 (s, 1 H, 3-H_b), 6.94 (t, $^3J_{5',6'}$ = 8.5 Hz, $^4J_{5',F}$ = 8.5 Hz, 1 H, 5'-H), 7.09 (ddd, $^3J_{6',5'}$ = 8.5 Hz, $^4J_{6',2'}$ = 2.2 Hz, $^5J_{6',F'}$ = 1.1 Hz, 1 H, 6'-H), 7.27 (dd, $^3J_{2',F}$ = 12.7 Hz, $^4J_{2',6'}$ = 2.2 Hz, 1 H, 2'-H); ^{13}C-NMR (151 MHz; DMSO-d6): δ 52.3 (Me), 115.7 (d, $^2J_{2',F}$ = 19.6 Hz, C-2'), 116.9 (d, $^3J_{5',F}$ = 2.3 Hz, C-5'), 124.8 (d, $^4J_{6',F}$ = 3.3 Hz, C-6'), 126.5 (C-3), 129.5 (d, $^3J_{1',F}$ = 6.7 Hz, C-1'), 139.7 (C-2), 143.7 (d, $^2J_{4',F}$ = 14.3 Hz, C-4'), 150.5 (d, $^1J_{3',F}$ = 236.9 Hz, C-3'), 167.2 (C-1); ^{19}F-NMR (CDCl₃; 565 MHz): δ −(9.49–9.57) (m, F); IR (ATR, film): $\tilde{v}$ (cm^{-1}) = 3372, 1696, 1630,1613, 1598, 1519, 1436, 1404, 1372, 1298, 1185, 1166, 1113, 983, 953, 937, 884, 827, 815, 789, 742, 722, 686; MS (EI, 70 eV): m/z = 196 (80) [M$^+$], 166 (5) [M-OCH₃$^+$], 137 (100) [M-C₂H₃O₂$^+$], 109 (17) [M-C₄H₇O₂$^+$]. GC (0.6 bar H₂, Hydrodex-β-TBDAc; 140 °C (0.5 min iso), 15 °C · min^{-1} up to 173 °C (10 min iso): t$_r$ = 7.13 minutes.

6.4.1.5   Enzymatic Synthesis of 3

1. A solution of enoate **2** (49 mg, 0.25 mmol) in 2-methyl-tetrahydrofuran (500 µL) was mixed with 20 mM KP$_i$ buffer (pH 7, 49.5 mL) in a 50 mL Falcon tube.
2. NADP$^+$ (3.94 mg, 0.005 mmol), glucose (180 mg, 1 mmol), GDH (1 mg lyophilized powder, ~5 U; *Evocatal*), and YqjM (2 mg lyophilized powder, ~5 U) were added and the reaction was shaken for 24 hours at 25 °C.
3. After complete conversion, as determined by GC, the reaction mixture was thoroughly extracted with MTBE (50 mL), the organic phase was dried over MgSO₄, filtered, and the solvent was removed under reduced pressure.
4. The crude product was purified by chromatography over silica using EtOAc/petroleum ether (20 : 80) as eluent to afford product **3** (33 mg, 0.17 mmol, 68%) as a colorless solid. R$_f$ = 0.46 (petroleum ether/EtOAc 70 : 30); ^{1}H-NMR (600 MHz; DMSO-d6): δ 1.35 (d, $^3J_{3,2}$ = 7.2 Hz, 3 H, 3-H), 3.59 (s, 3 H, Me), 3.71 (q, $^3J_{2,3}$ = 7.2 Hz, 1 H, 2-H), 6.87–6.92 (m, 2 H, 5'-H, 6'-H), 7.05 (dd, J = 1.3 Hz, $^3J_{2',F}$ = 12.2 Hz, 1 H, 2'-H), 9.78 (s, 1 H, O-H); ^{13}C-NMR (151 MHz; DMSO-d6): δ 18.4 (C-3), 43.3 (C-2), 51.7 (Me), 115.0 (d, $^2J_{2',F}$ = 18.7 Hz, C-2'), 117.6 (d, $^3J_{5',F}$ = 3.2 Hz, C-5'), 123.4 (d, $^4J_{6',F}$ = 3.1 Hz, C-6'), 131.8 (d, $^3J_{1',F}$ = 5.9 Hz, C-1'), 143.7 (d, $^2J_{4',F}$ = 12.03 Hz, C-4'), 150.7 (d, $^1J_{3',F}$ = 240.7 Hz, C-3'), 174.3 (C-1); IR (ATR, film): $\tilde{v}$ (cm^{-1}) = 3456, 3016, 2971, 2948, 1745, 1625, 1600, 1519, 1436, 1367, 1290, 1229, 1217, 1117, 1092, 1067, 922, 875, 799, 785; MS (EI, 70 eV): m/z = 198 (25) [M$^+$], 139(100) [M-C₂H₃O₂$^+$], 109 (9) [M-C₄H₇O₂$^+$]. GC (0.6 bar H₂, Hydrodex-β-3P; 140 °C (0.5 min iso), 15 °C · min^{-1} up to 173 °C (10 min iso): t$_r$[(S)-**3**] = 5.19 minutes, t$_r$[(R)-**3**] = 5.25 minutes.

6.4.2   Conclusion

In terms of catalyst yield, the biocatalyst YqjM is easy to produce from the heterologous expression in *Escherichia coli*, and both its purification and its handling are convenient. The conversion of unsaturated esters such as compound **2** described here can be readily implemented, resulting in valuable products within 24 hours. Due to the co-factor recycling system used, the conversion is driven to virtually 100%. In addition, the setup is applicable to various derivatives of enoate **2** [24], which are all simple to synthesize from cheap precursors using a Knoevenagel condensation (Table 6.1).

Table 6.1 *Tested substrate variety for both the Knoevenagel reaction and the biocatalytic reduction.*

No.	Product	Yield 2	Conv. 3	ee 3[a]
a	methyl 2-(phenyl)acrylate ($R^1 = R^2 = R^3 = H$)	66%	100%	>99% (*R*)
b	methyl 2 (*p*-tolyl)acrylate ($R^2 = Me$, $R^1 = R^3 = H$)	60%	73%	>99%
c	methyl 2-(4-iso-buytlphenyl)acrylate ($R^2 = iso$-butyl, $R^1 = R^3 = H$)	75%	42%	>99% (*R*)
d	methyl 2-(3-bromophenyl)acrylate ($R^3 = Br$, $R^1 = R^2 = H$)	59%	56%	>99%
e	methyl 2-(4-bromophenyl)acrylate ($R^2 = Br$, $R^1 = R^3 = H$)	35%	71%	>99%
f	methyl 2-(3-fluorophenyl)acrylate ($R^3 = F$, $R^1 = R^2 = H$)	31%	100%	>99%
g	methyl 2-(3-fluoro-4-hydroxyphenyl)acrylate ($R^3 = F$, $R^2 = OH$, $R^1 = H$)	40%	100%	>99%

[a] ees and conversions determined by GC. For compounds **3a** and **3c**, the configuration has been determined experimentally to be (*R*). In all other cases, the (*R*) configuration is most probable.

6.5 Productivity Improvement of the Bioreduction of α,β-Unsaturated Aldehydes by Coupling of the *In Situ* Substrate Feeding Product Removal (SFPR) Strategy with Isolated Enzymes

Elisabetta Brenna,[1] Francesco G. Gatti,[1] Daniela Monti,[2] Fabio Parmeggiani,[1] and Alessandro Sacchetti[1]

[1] *Department of Chemistry, Materials and Chemical Engineering "G. Natta", Politecnico di Milano, Italy*
[2] *Institute of Molecular Recognition Chemistry (CNR), Italy*

Even though baker's yeast-mediated reduction of α,β-unsaturated aldehydes and other activated alkenes has been investigated for many decades [26], its application in the large-scale synthesis of optically pure chemicals is hampered by several drawbacks: long reaction times, low substrate loadings, troublesome work-up, competitive side reactions, and so on. Overall, these factors contribute to give modest productivities, rarely higher than 0.5 g.L^{-1}.d^{-1}. So far, two strategies have been adopted to address some of these issues. One is the *in situ* substrate feeding product removal (SFPR) concept [27], which is based on the adsorption of substrate and products on a hydrophobic resin (e.g., XAD 1180N), ensuring very low concentrations in the aqueous phase and thus preventing toxic or inhibitory effects. Its application led to an improvement of yield, enantioselectivity and work-up practicality, but the productivities remained unsatisfactory. The second strategy is based on the use of isolated proteins as biocatalysts, rather than whole cells. The enzymes responsible for the C=C reduction

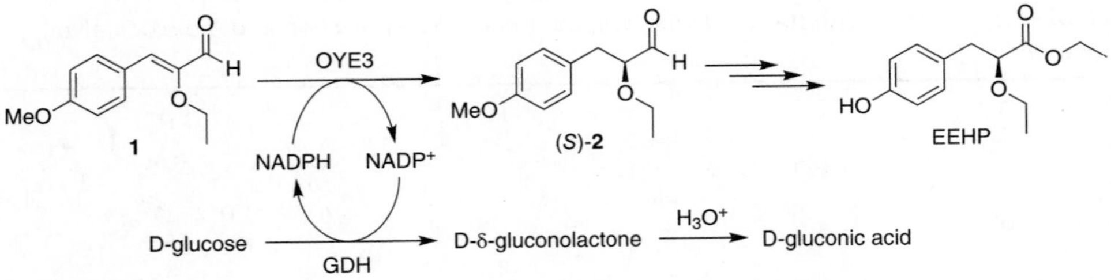

Scheme 6.8 *ER-mediated bioreduction of **1** (performed with the SFPR strategy in Procedure 1).*

activity are ene-reductases (ERs), most of which belong to the Old Yellow Enzyme superfamily (including OYE2 and OYE3, present in baker's yeast) [28]. In many cases, this approach allowed quantitative conversions to be obtained without side products. Nevertheless, the productivities were usually still low, because high substrate concentrations could not be employed.

In recent years, we have demonstrated that a substantial increase in productivity can be achieved by the combination of these two complementary strategies [29–33], as exemplified by the reduction of the α-substituted cinnamaldehyde derivative **1** (Scheme 6.8) [30]. The reduced product (*S*)-**2** is a precursor of the key pharmaceutical intermediate ethyl 2-ethoxy-3-(4-hydroxyphenyl)propanoate (EEHP), employed in the synthesis of several optically pure antidiabetic drugs (e.g., Tesaglitazar, Ragaglitazar) [34]. In order to keep the process economically feasible, the expensive NADPH co-factor required is employed in catalytic amounts using a regeneration system comprising glucose as a sacrificial substrate and GDH as a regeneration enzyme.

Another important feature of this methodology is that it prevents (or at least mitigates) the racemization of chiral α-substituted aldehydes, which occurs in water even at neutral pH [29,30]. Indeed, the *ee*s of the products obtained with isolated ERs are always higher using *in situ* SFPR than in homogeneous phase [29–33].

In several cases, the phenomenon is so severe that even the prompt recovery of the product on resin does not lead to high *ee* values. To overcome this problem, a cascade system has been devised, which involves coupling an alcohol dehydrogenase (ADH) to the ER. The ADH rapidly reduces the saturated aldehyde to the corresponding saturated alcohol, which does not undergo racemization. An example is provided by the reduction of the unsaturated aldehyde **3** to the saturated alcohol (*S*)-**5**, a precursor to the synthesis of the antidepressant robalzotan (Scheme 6.9) [33]. In this specific case, with isolated ERs, the *ee* was 8–11%, with the ER-SFPR system, the *ee* increased to 80–84%, and with the ER-ADH-SFPR cascade system, the *ee* reached 95–99% [29].

As an additional advantage, the two enzymes share the same regeneration system, because, although the co-factor preference is different (NADPH for the ER, NADH for the ADH), GDH is able to regenerate both.

In order to avoid the formation of the allylic alcohol (such as **6**) as a side product, it is necessary to select an ADH that is chemoselective for the saturated aldehyde over the unsaturated one. In the example described, the ADH from horse liver (HLADH) was found to be suitable for this purpose.

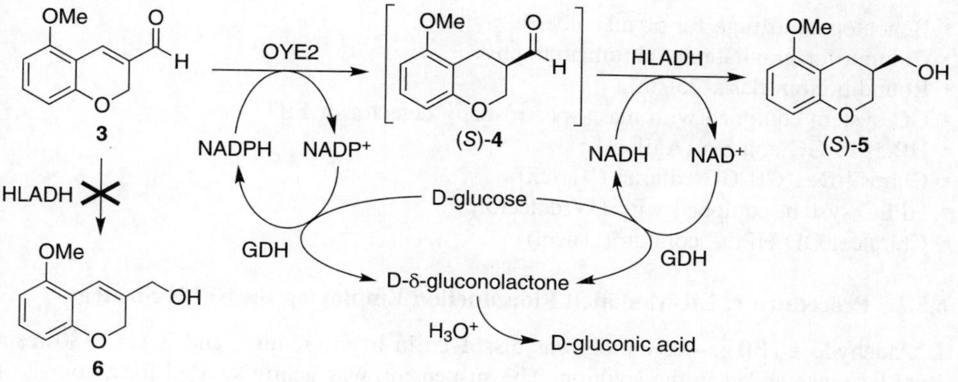

Scheme 6.9 *ER-ADH cascade bioreduction of* **3** *(performed with the SFPR strategy in Procedure 2).*

6.5.1 Materials and Equipment

- (Z)-2-Ethoxy-3-(4-methoxyphenyl)acrylaldehyde **1** (1.0 g, 4.85 mmol) (prepared according to reference [35])
- 5-Methoxy-2H-chromene-3-carbaldehyde **3** (250 mg, 1.25 mmol) (prepared according to reference [33])
- Amberlite XAD 1180N aromatic hydrophobic resin (Rohm & Haas)
- KP_i buffer (K_2HPO_4/KH_2PO_4, 100 mM, pH 7.0)
- D-glucose
- NAD^+
- $NADP^+$
- Recombinant OYE2 (1 mg.mL^{-1} solution in KP_i, prepared according to reference [30])
- Recombinant OYE3 (1 mg. mL^{-1} solution in KP_i, prepared according to reference [30])
- Recombinant GDH (100 U.mL^{-1} solution in KP_i, prepared according to reference [30])
- Recombinant HLADH (lyophilized powder, Sigma-Aldrich)
- Diethyl ether
- Dichloromethane
- *n*-Hexne or petroleum ether 40–60 °C
- Ethyl acetate
- Sodium sulfate anhydrous
- Silica for gravitational column chromatography (70–230 mesh)
- HPLC-grade isopropanol
- HPLC-grade *n*-hexane
- Large glass vial or glass bottle, with cap
- Rotary evaporator
- Water aspirator or vacuum pump
- Orbital shaker with temperature control
- Sintered glass Hirsch funnel, 5 cm diameter, porosity 0 (>165 µm)
- Side-armed filtration flask
- 50 mL Centrifuge tubes

- Benchtop centrifuge for 50 mL tubes
- Column for gravitational chromatography
- Round-bottom flask, 250 mL
- GC system equipped with mass-spectroscopy detector or FID
- HP-5MS GC column (Agilent)
- Chirasil-Dex CB GC column (Varian)
- HPLC system equipped with UV detector
- Chiralcel OD HPLC column (Daicel)

6.5.2 Procedure 1: ER-Mediated Bioreduction Employing the SFPR Strategy

1. Aldehyde **1** (1.0 g, 4.85 mmol) was dissolved in Et$_2$O (15 mL), and XAD 1180 resin (1.0 g) was added to the solution. The suspension was gently swirled for a couple of minutes to soak the beads thoroughly, then the solvent was slowly removed under reduced pressure.
2. The following components were added to a 500 mL glass vial: KP$_i$ buffer (26 mL, 50 mM, pH 7.0), glucose (3.5 g, 19.4 mmol), NADP$^+$ (2.6 mg, 3.3 μmol), OYE3 (5.7 mL, solution 1.0 mg.mL^{-1}), GDH (1.3 mL, solution 100 U.mL^{-1}). The homogeneous solution thus obtained (33 mL overall) contained 175 μg.mL^{-1} ER, 4 U.mL^{-1} GDH, 0.1 mM NADP$^+$, and 4 eq. glucose with respect to **1**.
3. The substrate adsorbed on the beads was added to the reaction mixture and the vial was capped and incubated in an orbital shaker (30 °C, 160 rpm). The shaking conditions (speed, tilt angle, etc.) were adjusted in order to keep the beads suspended in the aqueous phase at all times: having a portion of the beads sticking to the glass might be detrimental to the yield.
4. The evolution of the biotransformation was monitored by GC-MS, as described in Analytical Methods. After 12 hours, the mixture was filtered through a sintered-glass Hirsch funnel to separate the beads from the aqueous phase. The beads were washed with EtOAc (3 × 10 mL). The aqueous phase was transferred to a centrifuge tube and EtOAc (10 mL) was added. The tube was shaken vigorously and centrifuged (2000 rpm, 5 minutes) to separate the phases. The combined organic phases were dried over anhydrous Na$_2$SO$_4$ and concentrated under reduced pressure, to give (*S*)-**2** (0.99 g, 99% yield, *ee* 98%).
5. A sample of the product was dissolved in CH$_2$Cl$_2$ (ab. 1 mg.mL^{-1}) and analyzed by GC on a chiral stationary phase, as described in Analytical Methods.

6.5.3 Procedure 2: ER-ADH Cascade Bioreduction Employing the SFPR Strategy

1. Aldehyde **3** (250 mg, 1.32 mmol) was dissolved in Et$_2$O (10 mL), and XAD 1180 resin (750 mg) was added to the solution. The suspension was processed as described in Procedure 1, Step 1.
2. The following components were added to a 150 mL glass bottle: KP$_i$ buffer (81 mL, 50 mM, pH 7.0), glucose (900 mg, 5 mmol), NAD$^+$ (6.6 mg, 10 μmol), NADP$^+$ (7.4 mg, 10 μmol), OYE2 (15 mL, solution 1.0 mg.mL^{-1}), HLADH (10 mg), and GDH (4 mL, solution 100 U.mL^{-1}). The homogeneous solution thus obtained (100 mL overall) contained 150 μg.mL^{-1} ER, 100 μg.mL^{-1} ADH, 4 U.mL^{-1} GDH, 0.1 mM NAD$^+$, 0.1 mM NADP$^+$, and 4 eq. glucose with respect to **3**.

3. The reaction and the work-up were carried out as described in Procedure 1, Steps 3–4.
4. The residue was purified by chromatography over silica using *n*-hexane/EtOAc (9 : 1) as eluent to afford the starting aldehyde **3**, followed by the saturated alcohol (*S*)-**5**. The fractions containing the latter were collected and concentrated under reduced pressure (224 mg, 88% yield, *ee* 99%).
5. A sample of the purified alcohol was dissolved in MTBE (ab. 1 mg. mL^{-1}) and analyzed by HPLC on a chiral stationary phase, as described under Analytical Methods.

6.5.4 Analytical Methods

GC-MS analyses were performed on an Agilent HP 6890 gas chromatograph equipped with a 5973 mass detector and an Agilent HP-5-MS column (30 m × 0.25 mm × 0.25 μm), employing the following temperature program: 60 °C (1 minute)/6 °C.min^{-1}/150 °C (1 minute)/12 °C.min^{-1}/280 °C (5 minutes). GC retention times: **1** 21.59 minutes, **2** 19.31 minutes, **3** 22.33 minutes, **4** 21.25 minutes, **5** 22.74 minutes, **6** 23.25 minutes.

Chiral GC analyses were performed on a DANI HT 86.10 gas chromatograph equipped with an FID detector and a Varian Chirasil-Dex CB column (25 m × 0.25 mm), employing the following temperature program: 75 °C (1 minute)/3 °C.min^{-1}/119 °C (17 minutes)/30 °C.min^{-1}/180 °C (5 minutes). GC retention times: (*S*)-**2** 30.6 minutes, (*R*)-**2** 31.1 minutes.

Chiral HPLC analyses were performed on a Merck-Hitachi L-4250 chromatograph equipped with a UV detector (210 nm) and a Chiralcel OD column under isocratic elution at a flow rate of 0.6 mL.min^{-1}, with a mobile phase consisting of *n*-hexane/*i*-PrOH 97:3 (v/v). HPLC retention times: (*S*)-**5** 39.2 minutes, (*R*)-**5** 42.9 minutes.

6.5.5 Conclusion

Saturated aldehyde (*S*)-**2** was isolated in quantitative yield with *ee* 98%, according to Procedure 1. The productivity of the reaction was increased from 0.39 g.L^{-1}.d^{-1} with the original baker's yeast-based protocol [35] to 59.4 g.L^{-1}.d^{-1} with the ER-SFPR system [30].

Saturated alcohol (*S*)-**5** was obtained in 88% yield and *ee* 99%, according to Procedure 2. The productivity of the reaction was increased from 0.06 g.L^{-1}.d^{-1} with the original baker's yeast-based protocol to 17.4 g.L^{-1}.d^{-1} with the cascade ER-ADH-SFPR system [33].

The data obtained are summarized in Table 6.2, along with those from several similar case studies and the corresponding literature references. The coupling of enzymatic reduction with the *in situ* SFPR technology allowed a general productivity enhancement of two orders of magnitude.

In most of these examples, some of the parameters (such as resin/substrate ratio and reaction time) have been changed and optimized to achieve the highest yields and substrate loadings. Since the chiral center is established by the ER (and no DKR by the ADH is necessary), the only requirement for an effective suppression of the racemization is that the ADH be faster than the ER, which is often the case. Otherwise, it is possible to favor one reaction over the other by adjusting the ER/ADH ratio.

In summary, we demonstrated that isolated ERs and ADHs are compatible with the *in situ* SFPR technology, that the racemization of α-substituted aldehyde products can be effectively mitigated, and that a substantial increase in productivity can be achieved. Therefore, this approach opens a new perspective on the exploitation of recombinant ERs in preparative organic synthesis [29–33].

Table 6.2 *Bioreduction of α,β-unsaturated aldehydes employing the* in situ *SFPR strategy.*

Substrate	Product	Procedure	Biocatalyst	Yield (%)	ee (%)	Ref.
		1	OYE3	99	98	[30]
		1	OYE2	97	90	[29,31]
		1	OYE2	82	94	[29]
		2	OYE2 + HLADH	88	99	[29,33]
		2	OYE2 + HLADH	95	91	[33]
		2	OYE2 + HLADH	99	98	[32]
		2	OYE3 + HLADH	89	98	[31]

6.6 Reduction of Imines by Recombinant Whole-Cell *E. coli* Biocatalysts Expressing Imine Reductases (IREDs)

Friedemann Leipold, Shahed Hussain, and Nicholas J. Turner
School of Chemistry, Manchester Institute of Biotechnology, The University of Manchester, UK

Existing biocatalytic processes for enantiomerically-pure chiral amine synthesis based on transaminases, monoamine oxidases (MAOs), ammonia lyases, and lipases [36] are well established. Two emerging classes of enzymes have also been reported, including an engineered amino acid dehydrogenase (for the preparation of primary amines) [37] and imine reductases (IREDs) for the asymmetric reduction of imines to produce the

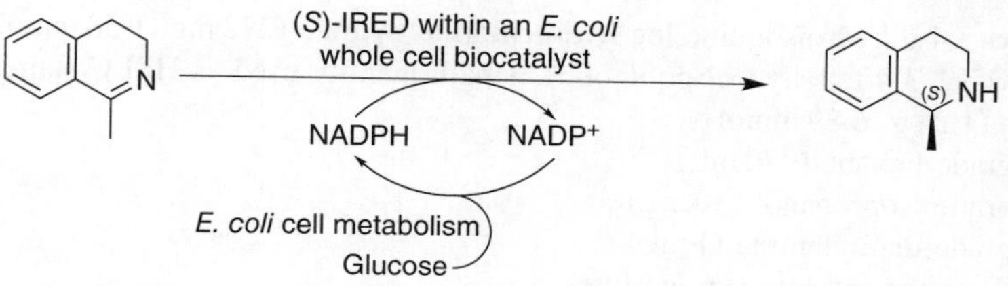

Scheme 6.10 *Biocatalytic synthesis of (S)-1-methyl-1,2,3,4-tetrahydroisoquinoline.*

corresponding secondary and tertiary amines [38,39]. The process has been applied successfully on a preparative scale.

Several IREDs have been described, originating mainly from the Gram-positive bacterial group of Actinobacteria, including two enantiocomplementary IREDs from *Streptomyces* sp. GF3546 and GF 3587 [40]. The codon-optimized genes of an (*S*)-selective IRED [(*S*)-IRED] from *Streptomyces* sp. GF3546 and an (*R*)-selective IRED [(*R*)-IRED] from *Streptomyces* sp. GF3587 were cloned into a pET-28a (+) plasmid and transformed in *E. coli* BL21(DE3). These *E. coli* recombinant whole-cell biocatalysts grow faster and express a higher level of IRED protein than *Streptomyces*. Although biocatalytic reductions catalyzed by purified IREDs have been demonstrated (using glucose-6-phosphate dehydrogenase as an NADPH-regeneration system on a small scale [41]), a non-growing *E. coli* whole-cell biocatalyst overcomes the need to add additional enzymes and cofactors by harnessing the cell's own metabolism for cofactor regeneration [42] when supplemented with glucose. The procedure can be used to reduce a wide range of imines, conceivably at any scale [38,39].

The reduction of 1-methyl-3,4-dihydroisoquinoline (Scheme 6.10) is taken as an example to illustrate the ease and efficiency of using the whole-cell IRED biocatalyst system. The experimental details are taken primarily from reference [38].

6.6.1 Biocatalytic Synthesis of (S)-1-Methyl-1,2,3,4-Tetrahydroisoquinoline

6.6.1.1 Materials and Equipment

- Expression plasmid featuring a T7 promoter such as pET-28a (+), containing a codon-optimized gene for (*S*)-IRED from *Streptomyces* sp. GF3546
- Chemocompetent or electrocompetent *E. coli* BL21(DE3) cells (in the case of electro-competent cells, an electroporator and suitable electroporation cuvettes are also required)
- LB broth miller (tryptone $10\,g.L^{-1}$, yeast extract $5\,g.L^{-1}$, NaCl $10\,g.L^{-1}$)
- LB agar miller (tryptone $10\,g.L^{-1}$, yeast extract $5\,g.L^{-1}$, NaCl $10\,g.L^{-1}$, agar $15\,g.L^{-1}$)
- Kanamycin ($30\,mg.mL^{-1}$ stock solution in water, filter-sterilized)
- IPTG (0.1 M stock solution in water, filter-sterilized)
- 60% v/v glycerol (solution in water, autoclaved)
- MTBE
- Sodium phosphate buffer (0.1 M, pH 7.0; prepared from sodium dihydrogenphosphate ($0.0372\,mol.L^{-1}$) and disodium hydrogenphosphate ($0.0627\,mol.L^{-1}$)
- Sodium hydroxide (10 M stock solution in water)
- Magnesium sulfate, anhydrous

- 1-Methyl-3,4-dihydroisoquinoline hydrochloride hydrate (272 mg, 1.38 mmol)
- 1-Methyl-1,2,3,4-tetrahydroisoquinoline (small quantity used as HPLC standard)
- Glucose (1.24 g, 6.89 mmol)
- HPLC-grade hexane (970 mL)
- HPLC-grade isopropanol (30 mL)
- HPLC-grade diethylamine (1 mL)
- 0.2 μm Cellulose acetate syringe filters
- 10 and 1 mL Syringes
- Petri dish
- Cryogenic vials, sterile
- 2 L Baffled Erlenmeyer flask; suitable bungs (cotton wool or foam)
- 50 mL Falcon tubes
- Autoclave
- Photometer capable of measuring absorbance at 600 nm
- Rotary shakers at 20, 30, and 37 °C
- Microcentrifuge capable of reaching 16 000× *g*
- Centrifuge capable of reaching 20 000× *g* while holding temperature at 4 °C
- Rotary evaporator
- HPLC system equipped with a UV detector
- CHIRALPAK IC HPLC column (5 μm, 250 × 4.6 mm; Daicel)

6.6.1.2 *Preparation of a Whole-Cell Biocatalyst*

1. A pET-28a (+) plasmid containing a gene of (*S*)-IRED from *Streptomyces* sp. GF3546 codon-optimized for expression in *E. coli* with a His$_6$ tag fused to the N-terminus was sourced (pET-His-SIRED).
2. The plasmid pET-His-SIRED was transformed into competent *E. coli* BL21(DE3) cells according to the manufacturer's instructions and the cells were spread onto LB agar plates with kanamycin (30 μg.mL^{-1}). A single colony was inoculated into 10 mL of LB medium (containing 30 μg.mL^{-1} kanamycin) in a 50 mL Falcon tube and grown overnight. 1 mL of the bacterial culture was transferred into a Cryogenic vial and 500 μL of a sterile 60% v/v glycerol solution was added. The tube was mixed gently by inverting and stored at −80 °C.
3. Crystals from a frozen glycerol stock of *E. coli* BL21(DE3)/pET-His-SIRED were inoculated into LB medium (10 mL containing 30 μg.mL^{-1} kanamycin) in a 50 mL Falcon tube and incubated overnight with shaking at 250 rpm on a rotary shaker at 37 °C.
4. A 1% inoculum derived from overnight pre-cultures was added into fresh LB medium (500 mL with kanamycin 30 μg.mL^{-1}) in a 2 L baffled Erlenmeyer flask. This culture was incubated with shaking at 250 rpm on a rotary shaker at 37 °C until an OD$_{600nm}$ of 0.6–0.8 was reached.
5. The protein expression was induced by addition of a 0.1 M IPTG solution (1 mL) to reach a final concentration of 0.2 mM. The culture was subsequently incubated with shaking at 250 rpm on a rotary shaker at 20 °C for a further 18–22 hours.
6. *E. coli* cells were transferred into a suitable centrifuge tube and pelleted by centrifugation at 4000× *g* for 15 minutes at 4 °C.
7. The *E. coli* cells were resuspended in sodium phosphate buffer (250 mL, 0.1 M, pH 7.0) and pelleted by centrifugation at 4000× *g* for 15 minutes at 4 °C. This step was repeated once to perform a total of two washing steps.

6.6.1.3 Whole-Cell Imine Reduction

1. *E. coli* cells were initially resuspended in sodium phosphate buffer (25 mL, 0.1 M, pH 7.0) and the OD_{600nm} was measured. Based on the measured OD_{600nm}, further sodium phosphate buffer was added to reach a final OD_{600nm} of 60. Note: Freshly harvested cells should be used for this biotransformation, as the use of frozen cell pellets significantly decreases conversion rates.
2. 1-Methyl-3,4-dihydroisoquinoline hydrochloride hydrate (272 mg, 1.38 mmol) and glucose (1.24 g, 6.89 mmol) were dissolved in sodium phosphate buffer (34.4 mL, 0.1 M, pH 7.0) and mixed with the previously prepared cell suspension (34.4 mL) to give the reaction mix.
3. The reaction was incubated at 30 °C with shaking at 250 rpm, and 500 µL samples were withdrawn at various time intervals for analysis. Samples were treated as described under Analytical Methods.
4. Full conversion was achieved after less than 24 hours. For recovery of the amine product, the reaction mixture was basified using 10 M NaOH (2.4 mL) until pH 12 was reached, then distributed into centrifuge tubes to about one-third of the maximum volume of each. The amine was extracted with MTBE (4 × 70 mL) and the phases were separated by centrifugation at 20 000× *g* for 10 minutes at 10 °C.
5. The pooled organic phases were dried over $MgSO_4$ and concentrated *in vacuo* to yield (*S*)-1-methyl-1,2,3,4-tetrahydroisoquinoline as a viscous, yellow liquid (176 mg, 87% yield, *ee* 98% (*S*)).

6.6.2 Analytical Methods

Sample aliquots of 500 µL from the biotransformation were basified with 30 µL 10 M NaOH and extracted twice using an equal volume of MTBE (2 × 500 µL). The mixtures were vortexed for 1 minute and microcentrifuged at 16 000× *g* for 5 minutes. The organic phase was transferred into a new microtube and dried over $MgSO_4$, and 10 µL of each sample was injected for HPLC analysis.

The HPLC system used a mobile phase consisting of hexane/isopropanol/diethylamine (97 : 03 : 0.1 v/v/v). Quantitation of standards and samples was achieved by isocratic elution over a CHIRALPAK IC (5 µm, 250 × 4.6 mm) column at a flow rate of 1 mL.min^{-1} at room temperature, and the absorbance was detected at 265 nm. HPLC retention times and for standards were as follows: 1-methyl-3,4-dihydroisoquinoline, 17.1 min; (*S*)-1-methyl-1,2,3,4-tetrahydroisoquinoline, 9.8 min; (*R*)-1-methyl-1,2,3,4-tetrahydroisoquinoline, 10.7 min. For calculation of conversions, the fact that the relative response of the imine is 9.103 times higher than that of the amine at 265 nm should be taken into account.

Other suitable imine substrates include 2-substituted pyrrolines, piperideines, and tetrahydroazepines, as well as dihydro-β-carbolines [38].

6.6.3 Conclusion

1-Methyl-3,4-dihydroisoquinoline was fully reduced by a resting *E. coli* biocatalyst containing (*S*)-IRED after overnight reaction to give (*S*)-1-methyl-1,2,3,4-tetrahydroisoquinoline. This reaction can also be applied to other substrates described for the IRED. This is an economic process, as glucose is added, avoiding the need for an external NADPH

Table 6.3 Whole-cell biotransformation of imines by (S)-IRED from Streptomyces *sp.*
GF3546 [38] and (R)-IRED from Streptomyces *sp. GF3587 [39].*

	(S)-IRED [38]			(R)-IRED [39]		
	Product config.	ee (%)	Conversion (%)	Product config.	ee (%)	Conversion (%)
	S	57	>98	R	>98	>98
	S	>98	>98	R	>98	>98
	R	>98	17	S	>98	20
	S	>98	>98	R	>98	>98
	n/a	n/a	>90	n/a	n/a	97
	S	98	>98	R	71	>98
	S	>98	>98			0

regeneration system. A comparable whole-cell biocatalyst system was also successfully applied to other IREDs, such as the (R)-IRED from *Streptomyces* sp. GF 3587. However, the applicability of this system may be limited to cyclic imines which do not contain ketone or aldehyde moieties as substrates, due to the possibility of background reactions from *E. coli's* endogenous dehydrogenases, which may reduce these functionalities to their corresponding alcohols. The broad substrate specificity of both the (S)-IRED and the (R)-IRED enables the wide application of these biocatalysts for the reduction of 5-, 6-, and 7-membered cyclic imines with a variety of substituents, as well as iminium ions. An exemplary selection of substrates, together with the conversions achieved, is displayed in Table 6.3. A more comprehensive chapter on biotransformations using IREDs was published recently [43].

References

1. Firn, R.D. and Jones, C.G. (2003) *Natural Product Reports*, **20**, 382–391.
2. Wheeler, M.H. and Stipanovic, R.D. (1985) *Archives of Microbiology*, **142**, 234–241.
3. Wheeler, M.H., Abramczyk, D., Puckhaber, L.S. *et al.* (2008) *Eukaryotic Cell*, **7**, 1699–1711.
4. Zhang, Y.L., Zhang, J., Jiang, N. *et al.* (2011) *Journal of the American Chemical Society*, **133**, 5931–5940.
5. Cai, Y.-S., Guo, Y.-W., and Krohn, K. (2010) *Natural Product Reports*, **27**, 1840–1870.
6. Schätzle, M.A., Flemming, S., Husain, S.M. *et al.* (2012) *Angewandte Chemie-International Edition in English*, **51**, 2643–2644.
7. Husain, S.M., Schätzle, M.A., Röhr, C. *et al.* (2012) *Organic Letters*, **14**, 3600–3603.
8. Schätzle, M.A., Husain, S.M., Ferlaino, S., and Müller, M. (2012) *Journal of the American Chemical Society*, **134**, 14742–14745.
9. (a) Bates, R.W. and Sa-Ei, K. (2002) *Tetrahedron*, **58**, 5957–5978; (b) Asano, N., Nash, R.J., Molyneux, R.J., and Fleet, G.W.J. (2000) *Tetrahedron: Asymmetry*, **11**, 1645–1680.
10. (a) Girling, P.R., Kiyoi, T., and Whitning, A. (2011) *Organic and Biomolecular Chemistry*, **9**, 3105–3121; (b) Buffat, M.G.P. (2004) *Tetrahedron*, **60**, 1701–1729; (c) Bailey, P.D., Millwood, P.A., and Smith, P.D. (1998) *Chemical Communications*, 633–640.
11. (a) Simon, R.C., Grischek, B., Zepeck, F. *et al.* (2012) *Angewandte Chemie-International Edition in English*, **51**, 6713–6716; (b) Simon, R.C., Zepeck, F., and Kroutil, W. (2013) *Chemistry - A European Journal*, **19**, 2859–2865; (c) Simon, R.C., Fuchs, C.S., Lechner, H., Zepeck, F., and Kroutil, W. (2013) *European Journal of Organic Chemistry*, 3397–3402.
12. (a) Mathew, S. and Yun, H. (2012) *ACS Catalysis*, **2**, 993–1001; (b) Malik, M.S., Park, E.-S., and Shin, J.-S. (2012) *Applied Microbiology and Biotechnology*, **94**, 1163–1171; (c) Koszelewski, D., Tauber, K., Faber, K., and Kroutil, W. (2010) *Trends in Biotechnology*, **28**, 324–332; (d) Hailes, H.C., Dalby, P.A., Lye, G.J., Baganz, F., Micheletti, M., Szita, N., and Ward, J.M. (2010) *Current Organic Chemistry*, **14**, 1883–1893; (e) Ward, J. and Wohlgemuth, R. (2010) *Current Organic Chemistry*, **14**, 1914–1927.
13. (a) Schneider, M.J., Montali, J.A., Hazen, D., and Stanton, C.E. (1991) *Journal of Natural Products*, **54**, 905–909; (b) Tawara, J.N., Blokhin, A., Foderaro, T.A., and Stermitz, F.R. (1993) *The Journal of Organic Chemistry*, **58**, 4813–4818.
14. Kaulmann, U., Smithies, K., Smith, M.E.B. *et al.* (2007) *Enzyme and Microbial Technology*, **41**, 628–637.
15. Yamada, Y., Iwasaki, A., and Kizaki, N. (2000) (Kaneka Corporation) EP 0987332 A1.
16. Mutti, F.G., Fuchs, C.S., and Kroutil, W. (2012) Synthesis of optically pure amines employing ω-transaminases, in *Practical Methods for Biocatalysis and Biotransformations*, vol. 2 (eds J. Whitall and P.W. Sutton), John Wiley & Sons, Ltd, Chichester, West Sussex (UK), pp. 69–74.
17. Klibanov, A.M. (2001) *Nature*, **409**, 241–246.
18. Mutti, F.G. and Kroutil, W. (2012) *Advanced Synthesis and Catalysis*, **354**, 3409–3413.
19. Truppo, M.D., Strotman, H., and Hughes, G. (2012) *ChemCatChem*, **4**, 1071–1074.
20. Mutti, F.G., Fuchs, C.S., Pressnitz, D. *et al.* (2011) *Advanced Synthesis and Catalysis*, **353**, 3227–3233.
21. Koszelewski, D., Göritzer, M., Clay, D. *et al.* (2010) *ChemCatChem*, **2**, 73–77.
22. Eriksen, J.L., Sagi, S.A., Smith, T.E. *et al.* (2003) *The Journal of Clinical Investigation*, **112**, 440–449.
23. Morihara, T., Chu, T., Ubeda, O. *et al.* (2002) *Journal of Neurochemistry*, **83**, 1009–1012.
24. Pietruszka, J. and Schölzel, M. (2012) *Advanced Synthesis and Catalysis*, **354**, 751–756.
25. Classen, T., Pietruszka, J., and Schuback, S.M. (2013) *ChemCatChem*, **5**, 711–713.
26. Csuk, R. and Glänzer, B.I. (1991) *Chemical Reviews*, **91**, 49–97.

27. Anderson, B.A., Hansen, M.M., Harkness, A.R. *et al.* (1995) *Journal of the American Chemical Society*, **117**, 12358–12359.
28. Toogood, H., Gardiner, J.M., and Scrutton, N.S. (2010) *ChemCatChem*, **2**, 892–914.
29. Brenna, E., Gatti, F.G., Monti, D. *et al.* (2012) *Chemical Communications*, **48**, 79–81.
30. Bechtold, M., Brenna, E., Femmer, C. *et al.* (2012) *Organic Process Research & Development*, **16**, 269–276.
31. Brenna, E., Gatti, F.G., Monti, D. *et al.* (2012) *ChemCatChem*, **4**, 653–659.
32. Parmeggiani, F. (2011) *Milestones in Chemistry (Suppl to Chimica Oggi/Chem Today)*, **29**, 13–15.
33. Brenna, E., Gatti, F.G., Malpezzi, L. *et al.* (2013) *The Journal of Organic Chemistry*, **78**, 4811–4822.
34. Linderberg, M.T., Moge, M., and Sivadasan, S. (2004) *Organic Process Research & Development*, **8**, 838–845.
35. Brenna, E., Fuganti, C., Gatti, F.G., and Parmeggiani, F. (2009) *Tetrahedron: Asymmetry*, **20**, 2694–2698.
36. Turner, N.J. and Truppo, M.D. (2010) *Chiral Amine Synthesis*, Wiley-VCH Verlag GmbH & Co. KGaA, pp. 431–459.
37. Abrahamson, M.J., Vázquez-Figueroa, E., Woodall, N.B. *et al.* (2012) *Angewandte Chemie International Edition*, **51**, 3969–3972.
38. Leipold, F., Hussain, S., Ghislieri, D., and Turner, N.J. (2013) *ChemCatChem*, **5**, 3505–3508.
39. Hussain, S., Leipold, F., Man, H. *et al.* (2015) *ChemCatChem*, **7**, 579–583.
40. Mitsukura, K., Suzuki, M., Tada, K. *et al.* (2011) *Organic & Biomolecular Chemistry*, **8**, 4533–4535.
41. Rodríguez-Mata, M., Frank, A., Wells, E. *et al.* (2013) *ChemBioChem*, **14**, 1372–1379.
42. Walton, A.Z. and Stewart, J.D. (2004) *Biotechnology Progress*, **20**, 403–411.
43. Leipold, F., Hussain, S., France, S.P., and Turner, N.J. (2015) *Science of Synthesis: Biocatalysis in Organic Synthesis 2*, vol. 2 (eds K. Faber, W.-D. Fessner, and N.J. Turner), Thieme Verlag, Stuttgart, New York, pp. 359–381.

7

Halogenation and Dehalogenation

7.1 Site-Directed Mutagenesis Changes the Regioselectivity of the Tryptophan 7-Halogenase PrnA

Alexander Lang,[1] Stefan Polnick,[1] Tristan Nicke,[1] Peter William,[1] Eugenio P. Patallo,[1] James H. Naismith,[2] and Karl-Heinz van Pée[1]

[1]General Biochemistry, Dresden University of Technology, Germany
[2]Centre for Biomolecular Science, University of St Andrews, UK

The occurrence of many halogenated metabolites suggests that biology possesses halogenating enzymes with a high degree of substrate specificity and the ability to introduce halogen atoms regioselectively.

The first halogenase shown to catalyze the regioselective chlorination or bromination of tryptophan was the tryptophan 7-halogenase PrnA, involved in pyrrolnitrin biosynthesis [1]. PrnA was identified as a flavin-dependent halogenase requiring a flavin reductase as a second enzyme component to produce FADH$_2$ from FAD and NADH (Scheme 7.1) [2]. When FADH$_2$ is bound by PrnA, it reacts with molecular oxygen to form a flavin hydroperoxide, which is attacked by a single chloride or bromide ion bound near the isoalloxazine ring, leading to the formation of hypohalous acid [3,4]. A lysine (K79) and a glutamate (E346) residue are located close to the substrate. Yeh et al. [3] suggested a reaction between HOCl and K79, resulting in the formation of chloroamine as the chlorinating intermediate. However, Flecks et al. [4,5] could demonstrate that E346 is also absolutely necessary for enzyme activity. Furthermore, Flecks et al. [5] suggested that a concerted interaction of hypochlorous acid with the lysine and the glutamate residue should increase the electrophilicity of the chlorine species and ensure the correct positioning of the chlorine species for regioselective incorporation into the indole ring of tryptophan. The regioselectivity of tryptophan halogenases is regulated by the orientation of the substrate in the active site. This was confirmed by comparison of the

Practical Methods for Biocatalysis and Biotransformations 3, First Edition.
Edited by John Whittall, Peter W. Sutton, and Wolfgang Kroutil.
© 2016 John Wiley & Sons, Ltd. Published 2016 by John Wiley & Sons, Ltd.

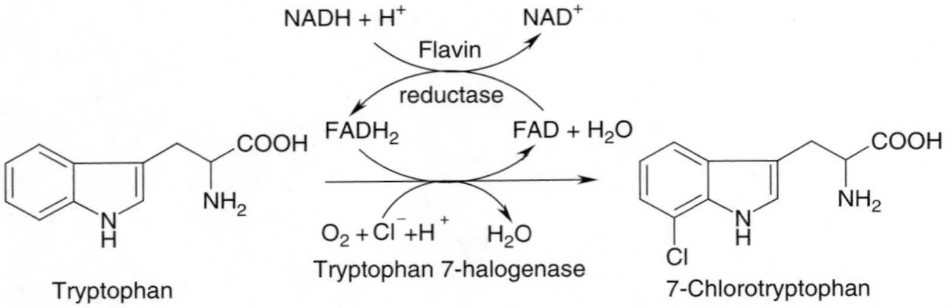

Scheme 7.1 *Reaction catalyzed by the two-component system of flavin-dependent halogenases.*

three-dimensional structure of the substrate (tryptophan) complexes of tryptophan 7-halogenase PrnA with that of the tryptophan 5-halogenase PyrH [6]. In PrnA, the 7-position of the indole ring points into the tunnel and is the only position accessible to the chlorine species. The other reactive positions of the indole ring are shielded by large aromatic amino acids (W455 and F103), which sandwich the substrate between them. In the tryptophan 5-halogenase PyrH, the 5-position of the indole ring is exactly at the same position as the 7-position in PrnA.

The obvious starting point in altering the regioselectivity of the chlorination of tryptophan was the large aromatic amino acid residues that sandwich and shield the indole ring, namely H101, F103, and W455. Each of these amino acids was exchanged with an alanine residue, and a double mutant W455A and F103A was constructed. Determination of the kinetic data of the purified His-tagged enzyme variants (Table 7.1) showed that PrnAF103A had quite similar kinetic properties like the His-tagged form of the wild-type enzyme. The HPLC-chromatogram of the reaction mixture of PrnAF103A showed 7-chlorotryptophan as the main product with a shoulder indicating the formation of a second product. The separation of the main product from the second, unknown product was more pronounced when using bromide rather than chloride as the halogenating substrate (Figure 7.1).

A comprehensive overview is given in Reference [7], from which most of the experimental details are taken.

7.1.1 Materials and Equipment

- *E. coli* TG1 pUC19*prnA* and pBSK containing *prnA* and its mutated variants (see Procedure)
- *Pseudomonas fluorescens* BL915 ΔORF 1–4 pCIBHis containing mutated *prnA* variants (see Procedure)
- LB medium for *E. coli*
- HNB medium for gene expression in *Pseudomonas* sp.
- Ampicillin (100 µg·mL^{-1}), tetracyclin (15 µg·mL^{-1}) for *E. coli*

Table 7.1 K_m and k_{cat} values of His-tagged wild-type PrnA and of some mutant forms [7].

Enzyme variant	K_m (µM)	k_{cat} (min^{-1})
HisPrnA	49.9 ± 5.2	6.79 ± 0.27
HisPrnAF103A	198.6 ± 27.3	3.99 ± 0.21
HisPrnAH101A	1785.1 ± 117.6	1.26 ± 0.05
HisPrnAW455A	1814.2 ± 382.1	1.05 ± 0.12
HisPrnAF103AW455A	inactive	inactive

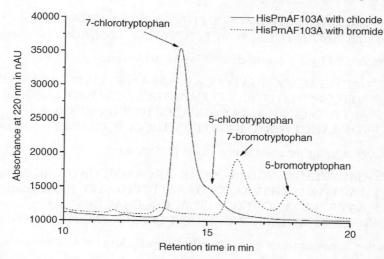

Figure 7.1   *HPLC analysis of the activity assay of HisPrnAF103A, using either chloride or bromide as the halogen source.*

- Ampicillin ($100\,\mu g \cdot mL^{-1}$), tetracyclin ($30\,\mu g \cdot mL^{-1}$), and kanamycin ($50\,\mu g \cdot mL^{-1}$) for *Pseudomonas* sp.
- Primers (see Procedure)
- Restriction enzymes
- Ni-chelating sepharose FF
- Enzyme assay components (see Procedure)
- HPLC-grade methanol
- HPLC-grade water
- TFA
- Laminar flow bench
- Rotary shakers at 30 and 37 °C
- Electroporator
- Centrifuge
- Chromatography column, pump, and UV detector for Ni-chelating chromatography
- HPLC system and UV detector
- Solid phase extraction column (Strata C 18-E, 1000 mg, Phenomenex)
- Petri dishes
- Stainless steel-capped DeLong flasks
- Microliter syringes

7.1.2   Procedure

1. The His-tagged PrnAH101A, PrnAF103A, and PrnAW455A single mutants were constructed by overlap extension PCR using pUC-*prnA* as template. Note: The mutagenic sites are underlined in the primers and restriction sites are in bold.

The primers used for the construction of PrnAH101A were:

primer a: 5′-GACTCTAGAGG**GGGATCCC**ATGAACAAGCCGATCAAG-3′ (sense)
primer b: 5′-CACGTTGCCGAACAA<u>GGC</u>GTAGAAGTGATCGTC-3′ (antisense)

primer c: 5′-GACGATCACTTCTAC<u>GCC</u>TTGTTCGGCAACGTG-3′ (sense)
primer d: 5′-ATC**AAGCTT**CTACAGGCTTTCCTGCGCTGCGAGCTT-3′ (antisense)

The primers used for the construction of PrnAF103A were:

primer a: 5′-GACTCTAGAGG**GGATCC**CATGAACAAGCCGATCAAGAAT-3′ (sense)
primer b: 5′-GTTCGGCACGTTGCC<u>GC</u>CCAAATGGTAGAAGT-3′ (antisense)
primer c: 5′-ACTTCTACCATTTG<u>GCC</u>GGCAACGTGCCGAAC-3′ (sense)
primer d: 5′-ATC**AAGCTT**CTACAGGCTTTCCTGCGCTGCGAGCTT-3′ (antisense)

The primers used for the construction of PrnW455A were:

primer a: 5′-GACTCTAGAGG**GGATCC**CATGAACAAGCCGATCAAG-3′ (sense)
primer b: 5′-TAGTTGCCGGTTCAAC<u>GCG</u>AAATTCTTGAATT-3′ (antisense)
primer c: 5′-AATTCAAGAATTT<u>CGC</u>GTTGAACGGCAACTA-3′ (sense)
primer d: ATC**AAGCTT**CTACAGGCTTTCCTGCGCTGCGAGCTT-3′ (antisense)

2. The fusion fragments were ligated into pBluescript II SK (+) and introduced into *E. coli* by electroporation.

3. For construction of the PrnAF103AW455A double mutant, the fragments containing the mutated genes were isolated from the respective pBluescript II SK (+) derivatives using *Eco*130I and *Bsh*TI. All mutants were confirmed by DNA sequencing. For expression, the halogenase genes were ligated into the *E. coli–Pseudomonas* shuttle vector pCIBHis and introduced into *Pseudomonas fluorescens* BL915 ΔORF 1–4 by conjugation.

4. His-tagged enzymes were purified using a Ni-chelating sepharose FF column.

5. One unit of halogenating activity is defined as 1 μmol product formed per minute. The reaction mixture was composed of enzyme-containing protein solution (50 μL), 10 mU FAD reductase, 1 μM FAD, 2.4 mM NADH, 12.5 mM MgCl$_2$, 100 U catalase, and 0.25 mM tryptophan in a total volume of 200 μL in 10 mM potassium phosphate buffer, pH 7.2. After incubation at 30 °C for 30 minutes, the reaction was stopped by boiling for 5 minutes. After removal of the precipitated protein, the assay mixture was analyzed by HPLC.

6. For determination of K_m and k_{cat} values, tryptophan concentrations were varied between 0.025 and 2 mM. Experiments were done in quadruplicate. The calculated rates were analyzed by linear and non-linear regression analysis methods. The presented K_m and k_{cat} values were originated from a hyperbola fit function (Michaelis–Menten equation) approximated by 30–50 cycles of 200 Levenberg–Marquardt iterations.

7. The reaction mixture for the large-scale preparation for the identification of the reaction products formed by the PrnAF103A variant comprised enzyme-containing protein solution (850 μL) in 10 mM potassium phosphate buffer, pH 7.2, 50 mU FAD reductase, 10 μM FAD, 2.4 mM NADH, 12.5 mM NaBr, and 0.25 mM tryptophan, in a total volume of 1000 μL.

8. After incubation at 30 °C for 4 hours, the reaction was stopped by boiling in a water bath for 5 minutes. The protein was removed by centrifugation and the resulting supernatant was loaded onto a solid phase extraction column (Strata C 18-E, 1000 mg, Phenomenex).

9. The column had been equilibrated with methanol and water. After washing with 10% methanol, the halogenated substances were eluted with 100% methanol. The eluates of 10 large-scale preparations were concentrated *in vacuo*.

10. The two halogenated species were separated and purified by HPLC (LiChrospher 100 RP-18, 5 µm, 250 × 4 mm, methanol/water 40 : 60, 0.1% v/v TFA, flow rate: 1.0 mL·min^{-1}).
11. The reaction products were identified by ^{1}H-NMR and ESI-MS.

7.1.3 Product ^{1}H-NMR and ESI-MS Data

^{1}H-NMR spectra were recorded with Avance 600, Bruker. ESI/APCI-MS data were acquired by LC/MS equipment HP 1100 – Bruker Esquire Ion Trap.

7-Br-Trp: ^{1}H NMR (600 MHz, D$_2$O) δ (in ppm) 3.17–3.22 (dd, ^3J = 15.4 Hz, ^2J = 8.0 Hz, 1H, β-CH$_2$), 3.33–3.37 (dd, ^3J = 15.4 Hz, ^2J = 4.9 Hz, 1H, β-CH$_2$), 3.93–3.96 (dd, $^{3a,b}J$ = 7.7 Hz, $^{3a,b}J$ = 5.0 Hz, 1H, α-CH), 6.98–7.00 (t, ^3J = 7.8 Hz, 1H, C5 indole), 7.27 (s, 1H, C2 indole), 7.36 (d, ^3J = 7.6 Hz, 1H, C6 indole), 7.60 (d, ^3J = 8.0 Hz, 1H, C4 indole); ESI-MS *m/z* 283.0, 285.0 [M + H$^+$], *m/z* 305.0, 307.0 [M + Na$^+$]

5-Br-Trp: ^{1}H NMR (600 MHz, D$_2$O) δ (in ppm) 3.16–3.24 (dd, ^3J = 15.4 Hz, ^2J = 8.0 Hz, 1H, β-CH$_2$), 3.32–3.36 (dd, ^3J = 15.4 Hz, ^2J = 4.9 Hz, 1H, β-CH$_2$), 3.92–3.95 (dd, $^{3a,b}J$ = 7.7 Hz, $^{3a,b}J$ = 5.0 Hz, 1H, α-CH), 7.18 (s, 1H, C2 indole), 7.18–7.20 (dd, ^3J = 10.0 Hz, ^4J = 1.5 Hz, 1H, C6 indole), 7.51 (d, ^3J = 8.6 Hz, 1H, C7 indole), 7.61 (d, ^4J = 1.3 Hz, 1H, C4 indole); ESI-MS *m/z* 283.0, 285.0 [M + H$^+$], *m/z* 305.0, 307.0 [M + Na$^+$]

7.1.4 Conclusion

This work is the first demonstration that the regioselectivity of tryptophan halogenases can be engineered by site-directed mutagenesis. In this example, a mixture of products halogenated at two different positions was obtained at a rate comparable to the native enzyme.

A large-scale reaction in the presence of bromide was performed and HPLC analysis showed that the two products were formed in a ratio of about 2 : 1. According to HPLC-MS analysis, both products were monobrominated tryptophan. ^{1}H-NMR analysis of the purified products revealed that the main product was 7-bromotryptophan and the new product was identified as 5-bromotryptophan. The exchange of the large amino acid F103 with the smaller alanine residue must allow the substrate to adopt a different conformation when bound to the protein.

Further experiments can now be designed to modify the regioselectivity and to change the substrate specificity of tryptophan halogenases to enhance the biotechnological potential of halogenases.

7.2 Controlling Enantioselectivity of Halohydrin Dehalogenase from *Arthrobacter* sp. Strain AD2, Revealed by Structure-Guided Directed Evolution

Lixia Tang and Xuechen Zhu
School of Life Science and Technology, University of Electronic Science and Technology of China, China

Halohydrin dehalogenases catalyze the dehalogenation of vicinal halohydrins and the epoxide ring-opening reaction with a series of anion nucleophiles, such as CN$^-$, N$_3^-$, NO$_2^-$, and OCN$^-$ [8–10]. HheC from *Agrobacterium radiobacter* AD1 shows great potential in the

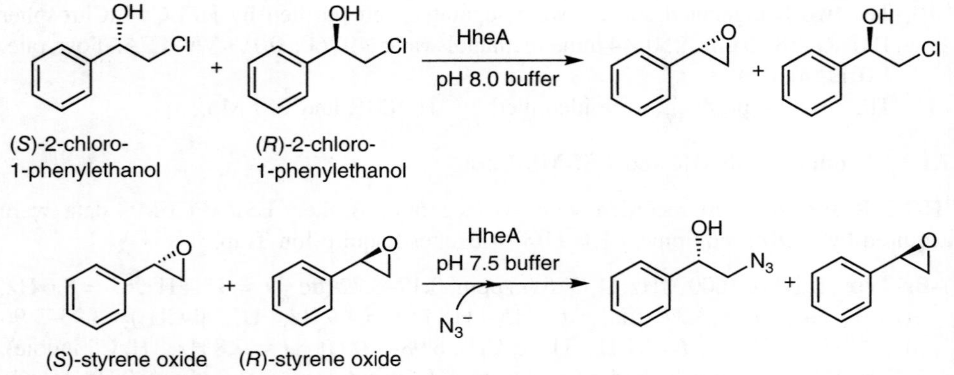

(S)-2-chloro-1-phenylethanol   **(R)-2-chloro-1-phenylethanol**

(S)-styrene oxide   (R)-styrene oxide

Scheme 7.2   *Sequential reaction catalyzed by HheA.*

preparation of (*R*)-enantiomers of epoxides and β-substituted alcohols, while HheA from *Arthrobacter* sp. AD2 shows low or modest (*S*)-enantioselectivity and low activity in most cases [11]. The availability of the crystal structure of the two enzymes has led to the identification of three key residues (V136, L141, N178) for controlling the enantio-selectivity of HheA by structure comparison. Through a saturation mutagenesis approach, two outstanding HheA mutants were obtained (Scheme 7.2). One shows an inverted enantioselectivity (from $E_S = 1.7$ to $E_R = 13$) toward 2-chloro-1-phenylethanol without compromising enzyme activity, while the other displays excellent enantioselectivity ($E_S > 200$) and a five- to six-fold enhanced specific activity toward (*S*)-2-chloro-1-phenylethanol [12]. This is an example to show that the active-site residues are crucial in modulating enzyme enantioselectivity. A comprehensive description is given in refer-ence [12], which is also the source of the following experimental details.

7.2.1   Materials and Equipment

- *E. coli* MC1061 (frozen glycerol stocks)
- Recombinant pBADHheA plasmid (frozen stocks)
- Phusion high-fidelity PCR kit (New England BioLabs)
- LB medium (10 g tryptone, 5 g yeast extract, and 10 g NaCl in each 1 L LB medium)
- LB agar medium (10 g tryptone, 5 g yeast extract, 10 g NaCl, and 12 g agar in each 1 L LB medium)
- Ampicillin (100 μg·mL^{-1} stock solution in water, filter-sterilized)
- L-Arabinose (0.2 g·mL^{-1} stock solution in water, filter-sterilized)
- Sterile 96-well microplates
- 1,3-Dicholoro-2-propanol (1,3-DCP) (5 mM solution: 10 μL in 20 mL solvent)
- Racemic 2-chloro-1-phenylethanol (*rac*-2-CPE) (5 mM solution: 13.2 μL in 20 mL water)
- (*R*)-2-Chloro-1-phenylethanol (*R*)-2-CPE) (2.5 mM solution: 6.6 μL in 20 mL water)
- (*S*)-2-Chloro-1-phenylethanol (*S*)-2-CPE) (2.5 mM solution: 6.6 μL in 20 mL water)
- Racemic styrene oxide (5 mM solution: 11.5 μL in 20 mL water)
- (*S*)-Styrene oxide (2.5 mM solution: 5.8 μL in 20 mL water)
- NaN$_3$ (200 mM stock solution in water)

- TIANprep Mini Plasmid Kit (Tiangen, China)
- Buffer A (10 mM Tris, 10% v/v glycerol, 70 µL β-mercaptoethanol, pH 7.5), Buffer B (10 mM Tris, 10% v/v glycerol, 70 µL β-mercaptoethanol, 0.5 M $(NH_4)_2SO_4$, pH 7.5) for AKTA purifier
- Tris-SO_4 buffer (200 mM Tris, pH 7.5 and pH 8.0; 50 mM Tris, pH 7.5) for enzymatic reactions
- Reagent I (30 mM $NH_4Fe(SO_4)_2$ in 1 M HNO_3), Reagent II (saturated solution of Hg $(SCN)_2$ in absolute ethanol) for steady-state kinetic parameter determination
- PCR amplifier (Bio-RAD)
- Electrophoresis system (GE)
- Centrifuge (Eppendorf)
- Refrigerated centrifuge holding in 4 °C
- Rotary shakers
- AKTA purifier system (GE)
- UV/Vis spectrophotometer
- Thermostatic water bath
- Gas chromatographer
- β-Dex 225 chiral GC column (30 m × 0.25 mm × 0.25 µm; Supelco)

7.2.2 Enzyme Engineering of HheA

1. Saturation mutagenesis libraries were constructed using a Phusion high-fidelity PCR kit (New England BioLabs), with a protocol of 3 minutes at 98 °C; 30 cycles of 10 seconds at 98 °C, 45 seconds at 50 °C, and 2 minutes at 72 °C; 10 minutes at 72 °C. The pBADHheA plasmid was used as the template in PCR amplification. The products of PCR were detected using electrophoresis and transformed into the chemical competent E. coli MC1061.
2. Saturation mutagenesis libraries were screened using a previously developed colorimetric activity assay on a 96-well plate format with whole cells [13]. Protein expression was induced by adding 0.002% L-arabinose (w/v) when the culture reached an OD_{600} of 0.1. After incubation at 37 °C for 15 hours, strains of E. coli MC1061 carrying recombinant plasmids were prepared as previously described. 5 mM 1,3-DCP was used in the first-round assay for activity, whereas 5 mM optical (R)-2-CPE and (S)-2-CPE were used in the second-round assay for enantioselectivity. 1000 colonies of the library containing two randomized sites and 90 colonies of the library containing a single randomized site were screened.
3. Positive variants were sorted for sequence analysis and the mutant plasmids were extracted from whole cells using TIANprep Mini Plasmid Kit (Tiangen, China).

7.2.3 Enyzme Expression and Purification

1. For preparation of proteins for kinetic resolution studies, strains of E. coli MC1061 carring pBADHheA and recombinant plasmids were cultivated in LB medium containing 100 µg·mL^{-1} ampicillin (20 mL) at 37 °C for 15 hours. 0.005% L-arabinose (w/v) was added when the culture reached an OD_{600} of 0.1, in order to induce the protein expression. Cells were harvested by centrifugation (5000× g, 50 minutes, 4 °C) and washed with 200 mM Tris-SO_4 buffer (pH 8.0 for dehalogenation reaction, pH 7.5 for ring-opening reaction), then the proteins were treated by ultrasonication. The supernatants were harvested by centrifugation (18 000× g, 60 minutes, 4 °C).

2. For the preparation of proteins for steady-state parameter measurement, *E. coli* MC1061 carrying pBADHheA and recombinant plasmids were cultivated in LB medium (1 L) at 37 °C for 15 hours. 0.005% L-arabinose (w/v) was added to induce protein expression. Cells were harvested by centrifugation (5000× *g*, 50 minutes) and washed with 10 mM Buffer A (pH 7.5). After crude extraction by ultrasonication and centrifugation, the proteins were purified by Q-Sepharose (20 mL, GE Health Care) column, as described previously [14].

3. Expression levels of crude extracts and the purities of purified proteins were estimated using SDS-PAGE. The concentrations of proteins were determined using Bradford's method.

7.2.4 Enzymatic Kinetic Resolution and Steady-State Kinetic Measurement

1. Kinetic resolution experiments were performed using crude extracts of both wild-type HheA and its variants in a sealed container at 30 °C. The kinetic resolution of 5 mM *rac*-2-CPE in Tris-SO₄ buffer (18 mL, 200 mM, pH 8.0) was started by addition of each crude extract (2 mL containing 5 mg protein in total). The kinetic resolution of 5 mM *rac*-styrene oxide used 5 mM NaN₃ as the nucleophile and was carried out in 200 mM Tris-SO₄ buffer (pH 7.5) with crude extract (2 mL with 25 mg protein for wild-type HheA, due to the low activity of the enzyme, and 5 mg for the variant).

2. The enzymatic conversions were monitored by periodically taking samples (0.5 mL each) from the reaction mixture. The samples were extracted with ethyl acetate (1 mL) containing mesitylene as an internal standard and were dried with Na_2SO_4. Samples were analyzed as described under Analytical Methods.

3. The kinetic parameters of wild-type and improved HheA toward both enantiomers of 2-CPE were determined in 50 mM Tris-SO₄ buffer (pH 8.0) at 30 °C using purified enzymes [15]. Less than 0.5% v/v dimethyl sulfoxide (DMSO) was used to facilitate the solubilization of the substrates. The total volume of the reaction system was 5 mL. A certain amount of enzyme was added to initialize the reaction. Samples (0.25 mL each) were taken periodically from the reaction mixture and treated and analyzed as described under Analytical Methods.

7.2.5 Analytical Methods

Each sample of the kinetic resolution reactions (2 µL) was analyzed by HPLC using a β-dex 225 chiral GC column (30 m × 0.25 mm × 0.25 µm; Supelco) under the following conditions: 6 minutes at 100 °C, 10 °C·min⁻¹ to 170 °C, 15 minutes at 170 °C. The peaks of (*R*)-2-CPE, (*S*)-2-CPE, and (*S*)-styrene oxide were determined by comparison to optically pure authentic standards. E values were calculated from the *ee* of the substrate ($ee_{(s)}$) and the product ($ee_{(p)}$) according to the formula $E = \ln((1 - ee_{(s)})/(1 + ee_{(s)}/ee_{(p)}))/\ln((1 + ee_{(s)})/(1 + ee_{(s)}/ee_{(p)}))$.

The kinetic parameters of wild-type and mutant HheA toward (*R*)-2-CPE and (*S*)-2-CPE were determined by monitoring chloride ion liberation by UV spectroscopy. A calibration curve was generated at 460 nm using aqueous NaCl solutions of known concentration. Samples of steady-state kinetic measurement were immediately mixed with 0.9 mL reagent I to stop the enzymatic reaction. Then reagent II (0.1 mL) was added and mixed well. After 15 minutes standing, the absorbance of the mixture at 460 nm was measured. The [Cl⁻] values of each sample were calculated based on the calibration curve. Initial rates were determined from the initial linear part of the reaction progress curves by varying the concentration of the substrate.

Kinetic data were fitted with the Michaelis–Menten equation using the Sigmaplot software program (version 10.0) to obtain the steady-state kinetic parameters. All k_{cat} values were defined as per monomer.

7.2.6 Conclusion

The improved HheAN178A mutant enzyme could transform the toxic vicinal halohydrin 2-CPE to reactive epoxide styrene oxide with a high enantioselectivity (E value > 200). It displayed $ee_{(S)}$ 98% and $ee_{(P)}$ 97% in the enzymatic kinetic resolution reaction. It also presented five- to six-fold activity improvement toward (S)-2-CPE in the enzymatic dehalogenation reaction. Moreover, the HheAN178A enzyme also catalyzed the nucleophilic substitution at β-C of the epoxide with high selectively (E value > 100). The HheAV136Y/L141G mutant enzyme presented a good R-type preference (E value = 13.3) and three-fold activity improvement toward 2-CPE.

The improved enantioselectivity and activity of the HheAN178A mutant enzyme enabled the wide application of this biocatalyst in the reaction of halohydrins, enantiomeric resolution of corresponding halohydrins and epoxides, and asymmetric synthesis of β-substituted alcohols.

7.3 Enzymatic Production of Chlorothymol and its Derivatives by Halogenation of the Phenolic Monoterpenes Thymol and Carvacrol with Chloroperoxidase

Thomas Krieg, Laura Getrey, Jens Schrader, and Dirk Holtmann
DECHEMA Research Institute, Germany

Halogenated natural products are widely distributed in nature, some of them being potent biologically active substances. Incorporation of halogen atoms into drugs is a common strategy by which to modify molecules in order to vary their bioactivities and specificities [16]. Chemical halogenation, however, often requires harsh reaction conditions and results in unwanted byproduct formation. Therefore, it is of great interest to investigate the biosynthesis of halogenated natural products and the biotechnological potential of halogenating enzymes [16].

Chlorothymol is a chlorinated phenolic antiseptic used as an ingredient in preparations for hand and skin disinfection and topical treatment of fungal infections. It has also been used in preparations for anorectal disorders, cold symptoms, and mouth disorders. Chlorothymol has a 75 times higher bactericidal potency, while also possessing low human toxicity, compared to thymol.

Among the various halogenating agents established, *in situ*-generated elementary halogens or hypohalogenites are the environmentally least harmful [17]. In this section, we describe the enzymatic generation of the reactive hypohalogenite by use of chloroperoxidase (CPO) from *Caldariomyces fumago* for the electrophilic chlorination of thymol and its derivative carvacrol (Scheme 7.3).

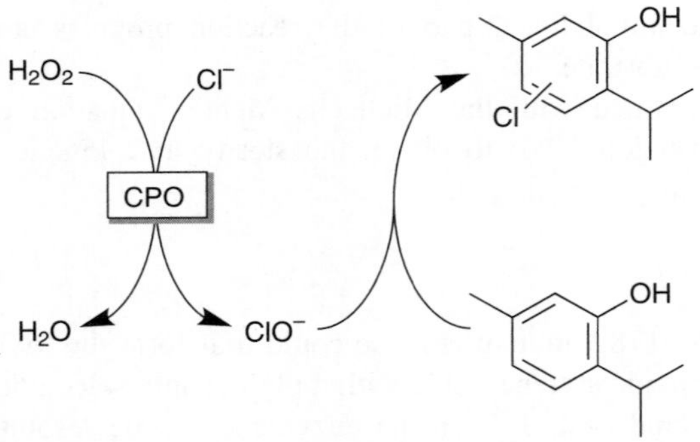

Scheme 7.3 *Hypothesized chemoenzymatic halogenation of thymol.*

7.3.1 Materials and Equipment

- 10 nM CPO (calculated with a molecular weight of 42 kDa) from *C. fumago* – commercially available or produced according to the protocol described previously [18]
- Photometer ($\lambda = 400$ nm)
- Cuvettes (PMMA)
- Citric acid buffer (100 mM), pH 3.5 (14.24 g.L^{-1} citric acid anhydrous, 7.61 g.L^{-1} sodium citrate dihydrate)
- Sodium chloride for chlorination or potassium bromide for bromination
- Thymol or carvacrol
- Hydrogen peroxide (3% solution)

7.3.2 Analytical Method

The concentration of CPO was determined photometrically at 400 nm ($\varepsilon = 75\,300$ M^{-1}. cm^{-1}). Reaction products were determined by GC/MS (GC17A; Shimadzu, Kyoto, Japan) using a Valcobond VB-5 column (l = 30 m; ID = 0.25 mm; VICI, Schenkon, Switzerland), with a temperature program of 5 minutes at 80 °C then 10 °C·min^{-1} to 200 °C (injector temperature: 250 °C, GC-MS interface temperature: 290 °C, scan range MS: 35–250 m/z). A sample volume of 950 µL was mixed with 50 µL of the internal standard (2 g.L^{-1} 1-octanol in ethanol) and subsequently extracted with 1 mL butyl acetate. The organic phase was dried with sodium sulfate and analyzed by GC-MS (retention times: thymol, 11.9 minutes; para-chlorothymol, 12.7 minutes; ortho-chlorothymol, 15.1 minutes).

7.3.3 Procedure 1: Chemoenzymatic Chlorination of Thymol to Chlorothymol

1. Halogenation of the substrates was investigated in 5 mL glass reactors. Thymol (2.5 mM, ~1.88 mg) and 10 nM CPO were added to 5 mL of citric acid buffer (100 mM, pH 3.5, containing 25 mM NaCl). The temperature was maintained at 30 °C using a water bath.
2. Hydrogen peroxide was dosed every 10 minutes to a final concentration of 166.8 µM (0.95 µL of a 3% solution) in the reaction mix to enable CPO activity.
3. After a reaction time of 120 minutes, maximum chlorothymol concentrations were obtained at a conversion of 73%, with product formation rates of up to 0.78 mM·h^{-1}. The ratio of *para-* to *ortho*-chlorothymol was 30:70 for all conversions measured.

Conversions of up to 90% were obtained with a thymol concentration of 1 mM and an addition of hydrogen peroxide up to 1 mM at doses of 66.7 µM (0.38 µL of a 3% solution) and a production rate of 0.59 mM·h^{-1}.

This procedure also works for the chlorination of carvacrol to chlorocarvacrol and the bromination of thymol and carvacrol to bromothymol and bromocarvacrol, respectively. For brominations, 25 mM potassium bromide was used in the buffer system instead of sodium chloride [19].

7.3.4 Procedure 2: Electrochemoenzymatic Chlorination of Thymol in a GDE-Based Reactor

7.3.4.1 Materials and Equipment

- Gas diffusion electrode (GDE) (GASKATEL, Kassel, Germany)
- Electrochemical flow-through reactor, as described previously [20]

7.3.4.2 Procedure

1. Briefly, a GDE with an apparent surface area of 5.5 cm^2 was used as the cathode and platinum was used as the anode. The reactor casing consisted of three parts, the middle part forming the flow channel, with an internal volume of 8 mL. The platinum anode and the gas diffusion cathode were assembled as side parts to the flow channel. Oxygen from the air diffused through a notch in the casing to the reverse side of the GDE. A current generator was used to apply suitable currents for the electro-generation of H$_2$O$_2$ by oxygen reduction.
2. The reactor was connected to a 50 mL reservoir and a flow of 30 mL·min^{-1} was pumped from the reservoir through the reactor cell. Experiments were carried out with a pretreated GDE. Pretreatment of the GDE was achieved by saturation with a 5 mM thymol solution for 24 hours at room temperature, to prevent adsorption of thymol to the PTFE layer of the electrode.
3. Thymol (2.5 mM) and 10 nM CPO were added to a volume of 50 mL citric acid buffer (50 mL, 100 mM, pH 3.5 containing 50 mM NaCl) and pumped through the electrochemical reactor. A current of 45 mA was applied to produce hydrogen peroxide, and chlorothymol was formed with a rate of 4.6 mM·h^{-1}. Conversions of 52% could be obtained.

7.3.5 Conclusion

Chemoenzymatic halogenations of thymol and carvacrol were investigated in this study. Hydrogen peroxide was produced with a gas diffusion cathode to enable CPO activity and to prevent rapid inactivation by the co-substrate. Compared to the current state of the art, using Cu^{II}-catalyzed *in situ* generation of hypochlorite from chloride and molecular oxygen [21], our method excels, especially in terms of catalyst performance compared to simple transition metal salts. Hence, the catalyst loading (0.0004–0.002 mol%) is orders of magnitude lower than in the chemical counterpart (12.5 mol%) but achieves comparable product yields. Additionally, the milder reaction conditions (30 °C and aqueous reaction media vs. 80 °C and acetic acid as solvent) presumably contribute to higher product qualities (and reduced downstream processing and purification efforts).

7.4 Halogenation of Non-Activated Fatty Acyl Groups by a Trifunctional Non-Heme Fe(II)-Dependent Halogenase

Sarah M. Pratter and Grit D. Straganz
Institute of Biotechnology and Biochemical Engineering and Institute of Biochemistry, Graz University of Technology, Austria

Oxidative halogenations are a potent synthetic tool for the synthesis of bioactive compounds in nature. Halogenated non-ribosomal lipopeptides are produced by a variety of prokaryotic and eukaryotic microbes, and the halogen substituents modulate their physiological properties, which range from antimicrobial and anti-infective to anticancer activities [22]. Consequently, halogenations in lipopeptide biosynthesis are of major interest for the study and engineering of pathways for the biosynthesis of pharmaceuticals.

The biochemical characterization of halogenases is an important prerequisite for the assessment of their synthetic potential and the required environment in the cell for efficient whole-cell catalysis. However, the fact that these enzymes are often embedded into the lipopeptide biosynthetic machinery complicates their study: arguably, the most powerful halogenases known to date are the non-heme Fe(II)-dependent halogenases, as they can selectively halogenate non-activated alkyl structures, whereby the substrate is generally covalently linked to a multi-domain protein complex (Scheme 7.4b). The substrate's transfer to and from the enzyme thereby depends on accessory proteins (Scheme 7.4a).

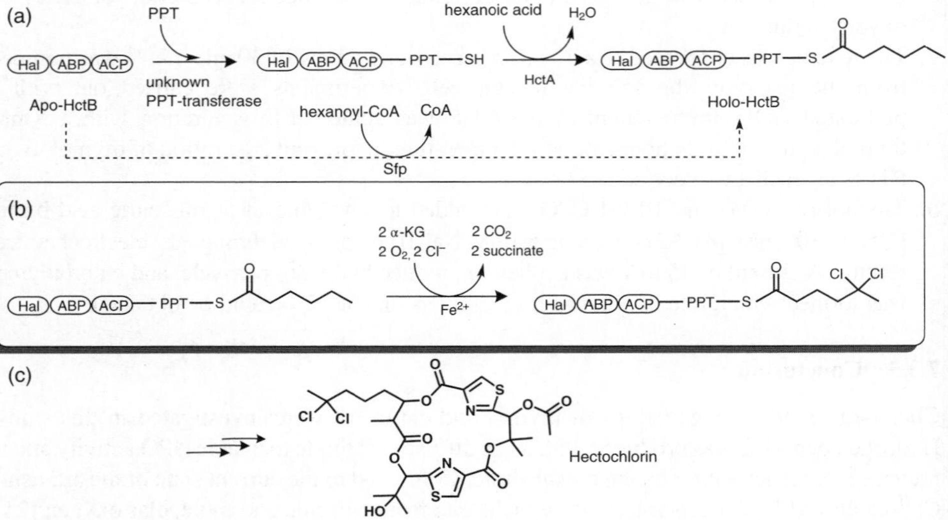

Scheme 7.4 *The primary steps in the biosynthesis of the cyanobacterial lipopeptide hectochlorin. (a) Pathway of constitution of the three-modular halogenase holo-HctB in vivo (arrows) and in vitro, as carried out in this study (dashed arrow). The three-domain structure of HctB, consisting of halogenase (Hal), acyl-Co-A binding protein domain (ABP), and acylcarrier protein domain (ACP), is shown. (b) Halogenation reaction of HctB, as suggested from genetic data and the hectochlorin structure [23]. (c) Structure of hectochlorin [23].*

Fatty acyl halogenases are one grouping of halogenases that are believed to generally convey halogenated acyl groups to non-ribosomal lipopeptide structures. The gene of one exponent, HctB, is found in the pathway of hectochlorin biosynthesis [23]. The putative hexanoyl-dichlorinating enzyme HctB (Scheme 7.4) consists of three domains, with the substrate-binding domain (ACP) and halogenase domain (Hal) present on one polypeptide chain. Scheme 7.4 depicts the biosynthesis of the substrate-carrying, acylated enzyme holo-HctB (panel A), its proposed enzymatic reaction (panel B), and the product of the hectochlorin biosynthetic pathway, hectochlorin (panel C). The three-domain structure of HctB is considered a general feature of fatty acyl halogenases and poses a special challenge to the proteins' functional expression and *in vitro* stability. In this section, we describe the procedure of purification and functional reconstitution of the α-ketoglutarate (α-KG)-dependent non-heme Fe(II) halogenase HctB, as well as the characterization of its substrate and reaction specificities.

7.4.1 General Materials and Equipment

- 2 Baffled flasks (300 mL)
- 14 Baffled flasks (1000 mL)
- Cotton plugs
- Autoclave
- Isopropyl β-D-1-thiogalactopyranoside (IPTG) 100 mg·mL^{-1} (Carl Roth)
- Incubation shaker (CERTOMAT BS-1, Satorius)
- Centrifuge (4500× g, 4 °C) and 500 mL centrifuge beakers
- French press and 35 mL French press cell (Aminco)
- Centrifuge (75 000× g, 4 °C) and 20 mL centrifuge beakers
- ÄKTAprime plus FPLC at 4 °C
- VivaSpin 20 ultrafiltration units
- Centrifuge for Sarsted tubes (e.g., Centrifuge 5804R, Eppendorf)
- HiPrep 26/10 desalting column (Amersham Biosciences)
- UV/Vis cuvette and spectrophotometer
- HEPES/NaOH buffer: 20 mM HEPES (Sigma), pH 7.5
- Roti-Quant concentrate for protein determination (Carl Roth)
- SDS-PAGE precast gels (GE Healthcare), equipment and buffers for PAGE electrophoresis, Comassie Blue protein staining solution, gel destaining solution

7.4.2 Expression and Purification of HctB

7.4.2.1 Materials and Equipment

- Oxygenase medium (OxM) (4.5 L, 0.2 g.L^{-1} NH$_4$Cl, 5 g.L^{-1} (NH$_4$)$_2$HPO$_4$, 4.5 g.L^{-1} KH$_2$PO$_4$, 5 g.L^{-1} Na$_2$HPO$_4$*2H$_2$O, 11 g.L^{-1} glucose hydrate, 10 g.L^{-1} yeast extract, 1.5 g.L^{-1} MgSO$_4$*H$_2$O, 1 g.L^{-1} ammonium ferric citrate, trace element solution SL6 (1 mL), and 50 mg.L^{-1} kanamycin) [24]
- Disruption buffer (50 mM Tris, 150 mM NaCl, 100 mM Na$_2$CO$_3$, 5% glycerol, pH 8.5)
- *E. coli* BL21 Gold (DE3) cells bearing the expression vector pKYB1_HctB-StrepII, which conveys kanamycin resistance [25]. StrepII-Tactin Superflow column, 5 mL bed volume (IBA BioTAGnology), Strep washing buffer (100 mM Tris, 150 mM NaCl,

pH 8.0), Strep elution buffer (100 mM Tris, 150 mM NaCl, 2 mM desthiobiotin (IBA), pH 8.0), and Strep regeneration buffer (100 mM Tris, 150 mM NaCl, 1 mM ethyl-enediaminetetraacetic acid (EDTA) (Sigma), 1 mM hydroxyl-azophenyl-benzoic acid (Sigma), pH 8.0))
- 3-(2-Pyridyl)-5,6-di(2-furyl)-1,2,4-triazine-5', 5''-disulfonic acid disodium salt, "FereneS" (Sigma)

7.4.2.2 Procedure

1. The *hctB* gene was synthesized based on the sequence of the second gene in the hectochlorin biosynthetic gene cluster (accession number AY974560). It had been codon-optimized and a streptavidin (II) affinity tag [26] had been fused to its C-terminus (Mr Gene GmbH, Regensburg, Germany) [25]. *E. coli* BL21 Gold (DE3) cells, transformed with the expression vector pKYB1_HctB-StrepII, were incubated in OxM medium [24] at 18 °C for 22 hours, as described previously [25].
2. Cells were harvested (4500× *g*, 20 minutes, 4 °C), resuspended in ice-cold disruption buffer, and passed through a French cell (approximately 24 000 psi) twice.
3. After centrifugation (75 000× *g*, 4 °C, 35 minutes), the supernatant was directly subjected to affinity purification using the Strep-Tactin Superflow column according to the manufacturer's procedure, using the respective EDTA-free buffers for washing and elution, as described previously.
4. The eluent was desalted by three cycles of ultrafiltration and backfilling with 20 mM HEPES buffer using a VivaSpin 20 ultrafiltration unit. The concentration of the resulting protein solution (1 mL) was determined by UV/Vis spectrometry using the method of Edelhoch and Pace [27,28], which is based on the absorbance of the protein solution at 280 nm and the extinction coefficient calculated from the HctB protein sequence ($\varepsilon = 84715\,M^{-1}.cm^{-1}$).
5. The Fe(II) content was assessed spectrophotometrically according to Hennessy *et al.* [29], by using an excess of FereneS (20 mM) and monitoring the evolution of a colored FereneS-Fe(II) complex at 492 nm over 20 hours. The second part of the resulting curve, a flat linear absorption increase between approximately 5 and 20 hours, was extrapolated to intersect the ordinate, and this absorption, together with the molar extinction coefficient of the complex ($35\,500\,M^{-1}.cm^{-1}$), was used to calculate the Fe(II) content [30].
6. Enzyme purity was estimated by SDS PAGE gel electrophoresis and Coomassie Blue staining using standard procedures [31]. Purified HctB samples were stored at −70 °C.

7.4.3 Expression and Purification of the Enzyme Sfp and the Thioesterase TycF

7.4.3.1 Materials and Equipment

- 2xYT broth medium (16 g.L^{-1} peptone, 10 g.L^{-1} yeast extract, 5 g.L^{-1} NaCl, and 50 mg. L^{-1} kanamycin (Sfp) or 100 mg.L^{-1} ampicillin (TycF))
- His-buffer A (50 mM Tris, 300 mM NaCl, pH 7.4)
- His-buffer B (50 mM Tris, 300 mM NaCl, 500 mM Imidazole, pH 7.4)
- PPTase buffer (50 mM Tris, 10 mM MgCl$_2$, 5 mM DTT, 5% glycerol, pH 6.9)
- HEPES/NaOH buffer (20 mM HEPES (Sigma), pH 7.5)
- *E. coli* BL21 Gold (DE3) cells bearing the expression vector pKYB1_Sfp-His [25], which conveys kanamycin resistance

- *E. coli* BL21 Gold (DE3) cells bearing the expression vector pET30a(+)_His-TycF [25], which conveys ampicillin resistance
- HisTrap FF column, 5 mL (GE Healthcare)

7.4.3.2 Procedure

1. The accessory enzyme Sfp from *Bacillus subtilis*, which can be used to transfer substituted phosphopanthetheinyl moieties to acyl and peptidyl carrier protein domains, was expressed from the pKYB1_Sfp-His plasmid in *E. coli* BL21 Gold (DE3) in 2xYT medium at 25 °C, as described previously [32].

2. The thioesterase TycF from *Bacillus brevis* hydrolyzes the acyl-group from an acyl-phosphopanthetheinyl moiety and can thus be used to free the reaction product from the enzyme–product complex. The pET30a(+)_His-TycF vector, containing the His-tagged TycF sequence [33], was transformed into *E. coli* BL21 Gold (DE3) cells, and His-TycF protein was expressed at 15 °C for 72 hours [33] (the pET30a(+) vector containing the His-tagged TycF sequence was a kind gift from the group of Prof. C.T. Walsh (Harvard Medical School)). Harvested cells expressing either of the His-tagged accessory proteins were harvested (4500× g, 20 minutes, 4 °C) and resuspended in PPTase buffer (Sfp) or 20 mM HEPES buffer (TycF), and crude cell lysates were prepared as described for HctB.

3. For the His-tag affinity chromatography, a 5 mL HisTrap FF column (GE Healthcare) was equilibrated with His-buffer A. After loading the crude cell lysate, unspecifically bound proteins were removed with 10% of His-buffer B. A stepwise gradient of His-buffer B was applied and His-tagged proteins were eluted between 20 and 50% buffer B.

4. Elution buffer was exchanged for PPTase buffer (Sfp) or 20 mM HEPES buffer (TycF) by size-exclusion chromatography using a HiPrep 26/10 desalting column. The resulting protein solutions were concentrated in VivaSpin 20 ultrafiltration units and the protein concentrations were assessed by comparison of a set of diluted protein samples with BSA standard solutions (0–1 mg·mL^{-1}) using the Roti-Quant protein assay (Carl Roth). Again, enzyme purity was estimated by gel electrophoresis and Coomassie Blue staining, using standard procedures [31]. Purified enzymes were stored at −70 °C.

7.4.4 Reconstitution of Holo-HctB

7.4.4.1 Materials and Equipment

- Fatty acyl CoAs, C6-C14 (Sigma)
- Thermomixer (Eppendorf)
- Microcentrifuge (Eppendorf)
- Transfer buffer (75 mM Tris, 10 mM MgCl$_2$, pH 7.5)
- HctB (~1 mM, 2 mL), purity >95%
- Accessory protein Sfp (~1 mM, 200 µL), purity >80%

7.4.4.2 Procedure

1. HctB, Sfp, and fatty acyl CoAs were mixed at a molar ratio of 10 : 1 : 20 in transfer buffer and incubated at 350 rpm and 24 °C for 60–90 minutes (see Scheme 7.4a, dashed arrow).

2. Precipitated protein was removed by centrifugation (16 800× g, 4 °C, 10 minutes), buffer was exchanged for 20 mM HEPES buffer pH 7.5, the holo-HctB solution was concentrated (VivaSpin 20 ultrafiltration units), and protein concentration was reassessed.

3. The procedure had been validated by MALDI-TOF mass spectrometry, and exhaustive hexanoyl transfer had been confirmed by a mass shift of ~56 780 to ~57 220 in the mass spectrum of HctB [25].

7.4.5 HctB Activity Assay and Product Analysis

7.4.5.1 *Materials and Equipment*

- $Fe(II)SO_4{}^*7H_2O$
- α-KG Stock solutions (0–100 mM, preferably 30 mM)
- NaCl Stock solutions (0–4 M, preferably 4 M)
- 20 mM HEPES/NaOH buffer, pH 7.5
- Solvent A: 1% formic acid in water
- Solvent B: 1% formic acid in acetonitrile
- Hexanoic acid (Sigma), 5-oxo-hexanoic acid (TCI), and 6-chloro hexanoic acid (TCI)
- 5 mL Air-tight V-Vial (Wheaton) capped with a screw-top septum (Supelco)
- N_2 stream (through a canula)
- Gas-tight syringe (optional: glove box)
- Temperature-controlled sealed cuvette with access channel for an O_2 sensor
- O_2 Sensor (Oxygen Microooptode, Presens)
- VivaSpin 20 ultrafiltration units (Satorius)
- Centrifuge for Sarsted tubes (e.g., Centrifuge 5804R, Eppendorf)
- Accessory protein TycF (~1 mM, 50–100 µL), purity >90%
- Thermomixer (Eppendorf)
- VivaSpin 500 ultrafiltration units (Satorius)
- Microcentrifuge (Eppendorf)
- 1200 Infinity Series HPLC system (Agilent Technologies)
- LiChroCART Purospher STAR HPLC-column (Merck Millipore; 308C, 0.8 mL·min^{-1}).

7.4.5.2 *Procedure*

1. A 15 mM anaerobic Fe(II) stock solution was prepared by transferring $Fe(II)SO_4{}^*7H_2O$ into a 5 mL air-tight screw-top vial equipped with a septum. After extensive N_2-purging, the salt was dissolved in degassed water, which had been prepared by bubbling through N_2. The transfer was performed either via an N_2-purged gas-tight syringe or in the glove box.

2. For HctB activity measurements, an aliquot of the respective holo-HctB was transferred to a sealed cuvette, which was equipped with an O_2 sensor and a magnetic stirrer and contained 20 mM HEPES buffer and the desired concentration of NaCl to give 100 µM of halogenase. Anaerobic Fe(II) stock solution (5 µL) was added, which resulted in a 1.5-fold excess of Fe(II) over HctB. The reaction was then started by addition of the respective α-KG stock solution (10–1000 µM end concentration).

3. The total reaction volume was 500 µL. Specific reaction velocities were in general determined from the measured O_2 consumption curves by dividing the initial rate of O_2 consumption by the HctB concentration. These specific activities were validated by fitting whole O_2 consumption curves to the equation $f = vy_0 + ae^{(-bx)}$ or $f = y_0 + ae^{(-bx)} + ce^{(-dx)}$, whereby y_0 is the final O_2 concentration after conversion, a and c are the amplitudes of the curve, and b and d are the respective reaction rates.

7.4.6 HPLC-MS

7.4.6.1 Sample Preparation

Holo-HctB samples from activity assays were pooled, buffer-exchanged into 20 mM HEPES buffer, and concentrated to 1–1.5 mM (VivaSpin 20 ultrafiltration unit). Acyl-moiety cleavage from the enzymes' ACP domains was accomplished by incubation (30 °C, 350 rpm, 16 hours) with thioesterase TycF (15–40 mM). Precipitated protein was removed by centrifugation (16 800× *g*, 4 °C, 10 minutes). The supernatant was subjected to a VivaSpin 500 centrifugal filter and the filtrate, which contained the cleaved carbonic acid, was collected [25].

7.4.6.2 Product Analysis

Reaction products and standard solutions of hexanoic acid, 5-oxo-hexanoic acid, and 6-chloro hexanoic acid were acidified with formic acid (1% v/v). HPLC analysis of the sample (20 μL injection volume) was performed using solvents A and B according to the following elution protocol: 10% B (10 minutes), followed by 10–50% B (30 minutes), 50–90% B (1 minute), 90% B (3 minutes), 90–10% B (1 minute), and 10% B (10 minutes). Separated components were analyzed by ESI-MS single-ion monitoring (SIM) in a negative ion-detection mode. Reaction products were identified based on their monoisotopic masses, and additionally, where standards were available, based on their retention times. The presence of chloride in the reaction products was confirmed by its prototypical isotope pattern $^{35}Cl:^{37}Cl$ of 3.1 : 1 for monochlorinated substances and $^{35}C:^{37}Cl$ $^{35}Cl:^{37}Cl$ of 9.8 : 6.3 : 1 for dichlorinated substances [25]. The method was validated by 2D-NMR of cleaved and uncleaved ^{13}C-labeled acyl-chains, which unambiguously gave the product structures and allowed quantification of the respective m/z signal intensities [25].

7.4.7 Results

HctB's activity toward its presumed native C6 substrate (Scheme 7.5) was measured at varying α-KG and NaCl conditions, and it was found that concentrations of >300 μM α-KG, >500 mM NaCl ensured saturating conditions [25], as summarized in Table 7.2.

The substrate spectrum was consequently assessed by measuring the reaction velocities of the respectively substituted holo-HctBs (C6–C14) under standardized conditions, and the

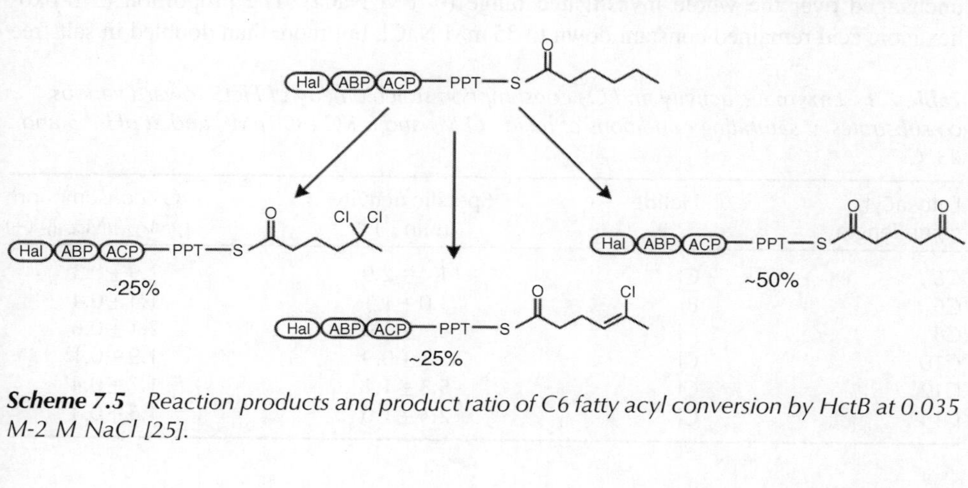

Scheme 7.5 *Reaction products and product ratio of C6 fatty acyl conversion by HctB at 0.035 M-2 M NaCl [25].*

Table 7.2 *Dependence of the enzymatic activity and O_2 consumption stoichiometry of HctB on co-substrate concentrations at pH 7.5 and 23 °C.*

Fatty acyl chain length	Co-substrate (mM)	Specific activity (min^{-1})	O_2 consumption $(\mu M_{O2}/\mu M_{HctB})$ (−)
Variation of Chloride ($c_{\alpha\text{-KG}}=0.3$ mM)			
C6	2000	14.9 ± 7.0	2.1 ± 0.4
C6	1000	11.4 ± 2.9	1.9 ± 0.2
C6	500	11.8 ± 3.5	1.8 ± 0.1
C6	100	10.7 ± 3.0	1.7 ± 0.1
C6	10	3.7 ± 0.5	1.3 ± 0.1
C6	1.0	0.9 ± 0.1	0.7 ± 0.2
C6	>0.10	0.4 ± 0.2	0.6 ± 0.4
Variation of α-KG ($c_{Cl^-}=1$ M)			
C6	0.05	8.1 ± 2.7	0.7 ± 0.1
C6	0.10	11.6 ± 4.1	1.1 ± 0.2
C6	0.20	15.1 ± 2.9	2.1 ± 0.1
C6	0.75	11.1 ± 4.0	1.9 ± 0.1

results suggested considerable promiscuity of HctB toward medium-chain fatty acyl moieties. Also, the potential of HctB to use bromide as a substrate was examined via the O_2 consumption assay, and bromide incorporation was verified via MALDI-TOF MS [25]. The specific activities and O_2 consumption rates are summarized in Table 7.3.

In order to determine the product spectrum of the native substrate C6 at varying NaCl concentrations (<1 mM–2 M), the respective O_2 consumption assays (10–15) were (i) pooled after reaction, (ii) exchanged into HEPES buffer, (iii) concentrated about 10-fold to yield an HctB product concentration of approximately 1.0–1.5 mM, and (iv) cleaved by reaction with TycF as described previously. Reaction samples were then subjected to HPLC-MS analysis. HPLC analysis, together with validation by 2D-NMR, determined the reaction products 5-oxo-hexanoic acid, 5,5-dichlorohexanoic acid, and 5-chloro-4-hexenoic acid in a ratio of 2 : 1 : 1 [25]. HPLC-MS analyses revealed that the relative concentrations of 5,5-dichloro-hexanoic acid and 5-chloro-4-hexenoic acid were roughly equal and remained unchanged over the whole investigated range (0–1 M NaCl). The proportion of 5-oxo-hexanoic acid remained constant down to 35 mM NaCl, but more than doubled in salt-free

Table 7.3 *Enzymatic activity and O_2 consumption stoichiometry of HctB toward various co-substrates at saturating conditions of halide (1 M) and α-KG (300 μM), and at pH 7.5 and 23 °C.*

Fatty acyl chain length	Halide	Specific activity (min^{-1})	O_2 consumption $(\mu M_{O2}/\mu M_{HctB})$ (−)
C6	Cl	11.4 ± 2.9	1.9 ± 0.3
C6	Br	3.0 ± 1.1	1.4 ± 0.4
C8	Cl	8.9 ± 2.0	2.1 ± 0.6
C10	Cl	5.1 ± 0.9	1.9 ± 0.1
C12	Cl	5.3 ± 1.1	1.7 ± 0.4
C14	Cl	2.4 ± 1.0	1.5 ± 0.3

sample (<1 mM NaCl), consistent with a ~2.5-fold relative increase in the oxo-compound compared to the halogenated products.

7.4.8 Conclusion

The results show that HctB possesses an intrinsic three-functionality, which includes an unprecedented ability to introduce a chlorovinyl functionality into a non-activated alkyl moiety [25]. The strong dependence of the reaction velocity on the Cl^- concentration, together with the constant product ratio over that concentration range, suggests some Cl^--induced switch in the mechanism, thus ensuring substantial halogenation at low halide concentrations, too. Furthermore, activity data imply considerable substrate promiscuity of HctB. The findings are in line with the general observation that biological lipopeptide pathways often employ promiscuous enzymes that generate a range of reactions products, which are then assembled into a variety of metabolites, thus increasing the structural diversity of the natural secondary metabolites [34]. The striking substrate and reaction promiscuity furthermore suggests that HctB is a useful target for the generation of considerable diversity in the biosynthesis of hectochlorin analogs by both feeding strategies and pathway engineering.

References

1. Hohaus, K., Altmann, A., Burd, W. *et al.* (1997) *Angewandte Chemie International Edition*, **36**, 2012–2013.
2. Keller, S., Wage, T., Hohaus, K. *et al.* (2000) *Angewandte Chemie International Edition*, **39**, 2300–2302.
3. Yeh, E., Blasiak, L.C., Koglin, A. *et al.* (2007) *Biochemistry*, **46**, 1284–1292.
4. Dong, C., Flecks, S., Unversucht, S. *et al.* (2005) *Science*, **309**, 2216–2219.
5. Flecks, S., Patallo, E.P., Zhu, X. *et al.* (2008) *Angewandte Chemie International Edition*, **47**, 9533–9536.
6. Zhu, X., De Laurentis, W., Leang, K. *et al.* (2009) *Journal of Molecular Biology*, **391**, 74–85.
7. Lang, A., Polnick, S., Nicke, T. *et al.* (2011) *Angewandte Chemie International Edition*, **50**, 2951–2953.
8. Hasnaoui, G., Lutje Spelberg, J.H., de Vries, E. *et al.* (2005) *Tetrahedron: Asymmetry*, **16**, 1685–1692.
9. Lutje Spelberg, J.H., van Hylckama Vlieg, J.E.T., Bosma, T. *et al.* (1999) *Tetrahedron: Asymmetry*, **10**, 2863–2870.
10. Majerić-Elenkov, M., Tang, L., Hauer, B., and Janssen, D.B. (2006) *Organic Letters*, **8**, 4227–4229.
11. Majerić Elenkov, M., Hoeffken, H.W., Tang, L. *et al.* (2007) *Advanced Synthesis and Catalysis*, **349**, 2279–2285.
12. Tang, L., Zhu, X., Zheng, H. *et al.* (2012) *Applied and Environment Microbiology*, **78**, 2631–2637.
13. Tang, L., Li, Y., and Wang, X. (2010) *Journal of Biotechnology*, **147**, 164–168.
14. van Hylckama Vlieg, J.E.T., Tang, L., Lutje Spelberg, J.H. *et al.* (2001) *Journal of Bacteriology*, **183**, 5058–5066.
15. Bergmann, J.G. and Sanik, J. (1957) *Analytical Chemistry*, **29**, 241–243.
16. Wagner, C., El Omari, M., and König, G.M. (2009) *Journal of Natural Products*, **72**, 540–553.
17. Podgoršek, A., Zupan, M., and Iskra, J. (2009) *Angewandte Chemie International Edition*, **48**, 8424–8450.

18. Kaup, B., Ehrich, K., Pescheck, M., and Schrader, J. (2008) *Biotechnology and Bioengineering*, **99**, 491–498.
19. Getrey, L., Krieg, T., Hollmann, F. *et al.* (2014) *Green Chemistry*, **16**, 1104–1108.
20. Krieg, T., Huttmann, S., Mangold, K.-M. *et al.* (2011) *Green Chemistry*, **13**, 2686–2689.
21. Menini, L. and Gusevskaya, E.V. (2006) *Chemical Communications*, 209–211.
22. Neumann, C.S., Fujimori, D.G., and Walsh, C.T. (2008) *Chemistry and Biology*, **15**, 99–109.
23. Ramaswamy, A.V.C., Sorrels, M., and Gerwick, W.H. (2007) *Journal of Natural Products*, **70**, 1977–1986.
24. Straganz, G.D., Slavica, A., Hofer, H. *et al.* (2005) *Biocatalysis and Biotransformation*, **23**, 261–269.
25. Pratter, S.M., Ivkovic, J., Birner-Gruenberger, R. *et al.* (2014) *Chembiochem*, **15**, 567–574.
26. Schmidt, T.G. and Skerra, A. (2007) *Nature Protocols*, **2**, 1528–1535.
27. Edelhoch, H. (1967) *Biochemistry*, **6**, 1948–1954.
28. Pace, C.N., Vajdos, F., Fee, L. *et al.* (1995) *Protein Science*, **11**, 2411–2423.
29. Hennessy, D.J.S., Reid, G.R., Smith, F.E., and Thompson, S.L. (1984) *Canadian Journal of Chemistry*, **62**, 721–724.
30. Straganz, G.D., Diebold, A.R., Egger, S. *et al.* (2010) *Biochemistry*, **49**, 996–1004.
31. Laemmli, U.K. (1970) *Nature*, **227**, 680–685.
32. Quadri, L.E.N., Weinreb, P.H., Lei, M. *et al.* (1998) *Biochemistry*, **37**, 1585–1595.
33. Yeh, E., Kohli, R.M., Bruner, S.D., and Walsh, C.T. (2004) *Chembiochem*, **5**, 1290–1293.
34. Firn, R.D. and Jones, C.G. (2003) *Natural Products Reports*, **20**, 382–391.

8

Cascade Reactions

8.1 Synthetic Cascades via a Combination of Artificial Metalloenzymes with Monoamine Oxidases (MAO-Ns)

Marc Dürrenberger,[1] Valentin Köhler,[1] Yvonne M. Wilson,[1] Diego Ghislieri,[2] Livia Knörr,[1] Nicholas J. Turner,[2] and Thomas R. Ward[1]

[1]*Department of Inorganic Chemistry, University of Basel, Switzerland*
[2]*School of Chemistry, Manchester Institute of Biotechnology, The University of Manchester, UK*

Despite the synthetic potential, the combination of native enzymes with transition metal catalysts to establish reaction cascades often suffers from mutual inhibition [1]. This drawback can be overcome by compartmentalization of the metal complex within a protein scaffold. In this context, the incorporation of the biotinylated transfer hydrogenation catalyst [Cp*Ir(biot-*p*-L)Cl] into streptavidin (Sav) variants affords a suitable procedure by which to combine the resulting artificial transfer hydrogenases (ATHases) with evolved monoamine oxidases (MAOs) from *Aspergillus niger* for the concurrent double stereoselective deracemization of amines (Scheme 8.1) [2]. 1-methyl-3,4-dihydroisoquinoline is reduced with moderate selectivity by the ATHase to the corresponding (*R*)-amine, whereas the minor (*S*)-amine is converted back to the initial imine by the highly (*S*)-selective MAO-N D9 variant [3]. Successive cycles of this process accumulate the (*R*)-product, typically at $ee > 99\%$. The presence of catalase prevents inactivation of the Ir center by H_2O_2 generated in stoichiometric amounts by the catalytic action of MAO-N. The presence of Sav is crucial, since it shields the organometallic moiety and prevents mutual inhibition of the catalysts, as illustrated by lower yields and by racemic product obtained in the absence of the host protein.

Practical Methods for Biocatalysis and Biotransformations 3, First Edition.
Edited by John Whittall, Peter W. Sutton, and Wolfgang Kroutil.
© 2016 John Wiley & Sons, Ltd. Published 2016 by John Wiley & Sons, Ltd.

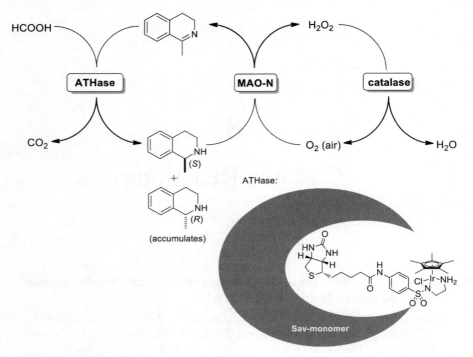

Scheme 8.1 *Highly stereoselective imine reduction through coupling of a moderately (R)-selective reduction step with a highly (S)-selective oxidation step. The incorporation of the Ir co-factor into streptavidin (Sav) enables the compatibility with MAO-N.*

8.1.1 Materials and Equipment

- Buffer A (0.6 M MOPS, 3 M sodium formate, adjusted to pH 7.8 with aq. KOH)
- Buffer B (0.6 M MOPS, adjusted to pH 7.8 with aq. KOH)
- [Cp*Ir(biot-*p*-L)Cl] stock solution (37.5 mM in DMF). For a detailed synthesis procedure for [Cp*Ir(biot-*p*-L)Cl], see reference [4]
- MAO-N stock solution (0.28 mg.mL^{-1} MAO-N-9 in buffer B). For a detailed preparation procedure, see reference [2]
- Sav-S112T (lyophilized solid). For a detailed preparation procedure for purified Sav mutants, see reference [5]
- Catalase (Sigma, lyophilized powder from bovine liver, 11 kU mg^{-1})
- 1-Methyl-3,4-dihydroisoquinoline hydrochloride stock solution (Acros, 1 M in water)
- Aqueous NaOH (10 N)
- *tert*-Butyl methyl ether (TBME)
- Brine
- HPLC-grade hexane
- HPLC-grade isopropanol
- HPLC-grade diethylamine
- 50 mL Falcon tube
- Incubator
- Separation funnel

- Rotary evaporator
- Normal phase HPLC system equipped with UV detection
- Daicel chiral phase IC column (250×4.6 mm, 5 μm)

8.1.2 Procedure

1. Sav-S112T (16.4 mg.mL^{-1}, 750 μM biotin binding sites, assuming three free binding sites per Sav tetramer) and catalase (12.5 kU.mL^{-1}) were dissolved in buffer A. [Cp*Ir (biot-*p*-L)Cl] stock solution was added (375 μM) and the mixture was gently agitated for about 30 seconds to achieve complete mixing.
2. The resulting ATHase stock solution (2.5 mL) was added to the MAO-N stock solution (2.5 mL) and placed in a Falcon tube. The reaction was started by addition of 1-methyl-3,4-dihydroisoquinoline stock solution (187.5 μL, 36.1 mM). and the mixture was incubated in a lying position at 37 °C and 250 rpm for 17 hours.
3. The reaction mixture was transferred into the separation funnel. After addition of water (6 mL) and aq. 10 M NaOH (1.2 mL), the solution was extracted with TBME (4×10 mL). The combined extracts were washed with brine. A gel-like layer, which formed upon extraction, was briefly centrifuged, and the resulting organic phase was added to the other extracts, which were then dried over anhydrous Na$_2$SO$_4$.
4. Evaporation of the solvent afforded (*R*)-1-methyl-1, 2,3,4-tetrahydroisoquinoline as a nearly colorless liquid (yield: 25.8 mg, 93.8%, containing 1.7% DMF).
5. The product *ee* was determined by chiral normal phase HPLC (Daicel IC 250×4.6 mm, 5 μm; hexane/i-PrOH/HNEt$_2$ = 97/3/0.06, 1 mL.min^{-1}, 25 °C, 265 nm; T$_R$, 11.0 minutes (*S*)-1-methyl-1,2,3,4-tetrahydroisoquinoline), 12.0 minutes (*R*)-1-methyl-1,2,3,4-tetra-hydroisoquinoline), 18.9 minutes (1-methyl-3,4-dihydroisoquinoline)). A product *ee* of > 99% is usually obtained.

8.1.3 Conclusion

The result demonstrates that the Sav incorporated biotinylated Ir complex [Cp*Ir(biot-*p*-L) Cl] is fully compatible with biocatalysts relying on FADH and heme co-factors. The procedure provides a versatile alternative to other common methods of circumventing incompatibility between enzymes and transition metal catalysts, such as immobilization, heterogeneous or biphasic reaction conditions, and encapsulation. Since the second coordination sphere provided by the host protein has a significant influence on the catalytic performance of the Ir center in terms of activity and selectivity, the reaction cascade can be further optimized by site-directed mutagenesis of streptavidin.

8.2 Amination of Primary Alcohols via a Redox-Neutral Biocascade

Johann H. Sattler,[1] Michael Fuchs,[1] Francesco G. Mutti,[2] and Wolfgang Kroutil[1]
[1]*Department of Chemistry, Organic and Bioorganic Chemistry, University of Graz, Austria*
[2]*School of Chemistry, Manchester Institute of Biotechnology, The University of Manchester, UK*

Diamines are common building blocks for polyamides (e.g., nylon), but nature provides mainly hydroxy functionalized compounds. Therefore, amination of, for example, 1,ω-diols

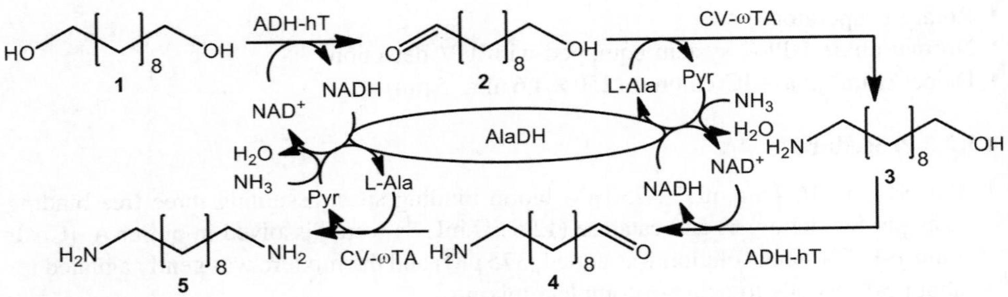

Scheme 8.2 *Double bioamination of long-chain n-alkane-1,ω-diols.*

is necessary in order to access novel 1,ω-diamine building blocks for polymers. To achieve this in an energy-efficient manner, an oxidation and reductive amination sequence has to take place twice. Thus, four sequential simultaneous steps are required in total to get access to the diamine. To avoid the need for additional redox reagents, a redox-neutral cascade has been designed (Scheme 8.2) [6–8]. The advantage of this artificial metabolism results from its self-sufficiency concerning redox equivalents: no external hydride source, such as glucose or formate, is required [9,10].

8.2.1 Materials and Equipment

- 1,10-Decanediol **1** (174.2 mg, 1 mmol)
- 10% v/v 1,2-Dimethoxyethane (DME) in water (2 mL)
- L-Alanine (445 mg, 5 mmol)
- NH$_4$Cl (320 mg)
- NAD$^+$ (10 mg)
- Pyridoxal 5′-phosphate (PLP) (2 mg, 8 μmol)
- Aqueous NaOH (6 M, 80 μL)
- Aqua bidest. (38 mL)
- Alcohol dehydrogenase (ADH) from *Bacillus stearothermophilus* as a cell-free heat-treated preparation (ADH-hT, 20 mg, 4 U)
- ω-Transaminase (ω-TA) from *Chromobacterium violaceum* (TA-CV) as a cell-free preparation (20 mg, 5 U)
- Purified alanine dehydrogenase from *Bacillus subtilis* (AlaDH 0.6 mg, 4 U)
- Round-bottom flask (50 mL)
- Orbital shaker
- DOWEX MAC-3 WAC from G Biosciences (15 mL)
- Gravity flow column (20 mL)
- Methanol (40 mL)
- 2% HCl in MeOH (90 mL)

8.2.2 Procedure

1. A solution of L-alanine (445 mg, 5 mmol) in H$_2$O (4 mL) and a solution of NH$_4$Cl (320 mg, 6 mmol) in H$_2$O (1.5 mL) were added to 1,10-decanediol **1** (174.2 mg, 1 mmol) dissolved in DME (2 mL, 10% v/v in water). The pH was adjusted to 8.5 by addition of a 6 M NaOH solution (~80 μL).

2. A solution of NAD$^+$ (10 mg, 12 μmol) in H$_2$O (0.5 mL), a solution of PLP (2 mg, 8 μmol) in H$_2$O (0.5 mL), H$_2$O (3.12 mL), the ADH solution (4 mL, 5 U), the transaminase solution (4 mL, 4 U), and the alanine dehydrogenase solution (0.3 mL, 5 U) were all added to the resulting mixture.

3. After shaking at 20 °C and 120 rpm (orbital shaker, Infors Unitron) for 22 h, the product was separated from the reaction mixture by a weak acid cation exchanger (using 15 mL of DOWEX MAC-3 resin packed into a 20 mL gravity flow column).

4. The pH of the reaction mixture was adjusted to 9.0 (NaOH 6 M), and the coagulated proteins were removed by filtration. The aqueous phase was loaded on to the column and successively washed with water (2 column volumes) and with MeOH (2 column volumes). The hydrochloride salt of the diamine was eluted with 2% HCl in MeOH, and six fractions of 15 mL were collected.

5. Fractions containing the diamine were combined, concentrated on a rotatory evaporator, and freeze-dried. The residue was washed with chloroform, and the organic phase was discarded. Further purification was achieved by taking up the residue in methanol and precipitating the product with diethylether to afford 1,10-diaminodecane dihydrochloride **5**.HCl (173 mg, 0.70 mmol, 70% yield).

8.2.3 Analytical Methods

^{1}H-NMR (D$_2$O, 300 MHz): δ 1.16–1.34 (m, 12 H, N-CH$_2$-CH$_2$-CH$_2$-CH$_2$-CH$_2$), 1.56 (m, 4 H, N-CH$_2$-CH$_2$), 2.89 (t, 4 H, $J_1 = J_2 = 7$ Hz, N-CH$_2$)

^{13}C-NMR (D$_2$O, 75 MHz): δ 25.5, 26.7, 28.1, 28.3, 39.6

The samples containing diamine **5** had to be derivatized for analysis. For derivatisation, ethyl(succinimidooxy)formate (60 mg, 320 μmol) dissolved in acetonitrile (500 μL) was added to the reaction mixture (50 mM substrate loading), followed by triethylamine (100 μl, 10% v/v in water), and the mixture was shaken at 45 °C and 500 rpm for 1 hour. Afterwards, the mixture was extracted three times with CH$_2$Cl$_2$ (500 μL). The combined organic phases were finally dried over Na$_2$SO$_4$.

Conversions were measured by GC-MS (column: Agilent J&W HP-5 column (30 m, 320 μm, 0.25 μm)).

GC program parameters:

Method: injector 250 °C
Flow: 1.8 mL.min^{-1}
Temperature program: 100 °C/hold 0.5 minutes; 10 °C.min^{-1} to 300 °C/hold 0 minutes
Retention times (minutes):

   1,10-Decanediol: 9.9 minutes
   Ethyl 10-hydroxydecylcarbamate: 15.2 minutes
   Diethyl decane-1,10-diyldicarbamate: 18.6 minutes

8.2.4 Conclusion

The redox neutral cascade requires ammonium as the amine source as its only stoichiometric reagent. Although alanine is required in excess, it is not consumed. To minimize the amount of salt needed, no common buffer salts like Tris or phosphate are used.

The network presented here is applicable to a variety of primary alcohols. For representative examples, see Table 8.1.

Table 8.1 Amination of primary alcohols employing a redox-neutral cascade.[a]

Entry	Alcohol	c (%)	Aldehyde (%)	Amine (%)
1	1-Hexanol	>99	<1	>99
2	1-Octanol	50	<1	50
3	1-Octanol	57[b]	<1	57
4	1-Decanol	2	<1	2
5	1-Decanol	25[b]	<1	25
6	1-Dodecanol	10[b]	<1	10
7	Benzyl alcohol	>99	13	87
8	Cinnamyl alcohol	>99	30	70
9	3-Phenyl-1-propanol	>99	<1	>99

[a] Reaction conditions: Substrate (50 mM), CV-ωTA (1 mg, 0.2 U) and ADH-hT (1 mg, 0.25 U), AlaDH (0.4 mg, 0.25 U), PLP (0.35 mM), NAD$^+$ (0.75 mM), ammonium chloride (275 mM), L-alanine (250 mM), pH 8.5, 24 hours, 20 °C.
[b] 1,2-dimethoxyethane (10% v v^{-1}) was added as co-solvent.

8.3 Biocatalytic Synthesis of a Diketobornane as a Building Block for Bifunctional Camphor Derivatives

Michael Hofer,[1] Harald Strittmatter,[1] and Volker Sieber[1,2]
[1] *Fraunhofer Institute for Interfacial Engineering and Biotechnology, Institute branch Straubing, BioCat – Bio-, Chemo- and Electrocatalysis, Germany*
[2] *Technical University Munich, Germany*

By combining enzymatic and chemical catalytic synthesis, the development of new routes to novel terpene-based bicyclic bifunctional derivatives is possible. The selective biocatalytic oxidation of camphor with the P450cam monooxygenase from *Pseudomonas putida* is already well known [11]. The further oxidation to the corresponding diketone is part of the natural camphor degradation pathway [12]. Recently, the dehydrogenase responsible for this step was cloned and used for the targeted and co-factor-efficient production of 2,5-diketobornane at lab scale [13]. This molecule was converted by chemical catalysis to the corresponding diol and diamine compounds (Scheme 8.3).

8.3.1 Procedure 1: Cell-Free Biocatalysis

8.3.1.1 Materials and Equipment

- *E. coli* Rosetta (DE3) pLysS pET22b-camD
- *E. coli* BL21(DE3) pET-PdR [14]
- *E. coli* BL21(DE3) pET-PdX [14]
- *E. coli* BL21(DE3) pET-P450cam [14]
- Luria–Bertani (LB) broth powder (20 g.L^{-1})
- LB agar powder (15 g.L^{-1})
- Ampicillin (100 μg.mL^{-1} stock solution in water, filter-sterilized)
- Isopropyl β-D-1-thiogalactopyranoside (IPTG) (1 M, filter-sterilized)
- ZnSO$_4$ (1 mM, filter-sterilized)

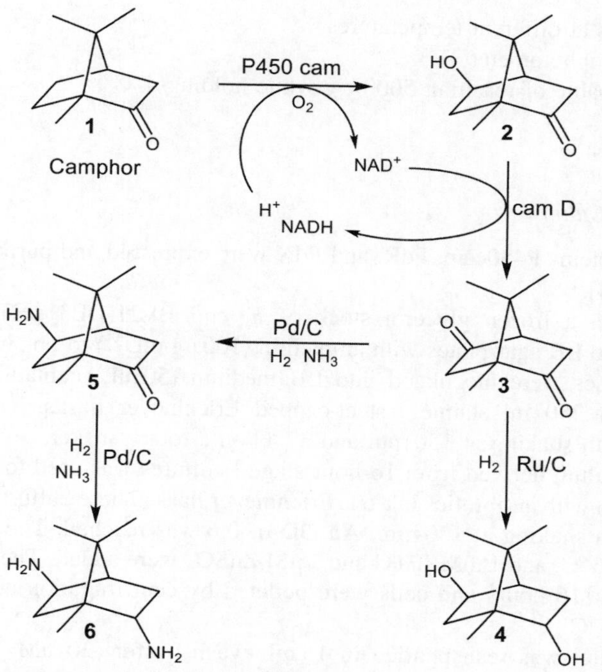

Scheme 8.3 *Synthesis of the diol-derivative* **4** *and, via a reduction intermediate with a monoamine function* **5**, *of diamine-derivative* **6**. *For the intermediate we assume the structure of molecule* **5**, *because of the lesser steric hindrance of this carbonyl group.*

- (1R)-(+)-Camphor
- Celite
- Tris HCl (20 mM, pH 7.4, containing 0.5 M NaCl and 20 mM imidazole)
- Tris HCl (20 mM, pH 7.4, containing 0.5 M NaCl and 500 mM imidazole)
- Sodium potassium phosphate (20 mM, pH 7.4)
- Potassium phosphate (20 mM, pH 7.4, containing 50 mM KCl, 100 µM ZnSO$_4$, and 250 µM β-nicotinamide adenine dinucleotide reduced (NADH))
- Disposable cuvettes
- HPLC-grade ethyl acetate
- HPLC-grade methylene chloride
- HPLC-grade methanol
- HiPrep 26/10 desalting column
- 0.2 µM Cellulose acetate membrane filter
- Ni-NTA column (HiTrap FF)
- 1 L Centrifuge tubes
- Sterile loop
- Petri dish
- Three stainless steel-capped Erlenmeyer flasks (5, 125, and 500 mL)

- Rotary shakers at different temperatures
- UV/Vis spectrophotometer
- Centrifuge capable of reaching 5000× *g* while holding 4 °C
- French press
- Rotary evaporator

8.3.1.2   Initial Culture

1. The three proteins P450cam, PdR, and PdX were expressed and purified as previously described [14].
2. Crystals from a frozen glycerol stock of *E. coli* BL21(DE3)/pET22b-camD were streaked on to LB agar plates with ampicillin (100 µg.mL^{-1}) to obtain single colonies.
3. Single colonies were inoculated into LB medium (50 mL) containing 100 µg.mL^{-1} ampicillin in 250 ml stainless steel-capped Erlenmeyer flasks. The cultures were incubated with shaking at 150 rpm and 37 °C on a rotary shaker.
4. A 10% inoculum derived from 16-hour stage I cultures was used to initiate fresh LB cultures (1 L) with antibiotics in a 5 L Erlenmeyer flask. These cultures were incubated at 37 °C with shaking at 150 rpm. An OD of 0.6 was reached. The temperature was reduced to 16 °C, and 1 mM IPTG and 1 µM ZnSO$_4$ were added. The culture was then incubated for 16 hours and cells were pelleted by centrifugation at 3000× *g* for 15 minutes at 4 °C.
5. The cell pellet was resuspended in 15 mL wash buffer (20 mM Tris HCl pH 7.4 containing 0.5 M NaCl and 20 mM imidazole) and disrupted by a French press. Disrupted cells were centrifuged at 20 000× *g* for 30 minutes at 4 °C. The supernatant was filtered through a 0.2 µM cellulose acetate membrane filter, applied to a Ni-NTA column, and eluted with 20 mM Tris HCl pH 7.4 containing 0.5 M NaCl and 500 mM imidazole. The purified protein was then passed through a gel filtration column (HiPrep 26/10 Desalting), equilibrated with 20 mM sodium potassium phosphate (pH 7.4).
6. Protein amounts were determined by Bradford for camD and photometric measurements for P450cam, PdX, and PdR [15].

8.3.1.3   Cell-Free Enzymatic Catalysis

1. Camphor (1 mM) was added to a solution of 20 mM potassium phosphate pH 7.4 (500 mL) containing KCl (50 mM), ZnSO$_4$ (100 µM), P450cam (0.5 µM), PdR (0.5 µM), PdX (2.5 µM), camD (44 nM), and NADH (250 µM) and shaken at 25 °C for 2 hours.
2. The aqueous solution containing 2,5-diketobornane **3** was filtered over celite and the resulting liquid was extracted with ethyl acetate (3 × 100 mL). The solvent was removed under reduced pressure and the residue was purified by chromatography over silica using 5% MeOH in methylene chloride.

8.3.1.4   Analytical Data

^{1}H-NMR (CDCl$_3$; 400 MHz): δ 0,95 (s, 3H); 1,05 (2 s, 6H); 2,00 (d, $^{2}J = 19,2$ Hz, 1H); 2,13 (dd, $^{2}J = 21,6$ Hz, 3,2 Hz, 1H); 2,29 (d, $^{2}J = 19,2$ Hz, 1H); 2,55–2,61 (m, 2H)
^{13}C-NMR (CDCl$_3$; 100 MHz): δ 214.3; 212.5; 57.6; 46.1; 42.7; 36.6; 19.5; 18.3; 9.0

8.3.2 Procedure 2: Chemical Catalysis for the Synthesis of Bifunctional Camphor Derivatives from 2,5-Diketobornane 3 [16]

8.3.2.1 Materials and Equipment

- 2,5-Diketobornane **3** (0.06 mmol)
- Ammonium chloride (65 equivalents)
- Pd/C (3 equiv.) for the synthesis of diamine **6** or Ru/C (0.4 equiv.) for the synthesis of diol **4**
- *t*-Butanol (5 g)
- Aqueous ammonia (0.6 g)
- Dichloromethane (10 ml)
- Deionized water (10 ml)
- NaOH (2 M, 1 mL)

8.3.2.2 Procedure

1. Diketobornane **3**, ammonium chloride, Pd/C or Ru/C, *t*-butanol, and aqueous ammonia were pressurized with 90 bar hydrogen gas. The reaction mixture was heated to 180 °C for 5 hours. The oil bath was removed and the reaction mixture was cooled overnight. The pressure was released.
2. The catalyst was removed by paper filtration and the resulting solid was washed with *t*-butanol in order to collect the adhesive product.
3. The solvent was removed under reduced pressure at 40 °C.
4. The residue was diluted with dichlormethane or alternatively dissolved in 2 M NaOH and extracted with dichlormethane and subsequently analyzed by GC-MS.

8.3.2.3 Analytical Methods

The product was analyzed after extraction following dilution with an equivalent volume of ethyl acetate containing 100 µM hexylbenzene as internal standard.

GC analysis was achieved on a BPX 5 capillary column (30 m × 0.25 mm × 0.25 µm film thickness, 5% phenyl polysilphenylene-siloxane). 1.0 µL of the sample was injected at 250 °C (split–splitless injector). The oven temperature was initially maintained at 50 °C for 1 minute, then raised to 100 °C at 3.33 °C.min^{-1}, then raised to 200 °C at 4 °C.min^{-1} and held for 10 minutes. The temperature program was adapted for analysis of the experiments on the reductive amination. The temperature was initially maintained at 50 °C for 1 minutes, then raised to 120 °C at 15 °C.min^{-1}, then raised to 170 °C at 5 °C.min^{-1}, then raised to 200 °C at 20 °C.min^{-1} and held for 7 minutes. The column flow (carrier gas: helium) was 1.69 mL. min^{-1}. The product was identified by comparing its retention time and mass spectrum to an authentic standard or by comparing its mass spectrum with the NIST08 library.

8.3.3 Conclusion

Combination of the enzymatic P450cam system with the 5-exo-hydroxycamphor dehydrogenase (camD) allows an efficient synthesis of 2,5-diketobornane. The redox co-factor NADH is regenerated within the process, so there is no requirement for an exogenous substrate like glucose or formate for co-factor regeneration, and oxidation takes place using cheap molecular oxygen as oxidant.

The transformation of the carbonyl groups of 2,5-diketobornane to amino functions (camphor derivative **6**) can be performed with palladium on carbon as a catalyst. By using Ru/C instead of Pd/C, diol **4** can be formed.

8.4 Three Enzyme-Catalyzed Redox Cascade for the Production of a Carvo-Lactone

Nikolin Oberleitner,[1] Christin Peters,[2] Jan Muschiol,[2] Florian Rudroff,[1,*] Marko D. Mihovilovic,[1] and Uwe T. Bornscheuer[2]
*corresponding author.
[1] *Institute of Applied Synthetic Chemistry, Vienna University of Technology, Austria*
[2] *Institute of Biochemistry, Department of Biotechnology and Enzyme Catalysis, University of Greifswald, Germany*

The concept of enzymatic multi-step one-pot reactions has become increasingly important to synthetic chemists in recent years [17]. Such cascades offer the advantages of decreasing the amounts of chemicals used for each reaction and reducing downstream processing efforts. Furthermore, toxic or unstable intermediates are almost completely avoided, which makes processes both safer and more ecological, due to waste reduction [17c]. For this reason, we established an *in vivo* enzymatic cascade reaction comprising three different enzymes, which were produced recombinantly in the same *E. coli* cell to produce potential monomers for biorenewable polyesters [18].

This three-step cascade (Scheme 8.4) allowed us to produce the normal lactone (**D**) from (1*S*,5*S*)-carveol (**A**) in a good yield. In the first reaction step, the alcohol dehydrogenase LK-ADH from *Lactobacillus kefir* [19] converted the unsaturated alcohol, (1*S*,5*S*)-carveol (**A**), to the corresponding ketone, (*S*)-carvone (**B**). Subsequently, the reduction of the 2,3-double bond of (*S*)-carvone (**B**) to (2*R*,5*S*)-dihydrocarvone (**C**) was carried out by the enoate reductase XenB from *Pseudomonas putida* [20]. In the last reaction, the Baeyer–Villiger monooxygenase CHMO from *Acinetobacter* sp. [21] converted the ketone (**C**) to the normal dihydrocarvo-lactone (**D**). With these three enzymes from different organisms – heterologously expressed in *E. coli* – the lactone was produced on preparative scale.

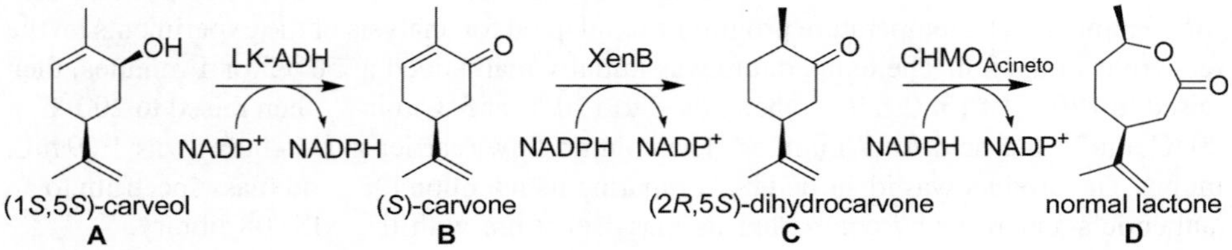

Scheme 8.4   Conversion of (1S,5S)-Carveol (A) to the normal dihydrocarvo-lactone (D).

8.4.1 Procedure 1: Recombinant Expression of the Three Enzymes in *E. coli* BL21(DE3)

8.4.1.1 Materials and Equipment

- Peptone from casein (5 g)
- Tryptone from casein (24 g)

- Yeast extract (50.5 g)
- NaCl (5 g)
- $K_2HPO_4 \cdot 3\,H_2O$ (32.8 g)
- KH_2PO_4 (4.6 g)
- Distilled water (dH_2O)
- Ampicillin (50 mg.mL^{-1} in dH_2O, filter-sterilized)
- Kanamycin (50 mg.mL^{-1} in dH_2O, filter-sterilized)
- IPTG (1 M in dH_2O, filter-sterilized)
- LB agar plate with colonies of *E. coli* BL21(DE3) harboring the expression vector pET28b(+), which bears the genes encoding CHMO and XenB, and the expression vector pET22b(+), with gene encoding LK-ADH [18a]
- Five 0.5 L Schott bottles with screw caps
- 0.1 L Erlenmeyer flask (baffled) with cotton cap
- Five 2 L Erlenmeyer flasks (baffled) with cotton caps
- Orbital shaker (InforsHT Multitron 2 Standard)
- Table autoclave (Tuttnauer 2540EL)
- Cooling centrifuge (min 4500 × *g*)

8.4.1.2 Procedure

1. Peptone (5 g), yeast extract (2.5 g), and NaCl (5 g) were dissolved in dH_2O (500 mL) and autoclaved (20 minutes, 121 °C) in a 0.5 L Schott bottle to give sterile lysogenic broth medium.
2. Tryptone (24 g), yeast extract (48 g), $K_2HPO_4 \cdot 3\,H_2O$ (32.8 g), and KH_2PO_4 (4.6 g) were dissolved in dH_2O (2 L). The solution was distributed to four 0.5 L Schott bottles and autoclaved (20 minutes, 121 °C) to give sterile terrific broth (TB) medium.
3. To prepare the pre-culture, sterile lysogenic broth medium (20 mL) was placed in a sterile baffled 0.1 L Erlenmeyer flask. Afterwards, ampicillin (40 μL) and kanamycin (20 μL) stock solutions were added to reach final concentrations of 100 and 50 μg.mL^{-1}, respectively. The solution was inoculated with a single colony of *E. coli* BL21(DE3), harboring pET28b(+)_CHMO_XenB and pET22b(+)_LK-ADH, and shaken overnight at 37 °C and 200 rpm.
4. The next day, the TB medium (2 L) was supplemented with ampicillin (4 mL) and kanamycin (2 mL) stock solutions to final concentrations of 100 and 50 μg.mL^{-1}, respectively and divided into 5×400 mL fractions in sterile baffled 2 L Erlenmeyer flasks with cotton caps. All were evenly inoculated from the lysogenic broth pre-culture to give an OD_{600} of 0.05.
5. The cells were grown at 37 °C and 200 rpm until an OD_{600} of 1 was reached. Then the cultivation was cooled to 25 °C and expression of the three recombinant enzymes was induced by the addition of 400 μL IPTG stock solution (0.1 mM final conc.) to every flask. The expression was performed for 7 hours at 25 °C.
6. The final OD_{600} was measured and the cells were harvested by centrifugation for 10 minutes at 4500× *g* and 4 °C. The amount of supernatant was estimated in a measuring flask, then the supernatant was discarded. The cells were handled further as described in Procedure 2.

8.4.2 Procedure 2: Preparation of Resting Cells

8.4.2.1 Materials and Equipment

- NaCl (1.25 g)
- KH$_2$PO$_4$ (7.5 g)
- Na$_2$HPO$_4$·7 H$_2$O (32 g)
- Distilled water (dH$_2$O)
- MgSO$_4$ (1 M in dH$_2$O, autoclaved for 20 minutes at 121 °C)
- CaCl$_2$ (1 M in dH$_2$O, autoclaved for 20 minutes at 121 °C)
- Two 0.5 L Schott bottles with screw caps

8.4.2.2 Procedure

1. Na$_2$HPO$_4$·7 H$_2$O (32 g), KH$_2$PO$_4$ (7.5 g), and NaCl (1.25 g) were dissolved in dH$_2$O. Then the volume was adjusted to 500 mL and the mixture was autoclaved (20 minutes, 121 °C) to give 5× nitrogen-free M9 salts.
2. 80 mL 5× nitrogen-free M9 salts, 800 µL MgSO$_4$ stock solution (2 mM final conc.), 40 µL CaCl$_2$ stock solution (0.1 mM final conc.), and 320 mL sterile dH$_2$O were combined to give 400 mL nitrogen-free M9 medium.
3. The cells from the enzyme expression were gently resuspended in nitrogen-free M9 medium to a calculated OD$_{600}$ of 100. The required amount of M9 medium was calculated from the final OD$_{600}$ and the determined volume of supernatant (Procedure 1, Step 6), using the following equation:

$$mL\,M9 = \frac{mL\;supernatant \times final\;OD_{600}}{100}$$

8.4.3 Procedure 3: Biocatalytic Conversion of (1S,5S)-Carveol

8.4.3.1 Materials and Equipment

- (1S,5S)-Carveol stock solution (0.5 g.mL^{-1} in high-purity ethanol)
- Glucose (20% w/v in dH$_2$O, filter-sterilized)
- Aqueous NaOH (3 N)
- Ethyl acetate
- Methyl benzoate
- Distilled water (dH$_2$O)
- Na$_2$SO$_4$, dry
- NaCl (saturated solution)
- Diethyl ether (Et$_2$O)
- Petrol ether (PL)
- Resting cells from Procedure 2
- 1 L Schott bottle with screw cap
- 1 L Separating funnel
- Two 1 L Erlenmeyer flasks
- Orbital shaker (InforsHT Multitron 2 Standard)
- pH Meter

Table 8.2 *GC method.*

Temperature program (r = °C/min^{-1})	Duration
100 °C 2 minutes − 2r → 110 °C − 1.5r → 120 °C 2 minutes − 1.5r → 125 °C − 5r → 150 °C − 1.5r → 165 °C − 2r → 180 °C 1 minute − 50r → 220 °C 1 minute	45 minutes

- Blood glucose meter (AKKU-CHEK Go or similar)
- GC system with FID detection (Thermo Finnigan Focus GC)
- Chiral GC column (BGB175: 30 m × 0.25 mm ID, 0.25 μm film)
- Ultrasonic wave generator (Bandelin Sonopuls HD 3200)
- Ultrasonic sonotrode (Bandelin KE 76)
- Cooling centrifuge (min 17 000 × *g*)
- Rotary evaporator (Rotavap)
- MPLC system with fraction collector (Büchi)
- TLC silica gel 60 F$_{254}$ aluminum plates (Merck)
- KMnO$_4$ dip reagent (1.0 g KMnO$_4$, 20.0 g K$_2$CO$_3$, 10.0 mL NaOH (5%), 150 mL H$_2$O)

8.4.3.2 Procedure

1. (1S,5S)-Carveol was chemically synthesized as recently reported elsewhere [18a].
2. The resting cells in nitrogen-free M9 medium (155.8 mL) obtained in Procedure 2 were supplemented with 8.2 mL glucose stock solution (10 g.L^{-1} final conc.) and placed in a 1 L Schott bottle with screw cap.
3. (1S,5S)-Carveol stock solution (200 μL) was added to the resting cells (0.61 mg.mL^{-1} final conc.). The reaction was performed at 24 °C and 160 rpm in an orbital shaker.
4. Glucose concentration was checked every hour using the AKKU-CHEK Go and maintained at 10 g.L^{-1} by the addition of glucose stock solution.
5. The pH was checked every 30 to 60 minutes using a pH meter and adjusted to 6.5–8.0 by the addition of 3 N NaOH.
6. For reaction monitoring by GC, a cell suspension (50 μL) was extracted with ethyl acetate (250 μL) supplemented with 1 mM methyl benzoate as a reference standard. Samples were analyzed by GC with a BGB175 column (see Table 8.2 for the GC method and Table 8.3 for retention times).
7. After 93% conversion, determined by GC, cells were centrifuged (10 minutes, 4600 × *g*, 4 °C). The aqueous layer was set aside and the cells were diluted with water and sonicated (10 seconds pulse (50% amplitude), 50 seconds cooling on ice, 10 cycles).

Table 8.3 *Retention times for GC analysis.*

Substance	Retention Time [min]
(1S,5S)-Carveol	14.50
(S)-Carvone	14.71
(2R,5S)-Dihydrocarvone	12.58
Normal dihydrocarvo-lactone	32.26
Methyl benzoate (standard)	6.82

8. The cell debris was removed by centrifugation (15 minutes, 17 000 × *g*, 4 °C). All aqueous layers were combined, saturated with NaCl, and extracted with Et$_2$O (200 mL) in a separating funnel. The organic phases were separated in Erlenmeyer flasks and the aqueous phase was extracted three more times with Et$_2$O (3 × 200 mL).
9. The combined organic layers were dried over Na$_2$SO$_4$ and filtrated, and the solvent was removed under reduced pressure.
10. The crude product was purified by chromatography over silica using LP/Et$_2$O (6 : 1) as eluent. Fractions were collected and checked for the desired substances via TLC. LP/Et$_2$O (10 : 1) was used as the TLC solvent, and staining was performed with KMnO$_4$ dip reagent. Fractions containing the desired product were combined and evaporated.
11. The desired dihydrocarvo-lactone, (4*S*,7*R*)-7-methyl-4-(prop-1-en-2-yl)oxepan-2-one (**D**), was obtained in 65 mg (60% yield, > 99% ee).

8.4.4 Conclusion

The procedure described in this section enabled the regio- and stereoselective production of the normal dihydrocarvo-lactone in high yield using standard chemical and microbiological laboratory equipment and methods. In principle, this procedure can be adapted for related substrates through the choice of suitable enzymes [18a].

8.5 Preparation of Homoallylic Alcohols via a Chemoenzymatic One-Pot Oxidation-Allylation Cascade

Michael Fuchs, Markus Schober, Mathias Pickl, Wolfgang Kroutil, and Kurt Faber
Department of Chemistry, Organic and Bioorganic Chemistry, University of Graz, Austria

Homoallylic alcohols serve as precursors for a variety of highly valuable targets, such as chiral chromanes [22], lactones [23], and urea derivatives [24]. Their preparation via the low-valent metal-mediated coupling of allylic bromides and aldehydes (Barbier-type coupling) [25] has been intensively studied, and among the metal reagents used, indium has emerged as a very mild (room-temperature) and water-compatible option [23,26]. A drawback of all these proctocols is the use of aldehyde precursors, which exhibit limited stability. Hence, the *in situ* formation of the aldehyde from the corresponding alcohol is desirable as it avoids the handling and storage of this sensitive reagent. Galactose oxidase from *Fusarium* NRRL 2903 is a copper-dependent oxidase that is able to oxidize benzylic and allylic alcohols to the corresponding aldehydes at the expense of molecular oxygen. Perfect selectivity, especially with regard to undesired overoxidation, is a salient feature of this green oxidation method. The side product – derived from molecular oxygen – is hydrogen peroxide. In order to avoid harming of the enzyme by high concentrations of this highly energetic compound, horse radish peroxidase in combination with a sacrificial substrate (2,2′-azino-bis(3-ethylbenzthiazoline-6-sulfonic acid) diammonium salt (ABTS)) converts the hydrogen peroxide to water and the stable and unreactive ABTS$^{+}\bullet$ radical. Both reactions – the bio-oxidation and the indium-mediated Barbier-type coupling – can be combined in a stepwise manner to a one-pot process, which gives homoallylic alcohols in high yields under mild and aqueous conditions from stable alcohol precursors. Additionally,

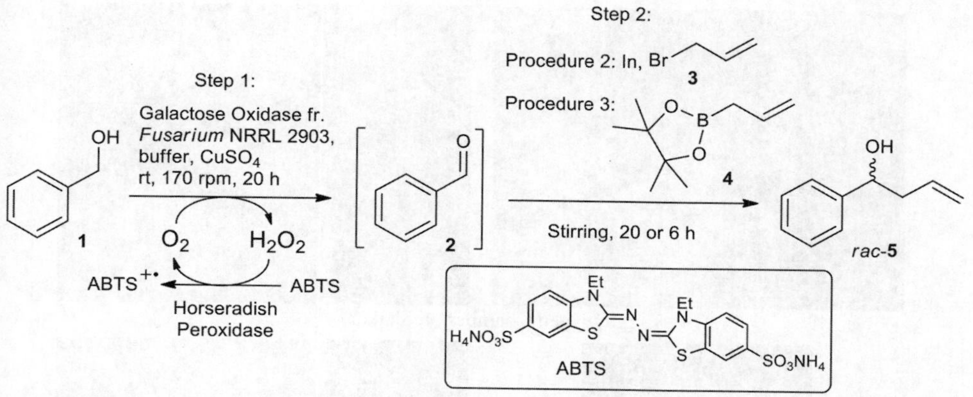

Scheme 8.5 *Preparation of homoallylic alcohols via a two-step bio-oxidation/indium- or boron-mediated allylation one-pot sequence.*

the exchange of allyl bromide/indium by allylboronic esters yields the same class of compounds, thereby avoiding the use of an additional metal mediator, as the boronic ester is activated by the polar solvent instead (see Scheme 8.5) [27].

8.5.1 Materials and Equipment

- *E. coli* BL21(DE3), harboring the plasmid GOX-M1 (frozen glycerol stock), referred to as *E. coli* BL21(DE3)/GOX-M1
- Yeast extract (Oxoid, # LP0021)
- Trypton without casein (Roth, # 8952.2)
- Sodium chloride (Roth, # 9265.1)
- Ampillicin sodium salt (Roth, # K029.2)
- IPTG (Roth, # 2316.4)
- Sodium phosphate buffer (100 mM, pH 7.0)
- Copper sulfate pentahydrate (Sigma Aldrich, # 209198)
- Horseradish peroxidase (Sigma Aldrich, # P8125)
- ABTS (Sigma Aldrich, # A1888)
- Benzylalcohol (Sigma Aldrich, # 402834)
- Benzaldehyde (Sigma Aldrich, # B1334)
- 1-Phenyl-3-buten-1-ol (Sigma Aldrich, # 535907)
- Allylbromide (Sigma # A29585)
- Indium (−100 mesh, Sigma Aldrich # 264032), pretreatment as described in Procedure 4 [28].
- Allylboronic acid pinacol ester (Sigma # 324647)
- GC-FID system
- Agilent Technologies HP5 GC-column (30 m × 0.32 mm × 0.25 μm film)
- Oxygen pressure chamber (see Figure 8.1)
- Oxygen gas (technical grade)
- Rotary shaker at 37 °C and room temperature

Oxygen Chamber on Shaker

 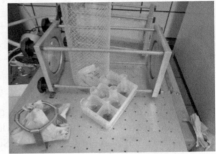

Screening plates for oxidation optimization   Screening plate for cascade reaction
                                                    screening (equipped with 6 x 10 mL flasks)

Figure 8.1 *Oxygen apparatus.*

8.5.1.1 *Optional*

- NMR
- SDS gel electrophoresis chamber, gel, and 2× Lämmli buffer

8.5.2 Procedure 1: Preparation of Biocatalyst

1. LB medium (20 mL, containing $100\,\mu g.mL^{-1}$ ampicillin) was inoculated with glycerol stock of *E. coli* BL21(DE3)/GOX-M1 ($100\,\mu L$). The culture was shaken at $37\,°C$ and 120 rpm overnight.
2. 1 L of LB medium (containing $100\,\mu g.mL^{-1}$ ampicillin and $250\,mg.mL^{-1}$ $CuSO_4x5H_2O$) was inoculated with 2 mL of the overnight culture and shaken at $37\,°C$ and 120 rpm until an OD_{600} of 0.6 was reached. A small sample ($100\,\mu L$) was withdrawn, the remaining cells were induced by the addition of IPTG (238 mg, 1.0 mmol in 1 mL of H_2O), and the culture was incubated at $20\,°C$ and 120 rpm overnight.
3. A small sample ($100\,\mu L$) was withdrawn to check overexpression via SDS gel electrophoresis by comparison with a sample from Step 2. The cells were harvested by centrifugation (8000 rpm, 10 minutes, $4\,°C$), washed with 0.9% aqueous NaCl solution, centrifuged again under the same conditions, and freeze-dried from sodium phosphate buffer (10 mL, 100 mM, pH 7.0).
4. The obtained freeze-dried cells were used for the biotransformations described in Procedures 2–4.

8.5.3 Procedure 2: Oxidation/Allylation Sequence Using Indium(0) and Allylbromide

1. Cells of *E. coli* BL21(DE3) (40 mg, lyophilized dry weight) containing the overexpressed enzyme were rehydrated in phosphate buffer (2 mL, 100 mM, pH 7.0, 10 mM $CuSO_4.5H_2O$) by shaking at 30 °C and 120 rpm in a horizontal position for 20 minutes.
2. The rehydrated cell suspension was transferred to a 10 mL round-bottom flask equipped with a stirring bar. Horseradish peroxidase (0.30 mg in 60 µL buffer), ABTS (0.30 mg, 0.5 µmol in 60 µL buffer), and benzylalcohol (**1**, 43 mg, 41 µL, 0.4 mmol) were added and the mixture was placed into an oxygen apparatus (see Figure 8.1).
3. The apparatus was flushed with oxygen (technical grade) for 1–2 minutes, closed, pressurized to 4 bar, and shaken at room temperature and 170 rpm for 20 hours.
4. After careful depressurization, indium powder (54 mg, 0.48 mmol, −100 mesh) and allylbromide (**3**, 70 µL, 96 mg, 0.8 mmol) were added. The flask was sealed with a glass stopper and stirred for 20 hours at room temperature.
5. The reaction mixture obtained was treated with saturated, aqueous NH_4Cl solution (10 mL) and extracted with EtOAc (3 × 20 mL). In order to facilitate phase separation, the extraction was performed in 50 mL Sarstedt tubes, which were centrifuged after each extraction step.
6. The combined organic phase was dried over Na_2SO_4, filtered, and concentrated under reduced pressure.
7. The product oil was purified via flash chromatography using silica gel (4 g) and petroleum ether/EtOAc (9 : 1) as eluent to give 1-phenyl-3-buten-1-ol (**5**) as a pale yellow oil (57 mg, 0.38 mmol, 96% yield over two steps).

8.5.3.1 Analytical Data

Data for **5**: ¹H-NMR (CDCl₃; 300 MHz): 7.39–7.28 (m, 5H), 5.90–5.77 (m, 1H), 5.22–5.15 (m, 2H), 4.78–4.74 (m, 1H), 2.57–2.51 (m, 2H); ¹³C-NMR (CDCl₃; 75 MHz): 143.9, 134.5, 128.4, 127.6, 125.8, 118.5, 73.3, 43.9; the conversion was determined using GC-FID analysis (N₂ carrier gas, 14.5 psi, 70 °C for 4 minutes, 10 °C.min⁻¹ to 180 °C, 20 °C.min⁻¹ to 280 °C; HP-5 column (30 m × 0.25 mm × 0.25 µm)]: t_{ret}[benzylalcohol (**1**)], 5.15 minutes; t_{ret}[benzaldehyde (**2**)], 3.54 minutes; t_{ret}[1-Phenyl-3-buten-1-ol (**5**)], 8.89 minutes)

8.5.4 Procedure 3: Oxidation/Allylation Sequence Using Allylboronic Acid Pinacol Ester

1. The oxidation was performed as described in Procedure 2, Steps 1–3, but using a different amount of benzylalcohol (**1**, 22 mg, 21 µL, 0.2 mmol).
2. The oxygen apparatus was depressurized carefully and the flask was removed, charged with allylboronic acid pinacol ester (**4**, 40 mg, 45 µL, 0.24 mmol), and closed with a glass stopper. The obtained suspension was stirred at room temperature for 6 hours.
3. The work-up was performed as described in Procedure 2, Steps 5–7, and gave 1-phenyl-3-buten-1-ol (**5**) as a pale yellow oil (28 mg, 0.19 mmol, 95% yield over two steps).

8.5.5 Procedure 4: Etching of Oxide Layer on Indium Metal [28]

1. An excess of indium was washed with hydrochloric acid (10 mL, 5 M) on a Büchner funnel.
2. After 5 minutes, vacuum was applied to remove the acid and the dissolved oxides, while the metal turned shiny.

3. The metal powder was washed with $H_2O_{dist.}$ (10 mL) and acetone (10 mL), quickly dried with air (~30 seconds), and used as such for the allylation reaction.

8.5.6 Conclusion

Homoallylic alcohols – important intermediates in organic synthesis – can be easily prepared via a one-pot biocatalytic oxidation/allylation cascade. Very stable alcohol precursors can be used, avoiding the isolation and storage of the aldehyde intermediate. This method represents one of the rare combinations of enzyme-catalyzed reactions with organometallic chemistry, yielding an easy and applicable route to obtaining these frequently used organic intermediates.

8.6 Cascade Biotransformations via Enantioselective Reduction, Oxidation, and Hydrolysis: Preparation of (R)-δ-Lactones from 2-Alkylidenecyclopentanones

Ji Liu and Zhi Li

Department of Chemical and Biomolecular Engineering, National University of Singapore, Singapore

Optically active δ-lactones are useful flavor and fragrance materials. They are produced chemically, due to the low concentrations found in nature [29,30]. It has been found that different enantiomers of δ-lactones have different odors [31]. The present method of chemical synthesis suffers from a number of drawbacks, including harsh conditions, toxic reagents, an expensive catalyst, and a multi-step preparation [31]. In comparison, enzymatic cascade reactions in one pot are a useful and green tool in sustainable chemical synthesis that avoids tedious isolation and purification of intermediates and reduces waste generation and production cost [32]. In this section, we report a reduction–oxidation–hydrolysis cascade biotransformation for the preparation of enantiopure δ-lactones from easily available 2-alkylidene-cyclopenta-nones using *Acinetobacter* sp. RS1 and *Escherichia coli* co-expressing cyclohexanone monooxygenase (CHMO) [33] and glucose dehydrogenase (GDH) [34] (Scheme 8.6).

Scheme 8.6 *Cascade synthesis of (R)-δ-lactones from 2-alkylidene-cyclopentanones.*

8.6.1 Procedure 1: Preparation of Microbial Catalysts

8.6.1.1 Materials and Equipment

- *Acinetobacter* sp. RS1 strain
- *E. coli* (pRSF-Duet-CHMO/pACYC-Duet-GDH) recombinant strain, hereafter referred as "*E. coli* (CHMO-GDH)"

- TB broth
- Kanamycin (50 mg.L^{-1} stock solution in water, filter-sterilized)
- Chloramphenicol (30 mg.L^{-1} stock solution in ethanol)
- IPTG (0.5 mM)
- Tris buffer (pH 8.0, 50 mM)

8.6.1.2 Procedure

1. *Acinetobacter* sp. RS1 was cultivated overnight with TB broth (30 °C, 250 rpm). Cells were harvested by centrifuge, rinsed three times, and suspended with Tris buffer (pH 8.0, 50 mM) to 12 g cdw.L^{-1}.
2. *E. coli* (CHMO-GDH) was cultivated with TB broth (30 °C, 250 rpm) containing kanamycin (50 µg.L^{-1}) and chloramphenicol (30 µg.L^{-1}). The expression of CHMO protein was started by adding IPTG to a final concentration of 0.2 µM. Cells were rinsed and suspended in Tris buffer to 15 g cdw.L^{-1}.
3. Both types of cell were freshly prepared to ensure good catalytic performance.

8.6.2 Procedure 2: Cascade Biotransformation of 2-Alkylidene-Cyclopentanones to Enantiopure (*R*)-δ-Lactones

8.6.2.1 Materials and Equipment

- 2-Alkylidene-cyclopentanones, 0.25 mmol
- Glucose, AR
- Ethyl acetate, AR
- HPLC-grade *n*-hexane
- 1 L Conical flask
- Orbital shaker
- Centrifuge
- Rotator evaporator
- Flash chromatography

8.6.2.1.1 Optional

- GC system with Agilent HP-5 capillary column (30 m × 322 µm × 0.25 µm)
- Normal phase HPLC system with UV detector and Daicel Chiralpak IA-3 column (250 × 4.6 mm)
- GC system with Macherey–Nagel Hydrodex β-TBDAc chiral column (25 m × 0.25 mm)

8.6.2.2 Procedure

1. 50 mL *Acinetobacter* cell suspension was added to a 1 L conical flask with 0.25 mmol 2-alkylidene-cyclopentanone substrate. The mixture was incubated at 30 °C and 250 rpm for 3 hours. The pH value of the reaction mixture was adjusted to 8 before the follow-up steps.
2. Optional: 1 mL reaction mixture was extracted with *n*-hexane. The resulting organic phase was analyzed by GC and normal phase chiral HPLC for conversion and product *ee*, respectively. The chromatography conditions and retention time is as shown in Table 8.4.
3. The *E. coli* (CHMO-GDH) cell suspension (150 mL) and 20 mg.L^{-1} of glucose were added to the reaction mixture. The reaction mixture was incubated at 30 °C and 250 rpm for 1.5 hours.

Table 8.4 *Analytical conditions for cascade biotransformation of 2-alkylidene-cyclopentanones to enantiopure (R)-δ-lactones.*

GC analysis with HP-5 column, oven temperature: 120 °C, 0 minutes; 20 °C.min^{-1} to 200 °C; 50 °C.min^{-1} to 250 °C; holding at 250 °C for 4 minutes			
Compound	Retention time (minutes)	Compound	Retention time (minutes)
1	4.2	2	4.7
3	3.8	4	4.3
5	5.2	6	5.8

Chiral GC with β-TBDAc column, oven 160 °C, 12 minutes; 4 °C.min^{-1} to 200 °C; 200 °C for 7 minutes			
(S)-5	23.6	(R)-5	23.8
(S)-6	27.4	(R)-6	27.6

Chiral HPLC, Chiralpak IA-3 column, λ = 300 nm, n-hexane/isopropanol = 97/3, oven temperature 25 °C			
(S)-3	8.6	(R)-3	9.0
(S)-4	8.5	(R)-4	8.8

4. The lactones produced were extracted with ethyl acetate, concentrated, and purified by chromatography using ethyl acetate/n-hexane(1 : 9) as eluent. Product *ee* was determined with chiral GC.

8.6.3 Conclusion

The reduction–oxidation–hydrolysis cascade reaction provides a simple and green synthesis of the valuable flavor and fragrance materials (R)-δ-lactones **5** and **6** from the easily available starting materials **1** and **2**. Such a one-pot biotransformation with high product *ee* and high yield is better than other reported methods, including two-step chemical synthesis [31]. Practically, the production yield was limited by the Baeyer–Villiger oxidation and the hydrolysis of δ-lactones **5** and **6** by the *Acinetobacter* cells.

8.7 One-Pot Tandem Enzymatic Reactions for Efficient Biocatalytic Synthesis of D-Fructose-6-Phosphate and Analogs

Israel Sánchez-Moreno,[1,2] Virgil Hélaine,[1,2] Nicolas Poupard,[1,2] Eduardo García-Junceda,[3] Roland Wohlgemuth,[4] Christine Guérard-Hélaine,[1,2] and Marielle Lemaire[1,2]

[1] *Clermont University, Blaise Pascal University, ICCF, Clermont-Ferrand, France*
[2] *CNRS, UMR 6296, France*
[3] *Department of Bioorganic Chemistry, Institute of General Organic Chemistry, CSIC, Spain*
[4] *Sigma-Aldrich, Research Specialties, Switzerland*

Phosphorylated chiral sugars are playing a vital role as central metabolites in various metabolic pathways, such as glycolysis, the non-oxidative pentose phosphate pathway,

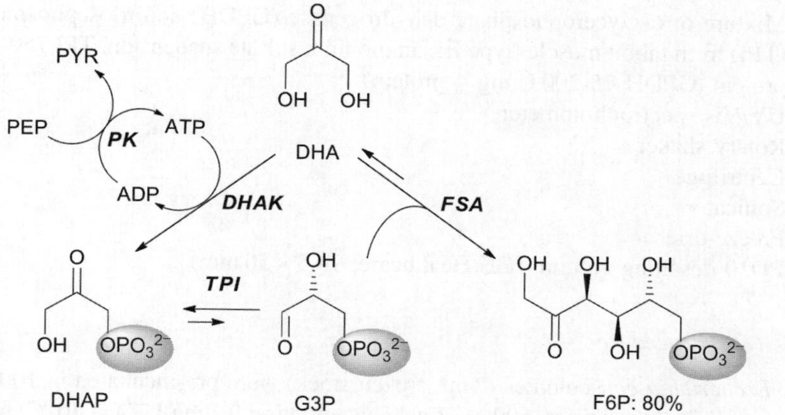

DHA: dihydroxyacetone; ATP: adenosine triphosphate; ADP: adenosine diphosphate; PEP: phosphoenolpyruvate; PYR: pyruvate
DHAP: dihydroxyacetone phosphate; G3P: D-glyceraldehyde-3-phosphate; F6P: D-fructose-6-phosphate
DHAK: dihydroxyacetone kinase; *PK*: pyruvate kinase; *TPI*: triose-phosphate-isomerase; *FSA*: D-fructose-6-phosphate aldolase

Scheme 8.7 *One-pot, two-step multi-enzymatic (3 + 3) reaction leading to D-fructose-6-phosphate.*

and the Calvin cycle. They are of particular interest in the study of various metabolic diseases [35]. Given the importance of such compounds, several tandem enzymatic methods have been implemented which combine a phosphorylation step (by the way of kinases [36] or phosphatases [37]) with a C—C bond formation step (using transketolase and/or fructose-1,6-*bis*phosphate aldolase [38] or a synthase [39]). In this section, we have decided to combine dihydroxyacetone kinase (DHAK) [40] with an ATP recycling system and fructose-6-phosphate aldolase [41] (Scheme 8.7). In this process, starting from achiral substrates, chirality is elegantly introduced using two enzymes: an isomerase, controlling one stereogenic center, and an aldolase, establishing two other contiguous stereocenters, resulting in the control of three configurations, with a cocktail of four biocatalysts [42].

8.7.1 FSA A129S

8.7.1.1 Materials and Equipment

- Stock of expression strain DH5α (*E. coli* K12), containing the plasmid pJF119-fsaA with the A129S mutation [43] (frozen at −80 °C)
- LB broth, Miller, from Difco
- Glycyl-glycine (Gly-Gly) buffer
- IPTG in water (0.5 M)
- Ampicillin solution in water (100 mg.mL^{-1})
- Bradford's reagent
- NADH disodium salt hydrate solution in water (12 mg.mL^{-1})
- D-Fructose-6-phosphate disodium salt hydrate solution in water (430 mg.mL^{-1})

- Mixture of α-glycerophosphate dehydrogenase (GPDH) and triosephosphate isomerase (TPI) from rabbit muscle (type III, ammonium sulfate suspension, TPI 750–2000 U.mg^{-1} protein, GPDH 75-200 U.mg^{-1} protein)
- UV/Vis spectrophotometer
- Rotary shaker
- Centrifuge
- Sonicator
- Freeze-drier
- PD10 desalting column (GE Healthcare, 14.5×50 mm)

8.7.1.2 Procedure

1. *Escherichia coli* colonies (2 mL frozen stock) were pre-incubated in 100 mL LB broth containing ampicillin (100 μL, final concentration 0.1 mg.L^{-1}) at 30 °C, with shaking at 250 rpm.
2. An aliquot of pre-inoculum (4 mL) was transferred in 400 mL LB with ampicillin (400 μL, final concentration 0.1 mg.mL^{-1}). The mixture was shaken at 250 rpm (37 °C).
3. When the culture reached an OD$_{600}$ of 0.8, protein expression was induced with IPTG (400 μL, final concentration 0.5 mM) and the temperature was dropped to 30 °C. The culture was incubated for a further period of 16 hours, with shaking at 250 rpm.
4. Cells were harvested by centrifugation for 15 minutes at 8000 rpm and 4 °C, washed twice with Tris buffer (15 mL, 50 mM, pH 7.5, containing 50 mM NaCl and 30 mM imidazole), and resuspended in Gly-Gly buffer (10 mL, 50 mM, pH 8.0).
5. The cell suspension was kept cold in ice and was disrupted by sonication (cycles of 8 seconds of sonication followed by 10 seconds of rest for 40 minutes). Cell debris was discarded by centrifugation at 8000 rpm for 20 minutes, and the clear supernatant was incubated at 70 °C for 30 minutes. The precipitate formed was separated by centrifugation (8000 rpm for 30 minutes) and the supernatant was freeze-dried.
6. The powder (70 mg) thus obtained was dissolved in water (2.5 mL) and loaded on to a PD-10 desalting column to remove the Gly-Gly buffer. Elution was carried out with distilled water and the protein content in each fraction (presence of protein) was monitored using the Bradford assay (10 μL from a fraction mixed with 200 μL Bradford's reagent). Aqueous fractions of the enzyme were pooled and used directly for the reaction.
7. The activity of D-fructose-6-phosphate aldolase (FSA) A129S was measured by a multi-enzymatic test, as depicted in Scheme 8.8:
   a. NADH solution (20 μL), D-fructose-6-phosphate (30 μL), GPDH/TPI (10 μL), the FSA sample from the PD10-column (5 μL), and 50 mM Gly-Gly buffer pH 8.0 (935 μL) were mixed, and the reaction was followed spectrophotometrically.
   b. The disappearance of NADH was directly correlated to D-fructose-6-phosphate retro-aldolization by FSA A129S.
   c. One unit (U) of enzyme activity is defined as the amount of enzyme that will catalyze the conversion of 1 μmol of D-fructose-6-phosphate per minute at pH 8.0 and 25 °C. The enzymatic activity of the pooled eluates was found to be 95 U.mL^{-1}.

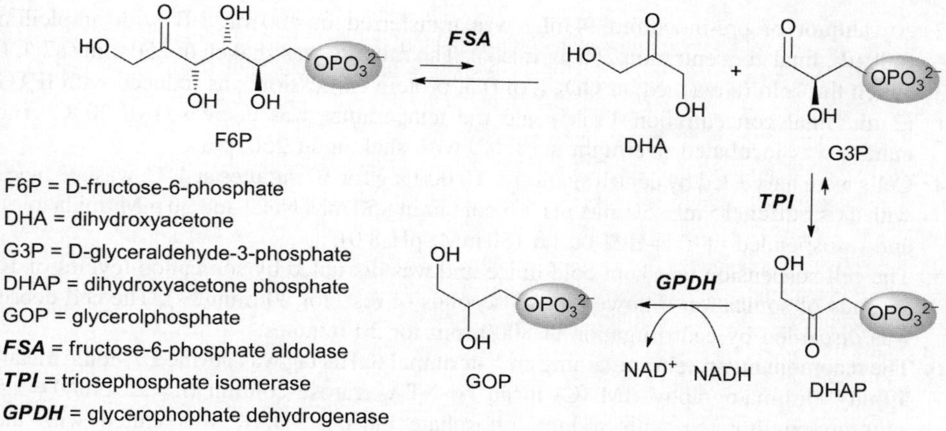
FSA
F6P
DHA
G3P
TPI
GPDH
NAD+
NADH
GOP
DHAP
F6P = D-fructose-6-phosphate
DHA = dihydroxyacetone
G3P = D-glyceraldehyde-3-phosphate
DHAP = dihydroxyacetone phosphate
GOP = glycerolphosphate
FSA = fructose-6-phosphate aldolase
TPI = triosephosphate isomerase
GPDH = glycerophophate dehydrogenase

Scheme 8.8 *Multi-enzymatic activity test for FSA.*

8.7.2 Dihydroxyacetone Kinase (DHAK)

8.7.2.1 Materials and Equipment

- Stock of expression strain BL21 (DE3), containing the plasmids pRSET-dhak (frozen at −80 °C)
- Mixture of GPDH and TPI from rabbit muscle (type III, ammonium sulfate suspension, TPI 750-2000 $U.mg^{-1}$ protein, GPDH 75-200 $U.mg^{-1}$ protein)
- LB broth, Miller, from Difco
- Sodium phosphate buffer (50 mM, pH 7.5)
- Tris-HCl buffer (50 mM, pH 8.0)
- Ampicillin solution in water (250 $\mu g.mL^{-1}$)
- IPTG in water (0.5 M)
- Ni-NTA-agarose resin (Quiagen)
- Bradford's reagent
- NADH disodium salt hydrate solution in water (20 mM)
- Adenosine 5′-triphosphate (ATP) disodium salt hydrate solution in water (50 mM)
- $MgSO_4$ solution in water (50 mM)
- 1,3-Dihydroxyacetone (DHA) dimer solution in water (1 M)
- Rotary shaker
- Spectrophotometer UV/Vis
- Centrifuge (10,000 × *g*)
- Sonicator
- Freeze-drier
- Dialysis tubing cellulose membrane (average flat width: 33 mm).

8.7.2.2 Procedure

1. *Escherichia coli* BL21 (DE3) colonies containing the plasmids pRSET-dhak (2 mL frozen stock) were pre-incubated in 100 mL LB broth containing ampicillin (100 μL, final concentration 250 $\mu g.mL^{-1}$) at 37 °C, with shaking at 250 rpm.

2. An aliquot of pre-inoculum (4 mL) was transferred in 400 mL LB with ampicillin (400 μL, final concentration 250 μg.mL^{-1}). The mixture was shaken at 250 rpm (37 °C).

3. When the culture reached an OD_{600} of 0.5, protein expression was induced with IPTG (2 mL, final concentration 1 mM) and the temperature was decreased to 30 °C. The culture was incubated overnight at 30 °C, with shaking at 250 rpm.

4. Cells were harvested by centrifugation at $10\,000 \times g$ for 10 minutes at 4 °C, washed twice with Tris buffer (15 mL, 50 mM pH 7.5 containing 50 mM NaCl and 30 mM imidazole), and resuspended in Tris-HCl buffer (50 mM, pH 8.0).

5. The cell suspension was kept cold in ice and was disrupted by sonication (cycles of 10 seconds of sonication followed by 10 seconds of rest, for 20 minutes). The cell debris was discarded by centrifugation at 8000 rpm for 20 minutes.

6. The recombinant protein containing an N-terminal 6xHis tag was purified by ionic metal-affinity chromatography (IMAC) in an Ni-NTA-agarose column (h = 2.5 cm/Ø = 2.5 cm) pre-equilibrated with sodium phosphate buffer. DHAK was eluted with the equilibrating buffer, supplemented with imidazole (0.5 M). The presence of enzyme in each fraction was monitored using the Bradford assay (aliquots of 10 μL in 200 μL of Bradford's reagent) and the DHAK activity assay.

7. In order to remove imidazole, a dialysis step was carried out. Fractions containing DHAK were pooled together in a dialysis membrane and incubated overnight at 4 °C in 5 L of mili-Q water (pH ~7.0) under gentle stirring. The dialyzed protein was then freeze-dried.

8. DHAK activity was measured spectrophotometrically by enzymatic quantification of the DHAP formed during the phosphorylation reaction of DHA, using GPDH (Scheme 8.8). This assay was run at room temperature, following NADH absorbance at 340 nm for 15 minutes in a reaction mixture of 1 mL containing Tris-HCl buffer (800 μL), NADH (20 μL), ATP (75 μL), $MgSO_4$ (100 μL), GPDH/TPI mixture (2 μL), DHAK sample (1–5 μL), and DHA (2.5 μL). One unit (U) is defined as the amount of enzyme able to catalyze the phosphorylation 1 μmol of DHA per minute at pH 7.5 and 25 °C.

8.7.3 Tandem Reaction (One-Pot, Two-Step Reaction)

8.7.3.1 Materials and Equipment

- Tris-HCl buffer 80 mM, pH 7.5 (10 mL)
- DHA dimer (72 mg, 0.8 mmol)
- Phosphoenolpyruvic acid (PEP) monosodium salt hydrate (122 mg, 0.64 mmol)
- $MgSO_4$ (77 mg)
- Adenosine 5′-triphosphate (ATP) disodium salt hydrate (28 mg, 52 μmol)
- DHAK, obtained as already described (11 U)
- FSA A129S, obtained as already described (50 U)
- Pyruvate kinase (PK) from rabbit muscle (type III, lyophilized powder (14 U)
- NADH solution in water (20 mM)
- Mixture of GPDH and TPI from rabbit muscle (type III, ammonium sulfate suspension, TPI 750–2000 U.mg^{-1} protein, GPDH 75–200 U.mg^{-1} protein)
- TPI from baker's yeast (type I, ammonium sulfate suspension, 400 U)
- 1 M HCl solution in water

- 1 M NaOH solution in water
- $BaCl_2$, $2H_2O$ (42 mg)
- Ethanol (100 mL)
- Acetone
- Centrifuge ($10\,000\times g$)
- Rotary evaporator
- Rotary shaker

8.7.3.2 Phosphorylation of DHA

1. The reaction was carried out at room temperature under gentle shaking (100–200 rpm).
2. DHA (72 mg, 0.8 mmol), PEP (122 mg, 0.64 mmol), and $MgSO_4$ (77 mg, 10 mM) were dissolved in Tris-HCl buffer (10 mL; 80 mM; pH 7.5). Water was added to a total reaction volume of 20 mL (final buffer concentration was 40 mM) and the pH of the mixture was adjusted to 7.5.
3. DHAK (11 U) and PK (14 U) were suspended in 1 mL of ultra-pure water (pH ~7.0) and added to the reaction. ATP was also added (28 mg, 52 µmol).
4. The amount of DHAP formed was quantified by the enzymatic assay described in Scheme 8.8 (reduction of DHAP is catalyzed by GPDH, with concomitant oxidation of NADH to NAD^+). The absorbance decrease at 340 nm was proportional to the DHAP concentration ($\varepsilon_{NADH} = 6220\,cm^{-1}.M^{-1}$). This spectrophotometric assay was run at room temperature for 10 minutes, in a final volume of 1 mL, containing Tris-HCl 50 mM, pH 8.0 (975 µL), NADH (20 µL), an aliquot of the reaction (2.5–5.0 µL), and a mixture of GPDH/TPI (2 µL).

8.7.3.3 Production of G3P and Aldol Reaction

1. The reaction was carried out at room temperature under gentle shaking (100–200 rpm).
2. When the DHAP accumulation in the previous procedure reached 95%, the mixture was centrifuged ($10\,000\times g$, 10 minutes, 4 °C) to discard any denatured or aggregated enzymes.
3. DHA (216 mg, 3 eq.) was added and the pH was adjusted to 8.0, with a 1 M NaOH solution.
4. Then FSA A129S (50 U) and TPI (400 U) were added and the reaction was allowed to proceed with gentle shaking. The reaction was monitored by DHAP assay. After 25 minutes, the reaction was evaluated to be incomplete; thus, 3 eq. of DHA were added. An assay 5 minutes later has revealed a total disappearance of DHAP.
5. The reaction was stopped by dropping the pH to 3.0 with a 1 M HCl solution, resulting in partial precipitation of the enzymes. The mixture was centrifuged ($10\,000 \times g$, 10 minutes, 4 °C) to eliminate the denatured or aggregated enzymes.

8.7.3.4 Purification of D-Fructose-6-Phosphate

1. The pH of the supernatant was adjusted to 6.0 with a 1 M NaOH solution, and two equivalents of $BaCl_2 \cdot 2H_2O$ (42 mg) were added.
2. The solution was centrifuged ($10\,000\times g$, 10 minutes, 4 °C) and the pellets were discarded.
3. After partial concentration under vacuum (rotary evaporator), five volumes of ethanol were added (100 mL) to precipitate the product and the suspension incubated overnight at 4 °C. Then the mixture centrifuged again ($10\,000\times g$, 10 minutes, 4 °C).

4. After removing the supernatant, the barium salt of the phosphorylated sugar was obtained as a white powder after one washing with ethanol and two washings with acetone, followed by drying.

8.7.4 Analytical Data

Fructose-6-phosphate barium salt: yield = 80% (202 mg)

^{1}H-NMR (400 MHz; D_2O + HCl): δ = 4.25 – 3.85 (m, 10H, H-3α, H-3β, H-4α, H-4β, H-5α, H-5β, 2H-6α, 2H-6β), 3.65 (d, 1HA, J = 12.1 Hz, H-1α), 3.61 (d, 1HB, J = 12.3 Hz, H-1α), 3.58 (d, 1HA, J = 12.1 Hz, H-1β), 3.53 (d, 1HB, J = 12.1 Hz, H-1β)

^{13}C-NMR (100 MHz; D_2O + HCl): δ 104.46 (C-2α), 101.63 (C-2β), 81.78 (C-3α), 79.79 (d, J = 7.9 Hz, C-5α), 79.21 (d, J = 8.2 Hz, C-5β), 75,75 (C-4α), 75,18 (C-3β), 74.21 (C-4β), 65.55 (d, J = 4.9 Hz, C-6β), 64.58 (d, J = 4.8 Hz, C-6α), 62.8 (C-1α), 62.67 (C-1β)

$[\alpha]_D^{20}$ = +3.5 (c = 1.5, HCl 1 N)

HRMS (ESI$^-$), m/z calculated for [$C_6H_{13}O_9P$-H]: 229.0113; found 229.0123

8.7.5 Conclusion

An enzymatic one-pot cascade reaction was successfully developed for the preparation of phosphorylated sugars of high biological interest. Indeed, by varying the structure of the FSA donor substrate, D-fructose-6-phosphate and its analogs (see Scheme 8.9 and Table 8.5) were obtained as their barium salts in a single precipitation step, directly from the reaction

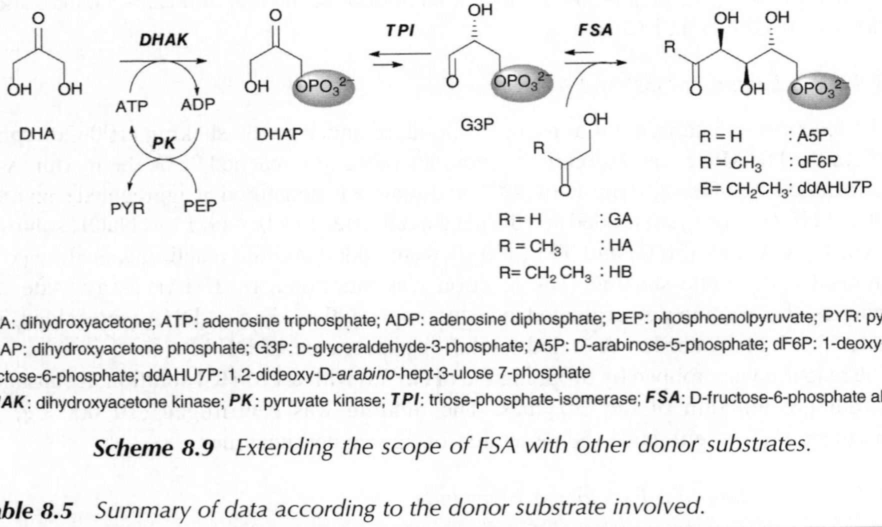

DHA: dihydroxyacetone; ATP: adenosine triphosphate; ADP: adenosine diphosphate; PEP: phosphoenolpyruvate; PYR: pyruvate
DHAP: dihydroxyacetone phosphate; G3P: D-glyceraldehyde-3-phosphate; A5P: D-arabinose-5-phosphate; dF6P: 1-deoxy-D-fructose-6-phosphate; ddAHU7P: 1,2-dideoxy-D-*arabino*-hept-3-ulose 7-phosphate
DHAK: dihydroxyacetone kinase; *PK*: pyruvate kinase; *TPI*: triose-phosphate-isomerase; *FSA*: D-fructose-6-phosphate aldolase

Scheme 8.9 *Extending the scope of FSA with other donor substrates.*

Table 8.5 *Summary of data according to the donor substrate involved.*

R	FSA source	FSA (U)	FSA donor (eq.)	Reaction time (minutes)	Product yield (%)
CH_2OH	FSA A129S	50	6	30	80
H	Wild-type	13	9	70	87
CH_3	Wild-type	13	9	55	95
CH_2CH_3	Wild-type	13	12	90	94

mixture, in high yields and purities. Thus, using wt-FSA [42] and replacing DHA with hydroxyacetone (HA) or hydroxybutanone (HB), the corresponding 1-deoxy-D-fructose 6-phosphate and 1,2-dideoxy-D-*arabino*-hept-3-ulose 7-phosphate were formed. In addition, a straightforward access to a phosphorylated aldose, namely D-arabinose 5-phosphate, was developed when using glycolaldehyde (GA). This straightforward multi-step enzymatic synthesis set-up offers versatile access to a broad molecular diversity of chiral phosphorylated carbohydrate metabolites [42].

8.8   Efficient One-Pot Tandem Biocatalytic Process for a Valuable Phosphorylated C8 D-Ketose: D-*Glycero*-D-*Altro*-2-Octulose 8-Phosphate

Christine Guérard-Hélaine,[1,2] Marine Debacker,[1,2] Pere Clapés,[3] Anna Szekrenyi,[3] Virgil Hélaine,[1,2] and Marielle Lemaire[1,2]

[1]*Clermont University, Blaise Pascal University, ICCF, Clermont-Ferrand, France*
[2]*CNRS, UMR 6296, France*
[3]*Biotransformation and Bioactive Molecules Group, Institute of Advanced Chemistry of Catalonia (IQAC-CSIC), Spain*

Since abnormal levels of phosphorylated sugars in mammals are directly correlated to a wide range of diseases [44], phosphorylated ketoses have been considered as important therapeutic targets. For instance, octulose 8-phosphates are involved in several pathways [45–48] and are interesting as precursors to and analogs of 3-deoxy-D-*manno*-oct-2-ulosonic acid (KDO), sialic acid, and related sugars [49,50]. An FSA A129S variant sample [51], containing traces of *E. coli* ribulose-5-phosphate epimerase (RPE) and ribose-5-phosphate isomerase (RPI) from the pentose phosphate pathway, is able to perform a one-pot multi-enzymatic cascade reaction involving two carboligation reactions, an isomerization, and an epimerization. The first carboligation with DHA and glycolaldehyde phosphate **1** leads to xylulose-5-phosphate **2**, which is then epimerized into ribulose-5-phosphate **3** (Scheme 8.10). The subsequent isomerization provides ribose-5-phosphate **4**, revealed as a good substrate for a second carboligation with DHA [52]. This last aldol

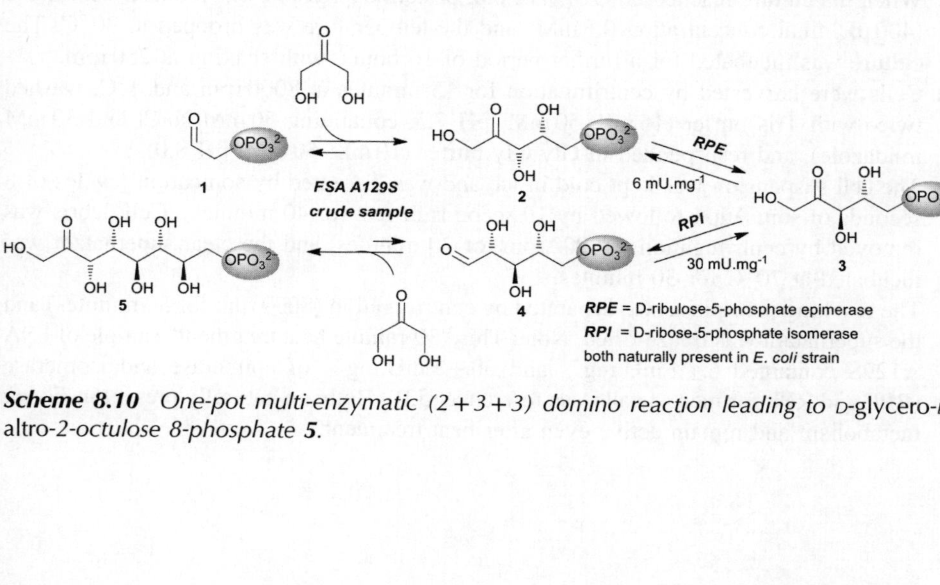

Scheme 8.10 *One-pot multi-enzymatic (2 + 3 + 3) domino reaction leading to D-glycero-D-altro-2-octulose 8-phosphate 5.*

reaction, driving the overall process, affords the desired D-*glycero*-D-*altro*-2-octulose 8-phosphate **5**. Five contiguous asymmetric centres were created using this route, with full control of the stereochemistry [52].

8.8.1 FSA A129S Partially Purified Production

8.8.1.1 Materials and Equipment

- Stock of expression strain DH5α (*E. coli* K12), containing the plasmid pJF119-fsaA, with the A129S mutation (frozen at $-80\,°C$)
- LB broth, Miller, from Difco
- Gly-Gly buffer (50 mM, pH 8.0)
- UV/Vis spectrophotometer
- Ampicillin solution in water (100 mg.mL^{-1})
- Bradford's reagent
- NADH disodium salt hydrate solution in water (12 mg.mL^{-1})
- D-Fructose-6-phosphate disodium salt hydrate solution in water (430 mg.mL^{-1})
- Mixture of GPDH and TPI from rabbit muscle (type III, ammonium sulfate suspension, TPI 750–2000 U.mg^{-1} protein, GDH 75–200 U.mg^{-1} protein)
- IPTG in water (0.5 M)
- Rotary shaker
- Centrifuge
- Sonicator
- Freeze-drier
- PD10 Desalting column (GE Healthcare, 14.5×50 mm)

8.8.1.2 Procedure

1. *Escherichia coli* colonies (2 mL frozen stock) were pre-incubated in 100 mL LB broth containing ampicillin (100 µL, final concentration 0.1 mg.L^{-1}) at 30 °C, with shaking at 250 rpm.
2. An aliquot of pre-inoculum (4 mL) was transferred into 400 mL LB with ampicillin (400 µL, final concentration 0.1 mg.mL^{-1}) and the mixture was shaken at 250 rpm (37 °C).
3. When the culture reached an OD$_{600}$ of 0.8, protein expression was induced with IPTG (400 µL, final concentration 0.5 mM) and the temperature was dropped to 30 °C. The culture was incubated for a further period of 16 hours, with shaking at 250 rpm.
4. Cells were harvested by centrifugation for 15 minutes at 8000 rpm and 4 °C, washed twice with Tris buffer (15 mL, 50 mM, pH 7.5, containing 50 mM NaCl and 30 mM imidazole), and resuspended in Gly-Gly buffer (10 mL, 50 mM, pH 8.0).
5. The cell suspension was kept cold in ice and was disrupted by sonication (cycles of 8 seconds of sonication followed by 10 seconds of rest for 40 minutes). Cell debris was removed by centrifugation at 8000 rpm for 20 minutes, and the clear supernatant was incubated at 70 °C for 30 minutes.
6. The resulting precipitate was separated by centrifugation (8000 rpm for 30 minutes) and the supernatant was freeze-dried. Note: This "30-minute heat treatment" sample of FSA A129S contained 6 ± 1 mU.mg^{-1} and 30 ± 3 mU.mg^{-1} of epimerase and isomerase (RPE and RPI; for more details, see reference [52]). Both are naturally present in *E. coli* metabolism and remain active even after heat treatment.

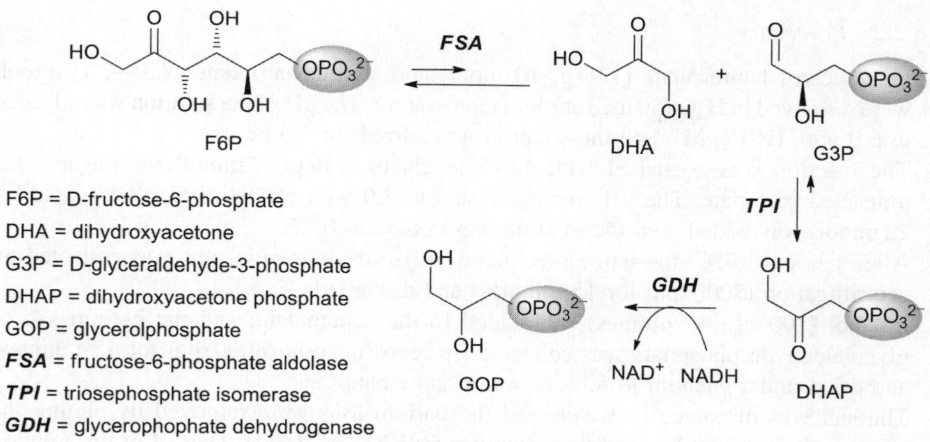

F6P = D-fructose-6-phosphate
DHA = dihydroxyacetone
G3P = D-glyceraldehyde-3-phosphate
DHAP = dihydroxyacetone phosphate
GOP = glycerolphosphate
FSA = fructose-6-phosphate aldolase
TPI = triosephosphate isomerase
GDH = glycerophophate dehydrogenase

Scheme 8.11 *Multi-enzymatic activity test for FSA.*

7. The powder (70 mg) thus obtained was dissolved in water (2.5 mL) and loaded onto a PD-10 desalting column to remove the Gly-Gly buffer. Elution was carried out with distilled water and the protein content in each fraction (presence of protein) was monitored using Bradford's test (10 μL from a fraction mixed with 200 μL of Bradford's reagent). Aqueous fractions of enzyme were pooled and directly used for the reaction.

8. Activity of FSA A129S was measured by a multi-enzymatic test, as depicted in Scheme 8.11.

   a. NADH solution (20 μL), D-fructose-6-phosphate (30 μL), GPDH/TPI (10 μL), an FSA sample from the PD10-column (5 μL), and 935 μL of 50 mM Gly-Gly buffer pH 8.0 were mixed. Reaction conversion was followed spectrophotometrically.

   b. The disappearance of NADH was directly correlated to D-fructose-6-phosphate retro-aldolization by FSA A129S.

   c. One unit (U) is defined as the amount of enzyme able to convert 1 μmol of D-fructose-6-phosphate per minute at pH 8.0. Enzymatic activity of the pooled eluates was found to be 95 U·mL^{-1}.

8.8.2 Glycolaldehyde Phosphate Synthesis

8.8.2.1 Materials and Equipment

- D,L-α-Glycerol phosphate magnesium salt hydrate (1.94 g, 10 mmol)
- Sodium periodate (2.35 g, 11 mmol)
- Distilled water
- 1 M HCl solution
- 1 M NaOH solution
- Ethylene glycol (110 μL)
- BaCl$_2$, 2H$_2$O (5.37 g)
- Ethanol (200 mL)
- Dowex 50WX8 ion-exchange resin (H$^+$ form) 100–200 mesh
- Centrifuge

8.8.2.2   Procedure

1. D,L-Glycerol-1-phosphate (1.94 g, 10 mmol) and sodium periodate (2.35 g, 11 mmol) were dissolved in H_2O (50 mL) at room temperature. The pH of the solution was adjusted to 6.0 with HCl (1 M), and the solution was stirred for 1.5 hours.
2. The reaction was quenched with ethylene glycol (110 μL 2 mmol) to consume the unreacted periodate. The pH was readjusted to 7.0 with NaOH (1 M), $BaCl_2$ (5.37 g, 22 mmol) was added, and the mixture was cooled to 0 °C.
3. After 1 hour at 0 °C, the white precipitate of barium inorganic salts was collected by centrifugation (8000 rpm for 15 minutes) and discarded.
4. Ethanol (200 mL, 4 volumes) was added to the supernatant, and the barium salt of glycolaldehyde phosphate was collected by centrifugation (8000 rpm for 15 minutes) and dried under vacuum to remove water and ethanol.
5. The salt was dissolved in water, and the barium ions were removed by eluting the solution through a column of resin (Dowex 50WX8, H^+ form). The pH of the solution was readjusted to 7.5 with NaOH (1 M), and the solution was stored in the freezer.

8.8.2.3   Analytical Data

^{1}H-NMR (400 MHz; D_2O): δ 5.18 (t, 1H, H-1), 3.88 (m, 2H, H-2)
^{13}C-NMR (100 MHz; D_2O): δ 90.7 (C-1), 68.2 (C-2)

8.8.3   Tandem Reaction

8.8.3.1   Materials and Equipment

- Freshly prepared 700 mM glycolaldehyde phosphate solution in water, pH 7.5 (235 μL, 0.164 mmol)
- DHA (32 mg, 0.355 mmol)
- Distilled water
- FSA A129S, obtained as already described (150 U)
- 1 M HCl solution in water
- 1 M NaOH solution in water
- $BaCl_2$, $2H_2O$ (42 mg)
- Ethanol (20 mL)
- Acetone
- Dowex 50WX8 ion-exchange resin (H^+ form) 100–200 mesh, exchanged with ammonium ion by passing a 1 M NH_4OH solution prior to use
- Centrifuge
- Rotatory evaporator
- Rotary shaker

8.8.3.2   Procedure

1. Glycolaldehyde phosphate (235 μL, 700 mM solution in water, pH 7.5) and DHA (32 mg, 2.1 eq.) were dissolved in water (3.4 mL), resulting in a 50 mM glycolaldehyde phosphate concentration.
2. The crude FSA A129S sample (150 U, containing residual amounts of RPE and RPI; see Scheme 8.10) was added and the resulting mixture was allowed to gently stir (100–200 rpm) at room temperature for 60 hours.

3. The reaction mixture was adjusted to pH 3 with HCl (1 M) and then to pH 6 with NaOH (1 M). The resultant suspension was centrifuged for 15 minutes at 14 000 rpm.

4. Barium chloride (42 mg, 2 eq.) was added to the supernatant, followed by ethanol (20 mL, 6 eq volumes), and the mixture was cooled at 4 °C for at least 1 hour.

5. The suspension was centrifuged for 15 minutes at 8000 rpm, and the precipitated product was washed twice with ethanol and acetone and concentrated under vacuum (rotatory evaporator).

6. The barium counter ion of the product was exchanged to ammonium using a Dowex 50WX8 ammonium-form ion-exchange resin in batch.

7. After removal of the resin by filtration, the filtrate was concentrated under vacuum (rotory evaporator) to afford the final product as its ammonium salt (38.5 mg, 66% yield).

8.8.3.3 Analytical Data

The ^{13}C-NMR spectrum was identical to that described in the literature [53]. Furanose, pyranose, and their α- and β-anomeric forms, together with the linear form, could co-exist for this compound. As noted by the authors of reference [53], due to overlapping and too-weak signals for minor product forms, only the resonances of the β-furanose structure could be fully assigned.

^{1}H-NMR (400 MHz; D$_2$O): δ 4.32 (t, 1H, $J = 7.8$ Hz, H-4β), 4.05 (d, 1H, $J = 7.9$ Hz, 3β), 4.00 (dd, 1H, $J = 3.8, 7.1$ Hz, H-5β), 3.93–3.91 (m, 2H, H-8β), 3.87–3.82 (m, 2H, H-6β, H-7β), 3.54 (d, 1H, $J = 12.2$ Hz, H-1aβ), 3.51 (d, 1H, $J = 12.1$ Hz, H-1bβ)

^{13}C-NMR (100 MHz; D$_2$O): δ (α: α-furanose; β: β-furanose; p: pyranose) 104.54 (C-2α), 101.49 (C-2β), 97.45 (C-2p), 80.34 (C-5β), 75.75 (C-3β), 74.71 (C-4β), 71.18 (C-6β), 71.09 (d, $J = 6.9$ Hz, C-7β), 65.46 (d, $J = 4.8$ Hz, 8β), 62.39 (C-1β)

HRMS (ESI$^-$); calculated for [C$_8$H$_{17}$O$_{11}$P-H]: 319.0430, found: 319.0416

8.8.4 Conclusion

This new one-pot cascade reaction provides an environmentally friendly way of accessing a valuable terminally *O*-phosphorylated D-ketose. This three-reaction sequence was accomplished with full control of five stereogenic centers, starting from achiral material. Moreover, the whole cascade reaction requires only one cloned and overexpressed enzyme; namely, FSA A129S. Epimerase and isomerase are wild-type activities that result directly from the expression strain *E. coli*.

8.9 Chemoenzymatic Synthesis of (*S*)-1,2,3,4-Tetrahydroisoquinoline-3-Carboxylic Acid by PAL-Mediated Amination and Pictet-Spengler Cyclization

Fabio Parmeggiani, Nicholas J. Weise, Syed T. Ahmed, and Nicholas J. Turner
School of Chemistry, Manchester Institute of Biotechnology, The University of Manchester, UK

The integration of chemical steps and biocatalytic reactions is becoming more and more popular in laboratory-scale synthesis, as well as in large-scale industrial chemical

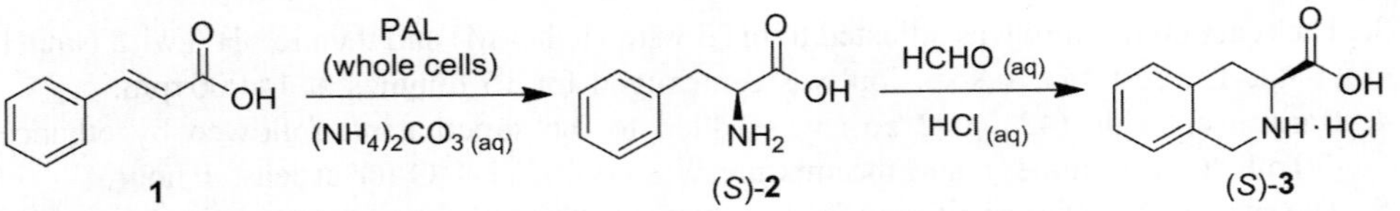

Scheme 8.12 *Two-step synthesis of (S)-1,2,3,4-tetrahydroisoquinoline-3-carboxylic acid by PAL-mediated amination and Pictet–Spengler cyclization.*

production. Indeed, it can provide great advantages, such as the reduction of the number of purification steps, the minimization of the amounts of chemicals used (e.g., solvents), and the simplification of the work-up or purification procedures. Clearly, these advantages match very well with the purposes and methods of green chemistry.

The hydroamination of substituted cinnamic acids catalyzed by phenylalanine ammonia lyases (PALs, E.C. 4.3.1.24) is one of the most atom-economic strategies to access optically pure α-amino acids in high yield, under environmentally friendly conditions, and starting from cheap and easily synthesized substrates [54].

Optically pure amino acids are among the chiral building blocks of choice for the synthesis of chiral compounds, and their structural moieties are ubiquitously present not only in numerous natural products and biologically active peptide-like molecules, but also in a large number of API intermediates and drug candidates. As an example, in this section we report the coupling of the amination of cinnamic acid **1** (mediated by AvPAL from *Anabaena variabilis* [55]) and Pictet–Spengler cyclization [56] on the crude product mixture, yielding (S)-1,2,3,4-tetrahydroisoquinoline-3-carboxylic acid (S)-**3**, a key intermediate in the pharmaceutical and fine chemical industry [57] (Scheme 8.12). This procedure has the advantage of achieving easier product purification.

8.9.1 Materials and Equipment

- *E. coli* BL21(DE3) cells expressing AvPAL (harvested from an autoinduction media culture)
- Cinnamic acid (500 mg, 3.38 mmol)
- Ammonium carbonate solution, 2.5 M in water (pH not adjusted)
- Conc. hydrochloric acid, 37%
- Sodium hydroxide
- Conc. formaldehyde solution, 40% (stab. with MeOH)
- Methanol
- Deionized water
- HPLC-grade water
- HPLC-grade methanol
- Orbital shaker with temperature control
- Heating mantle with magnetic stirring
- Lyophilizer (or centrifugal evaporator, e.g., Genevac)
- 500 mL Flask with stopper (or glass bottle)
- 50 mL Centrifuge tubes
- Benchtop centrifuge for 50 mL tubes

- Sintered-glass Hirsch funnel
- Side-armed filtration flask
- 50 mL Round-bottom flask
- Reflux condenser
- HPLC system equipped with UV detector

8.9.2 Procedure

1. Cinnamic acid (500 mg, 3.38 mmol) was dissolved in 2.5 M ammonium carbonate solution (250 mL). Wet *E. coli* BL21(DE3) cells expressing AvPAL (4.0 g wet pellet) were weighed and resuspended in the solution. The reaction mixture was incubated in an orbital shaker (30 °C, 250 rpm).
2. The course of the biotransformation was monitored by HPLC, as described in the Analytical Methods. After 12 hours, the suspension was centrifuged (4000 rpm, 20 minutes) to remove the cells.
3. The supernatant was lyophilized until all the water and most of the ammonium carbonate had been removed (alternatively, it is possible to remove water with a centrifugal evaporator, then heat the mixture at atmospheric pressure to sublime away the remaining ammonium carbonate).
4. The residue was resuspended in concentrated hydrochloric acid (4.0 mL) and heated to 60 °C under stirring for 10 minutes. After cooling, the white solid precipitate (unconverted cinnamic acid **1**) was removed by vacuum filtration and washed with concentrated hydrochloric acid (2.0 mL).
5. The combined acidic aqueous phase was transferred to a round-bottom flask equipped with a reflux condenser. Concentrated formaldehyde solution (1.0 mL) was added, and the mixture was stirred and heated under reflux for 5 hours. Two more portions of formaldehyde solution (1.0 mL each) were added after the first and second hour.
6. After cooling to room temperature, vacuum filtration, and washing with cold water (2.0 mL) and cold methanol (5.0 mL), the solid product (*S*)-**3** was obtained (377 mg, 53% yield, *ee* > 99%). Note: If required, the purity of the product (>95% by NMR) can be increased by recrystallization from boiling ethanol/water (1 : 1), neutralizing the hot solution with concentrated aqueous ammonia [57d].

8.9.3 Analytical Methods

Biotransformation samples (300 µL) were mixed with methanol (300 µL), thoroughly shaken, and centrifuged (13 000 rpm, 1 minute). The supernatant was transferred to a filter vial and used directly for HPLC analysis. Samples of the crude and recrystallized (*S*)-**3** were dissolved in the mobile phase for HPLC analysis.

For PAL reaction, the conversion can be monitored by reverse phase HPLC on a Zorbax Extend-C18 column (4.6 mm × 50 mm × 3.5 µm, Agilent), mobile phase NH_4OH 0.1 M/ MeOH 9:1, flow rate 1.0 mL.min^{-1}, temperature 40 °C, detection wavelength 210 nm. Retention times: **1** 2.26 minutes, **2** 5.77 minutes.

The *ee* of **2** and **3** can be determined by reverse phase HPLC on a Chirobiotic T column (4.6 mm × 250 mm × 5.0 µm, Astec), mobile phase MeOH/H_2O 60 : 40, flow rate 0.8 mL. min^{-1}, temperature 40 °C, detection wavelength 210 nm. Retention times: (*S*)-**2** 5.28 minutes, (*R*)-**2** 6.01 minutes, (*S*)-**3** 7.89 minutes, (*R*)-**3** 10.92 minutes.

8.9.4 Conclusion

Due to the simplicity of the reaction conditions and the 100% atom economy, the PAL-mediated hydroamination of cinnamic acids is a very valuable strategy for the synthesis of L-arylalanines in optically pure form, but the isolation and purification of the product from salts and unreacted starting material is often difficult, typically involving ion-exchange chromatography.

The Pictet–Spengler cyclization step, performed on the crude reaction product of the PAL amination (yield 88%), allows (*S*)-**3** to be obtained in >95% purity by NMR (purity can be increased to >99% by recrystallization) and *ee* > 99% by HPLC. The latter compound is a key chiral building block, used for the synthesis of various tetrahydroisoquinoline-based pharmaceuticals [57a–c], chiral ligands [57d], and conformationally restricted peptides [57e].

Due to the incompatibility of the reaction conditions, the two steps cannot be carried out as a one-pot process, but submitting the crude biotransformation mixture to the cyclization step enables a simpler purification of the final product.

The overall yield of the process is 53% from **1** (377 mg of (*S*)-**3**·HCl), with *ee* > 99%. The yield of the first step is not quantitative, because the conversion cannot be increased further when the equilibrium conditions have been reached (equilibrium conversion 60%). However, upon acidification of the mixture for the cyclization step, the cinnamic acid starting material separates spontaneously in >95% purity by NMR and in almost quantitative yield (in this example, 185 mg of unconverted **1** was isolated, 93% recovery yield). Therefore, in a large-scale process, it could be conveniently recycled in the following batch. Including this option, the overall yield of the two-steps process would be 82%, based on recovered starting material.

8.10 ω-TA/MAO Cascade for the Regio- and Stereoselective Synthesis of Chiral 2,5-Disubstituted Pyrrolidines

Elaine O'Reilly[1,2] and Nicholas J. Turner[1]
[1]*School of Chemistry, Manchester Institute of Biotechnology, The University of Manchester, UK*
[2]*School of Chemistry, University of Nottingham, UK*

The ability of complementary biocatalysts to operate under similar reaction conditions has enabled the combination of several enzymatic transformations in concurrent one-pot processes, allowing complex, optically pure architectures to be isolated from simple starting materials [58]. Such tandem biocatalytic processes offer highly efficient routes to target molecules by alleviating the need for costly intermediate purification steps, toxic chemicals, high temperatures, and protecting group manipulations [59]. Biocatalytic cascade processes for the synthesis of optically active 2,5-disubstituted pyrrolidines are particularly desirable, as such motifs are present in a wide variety of natural products and active pharmaceutical ingredients (APIs) [60]. This methodology exploits the complementary regio- and stereoselectivity of an ω-TA and an MAO-N cascade for the generation of optically active 2,5-pyrrolidines (**4a–d**), starting from achiral 1,4-diketones (**1a–d**) [61] (Scheme 8.13).

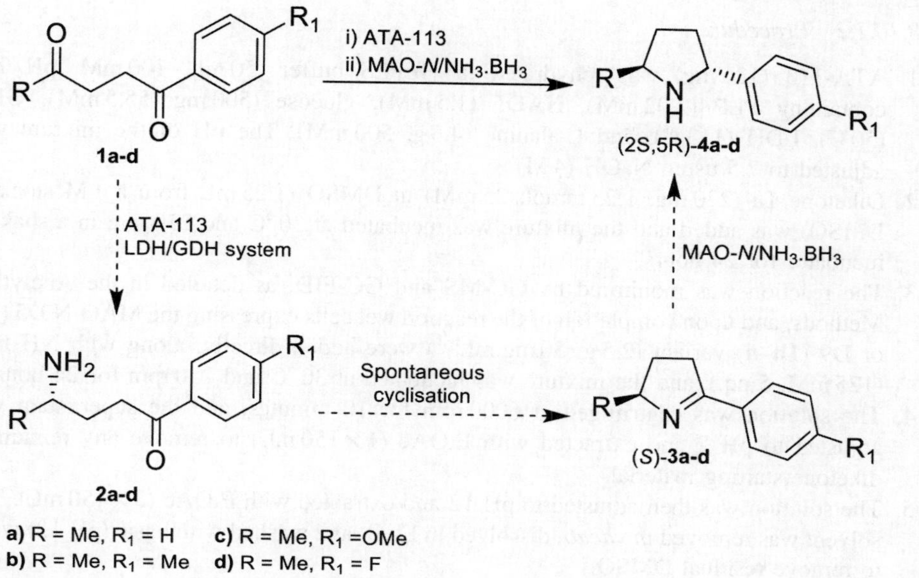

Scheme 8.13 One-pot ω-TA/MAO-N-medated synthesis of 2,5-disubstituted pyrrolidines **4a–d**, starting from **1a–d**. ATA113 is an (S)-selective transaminase purchased from Codexis in the form of a lyophilized cell extract.

8.10.1   One-Pot Asymmetric Synthesis of 4a on a Preparative Scale

8.10.1.1   Materials and Equipment

- (S)-Selective transaminase, ATA-113, from Codexis (2.5 mg.mL^{-1})
- *Escherichia coli* BL21(DE3) cells expressing MAO-N D5 and D9 variants (50 mg.mL^{-1})
- 4-(2-Hydroxyethyl)piperazine-1-ethanesulfonic acid (HEPES) buffer (100 mM, pH 7.5)
- PLP (2.02 mM)
- Nicotinamide adenine dinucleotide (NAD$^+$) (1.5 mM)
- Glucose (10 mg.mL^{-1}, 55.5 mM)
- *Lactate dehydrogenase* (LDH-103, from Codexis) (113 U) (displaces the equilibrium by removing pyruvate)
- Glucose dehydrogenase (GDH-901, from Codexis) (50 U) (co-factor recycling)
- L-Alanine (45 mg.mL^{-1}, 500 mM)
- Diketones (5 mM or 25 mM) (synthesized as in Reference [62])
- Dimethyl sulfoxide
- Ethyl acetate
- Cyclohexane
- NH$_3$BH$_3$ complex (25 mM, 5 eq.)
- Silica gel (used for preparative scale biotransformation)

8.10.1.2 Procedure

1. ATA-113 (125 mg) was rehydrated in HEPES buffer (50 mL, 100 mM, pH 7.5) containing PLP (2.02 mM), NAD$^+$ (1.5 mM), glucose (500 mg, 55.5 mM), GDH (50 U), LDH (113 U), and L-alanine (4.5 g, 500 mM). The pH of the mixture was adjusted to 7.5 using NaOH (4 M).

2. Diketone **1a** (220 mg, 1.25 mmol, 25 mM) in DMSO (1.25 mL from a 1 M stock in DMSO) was added and the mixture was incubated at 30 °C and 250 rpm in a shaking incubator for 24 hours.

3. The reaction was monitored by GC-MS and GC-FID, as detailed in the Analytical Methods, and upon completion of the reaction wet cells expressing the MAO-N D5 (**1a**) or D9 (**1b–d**) variant (2.5 g, 50 mg.mL^{-1}) were added directly, along with NH$_3$BH$_3$ (125 mM, 5 eq.), and the mixture was incubated at 30 °C and 250 rpm for 24 hours.

4. The solution was centrifuged at 6000 rpm for 10 minutes, and the supernatant was adjusted to pH 2 and extracted with EtOAc (1 × 150 mL) to remove any remaining diketone starting material.

5. The solution was then adjusted to pH 12 and extracted with EtOAc (3 × 150 mL). The solvent was removed *in vacuo*, dissolved in Et$_2$O, and washed with basic (pH 12) water to remove residual DMSO.

6. The organic layer was dried over MgSO$_4$ and filtered, and the solvent was removed *in vacuo*. The resulting pyrrolidine **4a** was filtered through a short pad of silica using cyclohexane/EtOAc (90:10) as the eluent and isolated as a clear oil (82% yield, *ee* > 99%, *de* > 99%). ^{1}H-NMR (400 MHz; CDCl$_3$): δ 7.43 – 7.28 (m, 4H), 7.23 (m, 1H), 4.38 (t, *J* = 7.6, 1H), 3.55 (m, 1H), 2.35 – 2.22 (m, 2H), 2.12 (m, 1H), 1.88–1.75 (m, 1H), 1.23 (d, *J* = 6.3 Hz, 3H).

8.10.1.3 Analytical Methods

GC-MS spectra were recorded on a Hewlett Packard HP 6890 equipped with an HP-1MS 30 m × 0.32 mm 025 μm column, an HP 5973 Mass Selective Detector, and an ATLAS GL FOCUS sampling robot. GC-FID analysis was performed on an Agilent 6850 equipped with a Gerstel MultiPurpose Sampler MPS2L and a Varian CP CHIRASIL-DEX CB 25 m × 0.25 mm DF = 0.25 column. The *ee* was determined by GC-FID and the absolute configuration was confirmed by comparing optical rotation values to literature values. The diastereomeric excess was determined by GC-FID using diastereomeric standards generated from the reduction of (*S*)-**3** and (*R*)-**3**.

8.10.2 Conclusion

(*S*)-Selective ATA-113 was highly regioselective, mediating the amination of **1a–d** exclusively on the methyl ketone. The resulting 1,4-amino ketone (*S*)-**2a–d** underwent spontaneous cyclization, providing (*S*)-**3a–d** in high conversion and excellent *ee* (>99%). Subsequent treatment of the pyrrolines with the MAO-N/NH$_3$BH$_3$ combination provided the corresponding pyrrolidines **4a–d** in excellent *de* (>99%) (Table 8.6). The complementary regioselectivities displayed by the ω-TA and MAO-N variants circumvent racemization of the C-2-(*S*)-centre installed by the TA and provide a method for accessing optically pure 2,5-pyrrolidines.

Table 8.6 *ATA113/MAO-N one-pot cascade for the synthesis of (2S,5R)-***4a–d***.*

Ketone	ω-TA	Mao-N	Conv. (%)	de (%)
1a[a]	ATA-113	D5	>99	>99 (2*S*,5*R*)-**4a**
1b[b]	ATA-113	D9	>99	>99 (2*S*,5*R*)-**4b**
1c[b]	ATA-113	D9	>99	>99 (2*S*,5*R*)-**4c**
1d[b]	ATA-113	D9	>99	>99 (2*S*,5*R*)-**4d**

[a] 25 mM substrate concentration.
[b] 5 mM substrate concentration.

References

1. Poizat, M., Arends, I.W.C.E., and Hollmann, F. (2010) *Journal of Molecular Catalysis B-Enzymatic*, **63**, 149–156.
2. Köhler, V., Wilson, Y.M., Dürrenberger, M. *et al.* (2013) *Nature Chemistry*, **5**, 93–99.
3. Ghislieri, D., Green, A.P., Pontini, M. *et al.* (2013) *Journal of the American Chemical Society*, **135**, 10863–10869.
4. Wilson, Y.M., Dürrenberger, M., and Ward, T.R. (2012) Organometallic chemistry in protein scaffolds, in *Protein Engineering Handbook*, vol. III (eds S. Lutz and U.T. Bornscheuer), Wiley VCH, Weinheim.
5. Köhler, V., Mao, J., Heinisch, T. *et al.* (2011) *Angewandte Chemie International Edition*, **50**, 10863–10866.
6. Kroutil, W., Fischereder, E.-M., Fuchs, C.S. *et al.* (2013) *Organic Process Research & Development*, **17**, 751.
7. Sattler, J.H., Fuchs, M., Tauber, K. *et al.* (2012) *Angewandte Chemie*, **124**, 9290.
8. Sattler, J.H., Fuchs, M., Tauber, K. *et al.* (2012) *Angewandte Chemie International Edition*, **51**, 9156–9159.
9. Fuchs, M., Tauber, K., Sattler, J. *et al.* (2012) *RSC Advances*, **2**, 6262–6265.
10. Schrittwieser, J.H., Sattler, J., Resch, V. *et al.* (2011) *Current Opinion in Chemical Biology*, **15**, 249–256.
11. Gould, P.V., Gelb, M.H., and Sligar, S.G. (1981) *Journal of Biological Chemistry*, **256**, 6686–6691.
12. Ryan, J.D. and Clark, D.S. (2008) *Biotechnology and Bioengineering*, **99**, 1311–1319.
13. Hofer, M., Strittmatter, H., and Sieber, V. (2013) *ChemCatChem*, **5**, 3351–3357.
14. Ichinose, H., Michizoe, J., Maruyama, T. *et al.* (2004) *Langmuir*, **20**, 5564–5568.
15. Verras, A., Alian, A., and de Montellano, P.R. (2006) *Protein Engineering Design and Selection*, **19**, 491–496.
16. Ikenaga, T., Matsushita, K., Shinozawa, J. *et al.* (2005) *Tetrahedron*, **61**, 2105–2109.
17. Oroz-Guinea, I. and García-Junceda, E. (2013) *Current Opinion in Chemical Biology*, **17**, 236–249; García-Junceda, E. (ed.) (2008) *Multi-Step Enzyme Catalysis*, Wiley-VCH; Ricca, E., Brucher, B., and Schrittwieser, J.H. (2011) *Advanced Synthesis and Catalysis*, **353**, 2239–2262; Pirie, C.M., de Mey, M., Prather, K.L.J., and Ajikumar, P.K. (2013) *ACS Chemical Biology*, **8**, 662–672; Guterl, J.-K. and Sieber, V. (2013) *Engineering in Life Sciences*, **13**, 4–18.
18. Oberleitner, N., Peters, C., Muschiol, J. *et al.* (2013) *ChemCatChem*, **5**, 3524–3528; Lowe, J.R., Martello, M.T., Tolman, W.B., and Hillmyer, M.A. (2011) *Polymer Chemistry*, **2**, 702–708.
19. Weckbecker, A. and Hummel, W. (2006) *Biocatalysis and Biotransformation*, **24**, 380–389.
20. Peters, C., Kölzsch, R., Kadow, M. *et al.* (2014) *ChemCatChem*, **6**, 1021–1027.

21. Chen, Y.C., Peoples, O.P., and Walsh, C.T. (1988) *Journal of Bacteriology*, **170**, 781–789.

22. Tietze, L.F., Sommer, K.M., Zinngrebe, J., and Stecker, F. (2005) *Angewandte Chemie International Edition*, **44**, 257–259; Tietze, L.F., Stecker, F., Zinngrebe, J., and Sommer, K.M. (2006) *Chemistry - A European Journal*, **12**, 8770–8776.

23. Li, C.J. and Chan, T.H. (1991) *Tetrahedron Letters*, **32**, 7017–7020.

24. Zhao, B., Peng, X., Cui, S., and Shi, Y. (2010) *Journal of the American Chemical Society*, **132**, 11009–11011.

25. Barbier, P. (1899) *Comptes Rendus de l'Académie des Sciences*, **128**, 110–111.

26. Li, C.J. (2005) *Chemical Reviews*, **105**, 3095–3165.

27. Fuchs, M., Schober, M., Pfeffer, J. *et al.* (2011) *Advanced Synthesis and Catalysis*, **353**, 2354–2358.

28. The Indium Corporation of America, Form No. 97752 R1: Etching Indium to Remove Oxides.

29. Wiebe, L. and Schmidt, T. (2010) (Danisco A/S), US Patent 7683187.

30. Bretler, G. and Dean, C. (2000) (Firmenich SA), US Patent 6025170.

31. Yamamoto, T., Ogura, M., Amano, A. *et al.* (2002) *Tetrahedron Letters*, **43**, 9081–9084.

32. Schrittwieser, J.H., Sattler, J., Resch, V. *et al.* (2011) *Current Opinion in Chemical Biology*, **15**, 249–256.

33. Mihovilovic, M.D., Chen, G., Wang, S.Z. *et al.* (2001) *Journal of Organic Chemistry*, **66**, 733–738.

34. Pham, S.Q., Gao, P., and Li, Z. (2013) *Biotechnology and Bioengineering*, **110**, 363–373.

35. Shaeri, J., Wright, I., Rathbone, E.B. *et al.* (2008) *Biotechnology and Bioengineering*, **101**, 761–767; Shaeri, J., Wohlgemuth, R., and Woodley, J.M. (2006) *Organic Process Research and Development*, **10**, 605–610; Wohlgemuth, R., Smith, M.E.B., Dalby, P., and Woodley, J.M. (2009) *Encyclopedia of Industrial Biotechnology* (ed. M. Flickinger), John Wiley & Sons, Hoboken; Wohlgemuth, R. (2009) *Journal of Molecular Catalysis B-Enzymatic*, **61**, 23–29; Charmantray, F., Hélaine, V., Légeret, B., and Hecquet, L. (2009) *Journal of Molecular Catalysis B-Enzymatic*, **57**, 6–9; Guérard, C., Alphand, V., Archelas, A. *et al.* (1999) *European Journal of Organic Chemistry*, 3399–3402; Solovjeva, O.N. and Kochetov, G.A. (2008) *Journal of Molecular Catalysis B-Enzymatic*, **54**, 90–92; Schörken, U. and Sprenger, G.A. (1998) *Biochimica et Biophysica Acta*, **1385**, 229–243, and references cited therein; Reimer, L.M., Conley, D.L., Pompliano, D.L., and Frost, J.W. (1986) *Journal of the American Chemical Society*, **108**, 8010–8015; Gauss, D., Schönenberger, B., and Wohlgemuth, R. (2014) *Carbohydrate Research*, **389**, 18–24; Wohlgemuth, R. (2009) *Biotechnology Journal*, **4**, 1253–1265.

36. Ahn, M., Cochrane, F.C., Patchett, M.L., and Parker, E.J. (2008) *Bioorganic and Medicinal Chemistry*, **16**, 9830–9836; Bednarski, M.D., Crans, D.C., DiCosimo, R. *et al.* (1988) *Tetrahedron Letters*, **29**, 427–430.

37. van Herk, T., Hartog, A.F., van der Burg, A.M., and Wever, R. (2005) *Advanced Synthesis and Catalysis*, **347**, 1155–1162, and references cited therein; van Herk, T., Hartog, A.F., Schoemaker, H.E., and Wever, R. (2006) *Journal of Organic Chemistry*, **71**, 6244–6247; van Herk, T., Hartog, A.F., Babich, L. *et al.* (2009) *Chembiochem*, **10**, 2230–2235.

38. Fessner, W.-D. and Walter, C. (1992) *Angewandte Chemie. International Ed. In English*, **31**, 614–616; Zimmermann, T., Schneider, A., Schörken, U. *et al.* (1999) *Tetrahedron: Asymmetry*, **10**, 1643–1646; Ricca, E., Brucher, B., and Schrittwieser, J.H. (2011) *Advanced Synthesis and Catalysis*, **353**, 2239–2262.

39. Li, H., Tian, J., Wang, H. *et al.* (2010) *Helvetica Chimica Acta*, **93**, 1745–1750; Zhou, Y.-F., Cui, Z., Li, H. *et al.* (2010) *Bioorganic Chemistry*, **38**, 120–123; Hecht, S., Kis, K., Eisenreich, W. *et al.* (2001) *Journal of Organic Chemistry*, **66**, 3948–3952; Gao, W., Raschke, M., Alpermann, H., and Zenk, M.H. (2003) *Helvetica Chimica Acta*, **86**, 3568–3577; Taylor, S.V., Vu, L.D., Begley, T.P. *et al.* (1998) *Journal of Organic Chemistry*, **63**, 2375–2377.

40. Sanchez-Moreno, I., Iturrate, L., Martin-Hoyos, R. *et al.* (2009) *Chembiochem*, **10**, 225–229; Sanchez-Moreno, I., Iturrate, L., Doyagueez, E.G. *et al.* (2009) *Advanced Synthesis and Catalysis*,

351, 2967–2975; Sanchez-Moreno, I., Francisco Garcia-Garcia, J., Bastida, A., and Garcia-Junceda, E. (2004) *Chemical Communications*, 1634–1635.

41. Clapés, P., Fessner, W.-D., Sprenger, G.A., and Samland, A.K. (2010) *Current Opinion in Chemical Biology*, **14**, 154–167.

42. Sánchez-Moreno, I., Hélaine, V., Poupard, N. *et al.* (2012) *Advanced Synthesis and Catalysis*, **354**, 1725–1730.

43. Castillo, J.A., Guérard-Hélaine, C., Gutiérrez, M. *et al.* (2010) *Advanced Synthesis and Catalysis*, **352**, 1039–1046.

44. Desvergnes, S., Courtiol-Legourd, S., Daher, R. *et al.* (2012) *Bioorganic and Medicinal Chemistry*, **20**, 1511–1520; Li, T.J. and Rosazza, P.N. (1997) *Journal of Bacteriology*, **179**, 3482–3487.

45. Horecker, B.L., Paoletti, F., and Williams, J.F. (1982) *Annals of the New York Academy of Sciences*, **378**, 215–224.

46. McIntyre, L.M., Thorburn, D.R., Bubb, W.A., and Kuchel, P.W. (1989) *European Journal of Biochemistry*, **180**, 399–420.

47. Williams, J. and MacLeod, J. (2006) *Photosynthesis Research*, **90**, 125–148.

48. Flanigan, I., MacLeod, J., and Williams, J. (2006) *Photosynthesis Research*, **90**, 149–159.

49. Unger, F.M. (1981) *Adv. Carbohydr. Chem. Biochem.*, 38 (eds R.S. Tipson and H. Derek), Academic Press, pp. 323–388.

50. Schauer, R. (1985) *Trends in Biochemical Sciences*, **10**, 357–360.

51. Castillo, J.A., Guérard-Hélaine, C., Gutiérrez, M. *et al.* (2010) *Advanced Synthesis and Catalysis*, **352**, 1039–1046.

52. Guérard-Hélaine, C., Debacker, M., Clapés, P. *et al.* (2014) *Green Chemistry*, **16**, 1109–1113.

53. Kapuscinski, M., Franke, F.P., Flanigan, I. *et al.* (1985) *Carbohydrate Research*, **140**, 69–79.

54. Turner, N.J. (2011) *Current Opinion in Chemical Biology*, **15**, 234–240; Heberling, M.W., Wu, B., Bartsch, S., and Janssen, D.B. (2013) *Current Opinion in Chemical Biology*, **17**, 250–260; Gloge, A., Zoń, J., Kövári, A. *et al.* (2000) *Chemistry - A European Journal*, **6**, 3386–3390; Paizs, C., Katona, A., and Rétey, J. (2006) *Chemistry - A European Journal*, **6**, 3386–3390.

55. Moffitt, M.C., Louie, G.V., Bowman, M.E. *et al.* (2007) *Biochemistry*, **46**, 1004–1012; Lovelock, S.L., Lloyd, R.C., and Turner, N.J. (2014) *Angewandte Chemie International Edition*, **53**, 4652–4656.

56. Cox, E.D. and Cook, J.M. (1995) *Chemical Reviews*, **95**, 1797–1842; Whaley, W.M. and Govindachari, T.R. (1951) *Organic Reactions*, **6**, 151–190.

57. Bingham, M.J., Dunbar, N.A., Huggett, M.J., and Wishart, G. (2011) Heterocyclic compounds as antagonists of the orexin receptors. PCT International Application WO2011061318 A1; Hirth, B.H., Quiao, S., Cuff, L.M. *et al.* (2005) *Bioorganic and Medicinal Chemistry Letters*, **15**, 2087–2091; Kothakonda, K.K. and Bose, D.S. (2004) *Bioorganic and Medicinal Chemistry Letters*, **14**, 4371–4373; Toselli, N., Fortrie, R., Martin, D., and Buono, G. (2010) *Tetrahedron: Asymmetry*, **21**, 1238–1245; Kazmierski, W.M., Yamamura, H.I., and Hruby, V.J. (1991) *Journal of the American Chemical Society*, **113**, 2275–2283.

58. Köhler, V. and Turner, N.J. (2015) *Chemical Communications*, **51**, 450–464; O'Reilly, E., Iglesias, C., and Turner, N.J. (2014) *ChemCatChem*, **6**, 992–995.

59. Reetz, M.T. (2013) *Journal of the American Chemical Society*, **45**, 12480–12496; Turner, N.J. and O'Reilly, E. (2013) *Nature Chemical Biology*, **9**, 285–288; Schrittwieser, J.H., Sattler, J., Resch, V. *et al.* (2011) *Current Opinion in Chemical Biology*, **15**, 249–256; Oroz-Guinea, I. and García-Junceda, E. (2013) *Current Opinion in Chemical Biology*, **17**, 236–249; Ricca, E., Brucher, B., and Schrittwieser, J.H. (2011) *Advanced Synthesis and Catalysis*, **353**, 2239–2262; Sattler, J.H., Fuchs, M., Tauber, K. *et al.* (2012) *Angewandte Chemie International Edition*, **51**, 9156–9159.

60. Brenneman, J.B. and Martin, S.F. (2004) *Organic Letters*, **6**, 1329–1331; Xu, F. (2013) *Organic Letters*, **15**, 1324–1345; Hanessian, S., Bayrakdarian, M., and Luo, X. (2002) *Journal of the*

American Chemical Society, **124**, 4716–4721; Aggarwal, V.K., Astle, C.J., and Rogers-Evans, M. (2004) *Organic Letters*, **6**, 1469–1471; Zhang, S., Xu, L., Miao, L. *et al.* (2007) *Journal of Organic Chemistry*, **72**, 3133–3136; Goti, A., Cicchi, S., Mannucci, V. *et al.* (2003) *Organic Letters*, **5**, 4235–4238; Trost, B.M., Horne, D.B., and Woltering, M.J. (2003) *Angewandte Chemie International Edition*, **42**, 5987–5990; Severino, E.A. and Correia, C.R.D. (2000) *Organic Letters*, **2**, 3039–3042.

61. O'Reilly, E., Iglesias, C., Ghislieri, D. *et al.* (2014) *Angewandte Chemie International Edition*, **53**, 2447–2450.

62. Shen, Z.-L., Goh, K.K.K., Cheong, H.-L. *et al.* (2010) *Journal of the American Chemical Society*, **132**, 15852–15855.

9

Biocatalysis for Industrial Process Development

9.1 Efficient Synthesis of (S)-1-(5-Fluoropyrimidin-2-yl)ethylamine Hydrochloride Salt Using an ω-Transaminase Biocatalyst in a Two-Phase System

Rebecca E. Meadows,[1] Keith R. Mulholland,[1] Martin Schürmann,[2]
Michael Golden,[1] Hans Kierkels,[2] Elise Meulenbroeks,[2] Daniel Mink,[2]
Oliver May,[2] Christopher Squire,[1] Harrie Straatman,[2] and Andrew S. Wells[1]
[1]AstraZeneca, Chemical Development, UK
[2]DSM Innovative Synthesis BV, The Netherlands

(S)-1-(5-Fluoropyrimidin-2-yl)ethylamine is an intermediate in the synthesis of AZD1480, a compound that was in development at AstraZeneca for the treatment of Myelofibrosis and Idiopathic Rubra Vera. Two main synthetic approaches to this compound were developed using asymmetric hydrogenation and the use of ω-transaminases. A biphasic reaction mixture was used to remove the byproduct acetophenone from the aqueous layer, thereby driving the reaction to completion.

The two procedures are described. The first employs a commercial formulation of the transaminase from *Vibrio fluvalis* and the second the Codexis transaminase TA-P1-A06. The first process worked well, although filtration of the enzyme on scale-up was slow due to the high loading required (2.8 mL·gL^{-1} ketone). The second process used only 2.5 wt-% TA-P1-A06 and did not display slow filtration on work-up.

Practical Methods for Biocatalysis and Biotransformations 3, First Edition.
Edited by John Whittall, Peter W. Sutton, and Wolfgang Kroutil.
© 2016 John Wiley & Sons, Ltd. Published 2016 by John Wiley & Sons, Ltd.

9.1.1 Procedure 1: Preparation of (*S*)-1-(5-Fluoropyrimidin-2-yl)ethylamine Hydrochloride Using Transaminase from *Vibrio fluvalis* [1]

9.1.1.1 Materials and Equipment

- (*S*)-1-Phenylethylamine (68.0 mL)
- Potassium phosphate-monobasic (13.21 g)
- Acetic acid (26.0 mL)
- Pyridoxal phosphate (0.643 g)
- 2-Acetyl-5-fluoropyrimidine (68.0 g)
- ω-Transaminase from *Vibrio fluvalis* (Codexis Inc) (182 mL)
- Toluene (340 mL)
- Potassium carbonate (6.7 g)
- Harborlite® (12.0 g)
- Water (136 mL)
- Toluene (136 mL)
- Potassium carbonate (134.1 g)
- Di-*tert*-butyldicarbonate (116.5 g)
- 2-Methyltetrahydrofuran (544 mL)
- Harborlite (12.0 g)
- Water (136 mL)
- 2-Methyltetrahydrofuran (136 mL)
- 2-Methyltetrahydrofuran (272 mL)
- 5.6 M HCl in isopropanol (238 mL)
- Heptane (204 mL)
- 2-Methyltetrahydrofuran (204 mL)

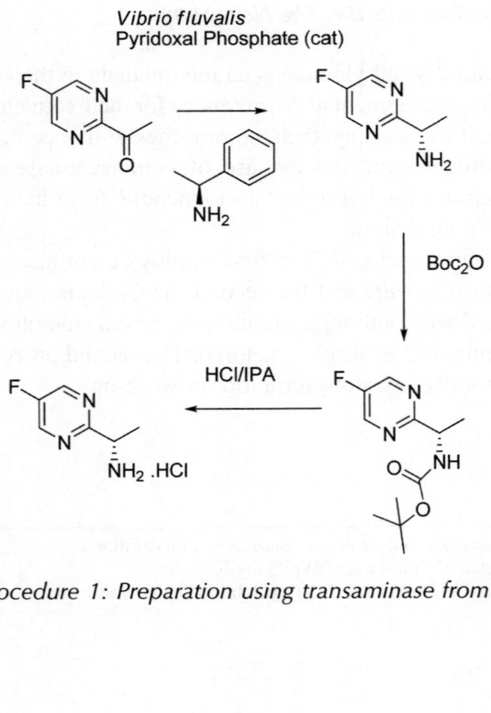

Figure 9.1 *Procedure 1: Preparation using transaminase from* Vibrio fluvalis.

- 2-L Jacketed vessel
- Overhead stirrer
- Thermometer
- Distillation apparatus

9.1.1.2 Procedure

1. (*S*)-1-Phenylethylamine (68.0 mL, 0.53 mol) was added to a solution of monobasic potassium phosphate (13.21 g, 0.097 mol) in water (1020 mL). The pH of the solution was adjusted to 7.5 by the addition of acetic acid (26.0 mL, 0.45 mol).
2. Pyridoxal phosphate (0.643 g, 0.0024 mol) was added, followed by 2-acetyl-5-fluoro-pyrimidine (68.0 g, 0.49 mol), a buffered solution of *Vibrio fluvalis* transaminase enzyme (182 ml, 35.3 KU) and toluene (340 mL).The reaction mixture was adjusted to pH 7.5 with potassium carbonate (6.71 g, 0.053 mol) then held at 29 °C for 18 h with stirring.
3. Harborlite (12.0 g) was added and the reaction mixture was stirred for a further 15 min before being filtered. The filter bed was washed with a mixture of water (136 mL) and toluene (136 mL). The organic layer was discarded.
4. Potassium carbonate (134.14 g, 0.98 mol) was added to the aqueous phase followed by a solution of di-*tert*-butyl dicarbonate (116.5 g, 0.53 mol) in 2-methyltetrahydrofuran (544 mL). The biphasic mixture was stirred at 40 °C for 2 h.
5. Harborlite (12.0 g) was added and the reaction mixture stirred for a further 15 min. The mixture was filtered. The filter bed was washed with a mixture of water (136 mL) and 2-methyltetrahydrofuran (136 mL). The aqueous layer was extracted with 2-methyltetra-hydrofuran (272 mL). The organic layers were combined and distilled at atmospheric pressure until the internal temperature reached 98 °C (~760 mL distillate was collected).
6. The residual solution was treated with 5.55 M hydrochloric acid in isopropanol (238 mL, 1.32 mol). The reaction mixture was heated to 40 °C for 18 h. The amine hydrochloride precipitated over this period. Heptane (204 mL) was added over 15 min and the resulting suspension stirred at 20 °C for 16 h.
7. The reaction mixture was filtered and the isolated amine hydrochloride washed with 2-methyltetrahydrofuran (102 mL) before being dried in the vacuum oven to give the title compound (68.2 g @ 98.4% w/w, 77.1%) ^{1}H NMR (400 MHz, CD$_3$SOCD$_3$) δ: 9.02 (d, 2H), 4.55 (m, 1H), 1.58 (d, 3H). LCMS: 142 [M + H]$^+$. ^{1}H NMR Assay versus maleic acid internal standard 98.4% w/w. Enantiomeric excess was determined by chiral HPLC (CrownPak CR+, aqueous perchloric acid, >99% enantiomeric excess [ee]).

9.1.2 Procedure 2: Preparation of (*S*)-1-(5-Fluoropyrimidin-2-yl)ethylamine Hydrochloride Using TA-P1-A06 Transaminase Enzyme [2]

9.1.2.1 Materials and Equipment

- (*S*)-1-Phenylethylamine (18.9 mL)
- Monobasic potassium phosphate (3.6 g)
- Water (300 mL)
- Acetic acid (7.0 mL)
- Pyridoxal phosphate (0.16 g)
- 1-(5-Fluoropyrimidin-2-yl)ethanone (20.0 g, 0.13 mol)

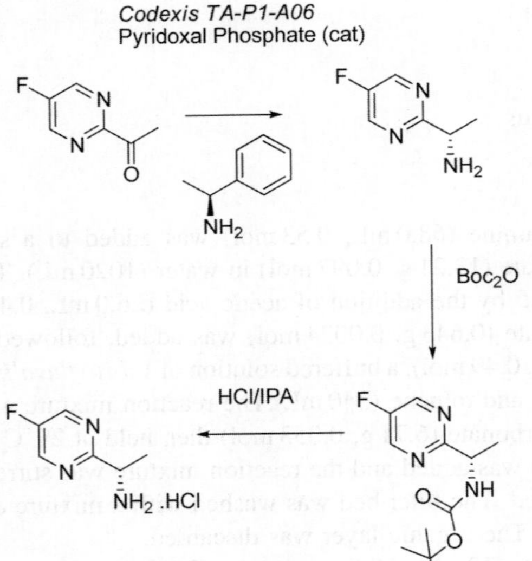

Figure 9.2 Procedure 2: Preparation using TA-P1-A06 transaminase enzyme.

- TA-P1-A06 Transaminase enzyme (Codexis Inc) (0.47 g)
- Toluene (100 mL)
- Potassium carbonate (1.84 g)
- Celite® (4.0 g)
- Toluene (40 mL)
- Water (100 mL)
- Potassium carbonate (36.8 g)
- Di-*tert*-butyldicarbonate (31.9 g)
- 2-Methyltetrahydrofuran (160 mL)
- 2-Methyltetrahydrofuran (58 mL)
- 5 M HCl in isopropanol (72.0 mL)
- Heptane (60 mL)
- 2-Methyltetrahydrofuran (40 mL)
- 2-L Jacketed vessel
- Overhead stirrer
- Thermometer
- Distillation apparatus

9.1.2.2 Procedure

1. (*S*)-1-Phenylethylamine (18.9 mL, 0.15 mol) was added to a solution of monobasic potassium phosphate (3.6 g, 0.27 mol) in water (300 mL). The pH of the solution was adjusted to 7.5 by the addition of acetic acid (7.0 mL, 0.12 mol).
2. Pyridoxal phosphate (0.16 g, 0.0007 mol) was added, followed by 1-(5-fluoropyrimidin-2-yl)ethanone (20.0 g, 0.13 mol), TA-P1-A06 enzyme (0.47 g) and toluene (100 mL).

The reaction mixture was adjusted to pH 7.5 with potassium carbonate (1.84 g, 0.013 mol) then held at 29 °C for 18 h.

3. Celite (4.0 g) was added and the reaction mixture was filtered. A mixture of toluene (40 mL) and water (100 mL) was added to the reaction vessel, stirred and discharged as a wash to the filter bed. The aqueous phase was separated and the organic phase discarded.
4. Potassium carbonate (36.8 g, 0.27 mol) was added to the aqueous phase followed by a solution of di-*tert*-butyl dicarbonate (31.9 g, 0.15 mol) in 2-methyltetrahydrofuran (160 mL). The mixture was heated to 40 °C for 18 h and the resulting biphasic mixture filtered. The organic layer was separated and evaporated to dryness.
5. The residue was dissolved in 2-methyltetrahydrofuran (58 mL) and a solution of 5 M hydrochloric acid in isopropanol (72.0 mL, 0.36 mol) was added. The reaction mixture was heated to 40 °C for 24 h. An off-white suspension was formed. The reaction mixture was cooled to 25 °C and heptane (60 mL) was added.
6. After 1 h the reaction mixture was filtered. The isolated solid was washed with 2-methyltetrahydrofuran (40 mL) and dried *in vacuo* to give the title compound as a monohydrochloride salt (16.1 g, 68%). ^{1}H NMR (400 MHz) δ: 9.02 (d, 2H), 4.55 (m, 1H), 1.58 (d, 3H). HRMS calculated for $C_6H_9FN_3$: 142.0775; HRMS found $[M + H]+$: 142.0774. Enantiomeric excess was determined by chiral HPLC (Chiralpak AD3 5 × 0.46 cm, >99.5% ee).

9.1.3 Conclusion

Two procedures have been developed for the synthesis of (*S*)-1-(5-fluoropyrimidin-2-yl) ethylamine hydrochloride using commercially available transaminase enzymes.

9.2 Preparative-scale Production of a Chiral, Bicyclic Proline Analog Intermediate for Boceprevir

Jack Liang, Jim Lalonde, and Gjalt Huisman
Codexis Inc, USA

Boceprevir (Victrelis®) is a peptidomimetic inhibitor of the NS3 serine protease for the treatment of Hepatitis C. It is composed of four moieties: a racemic β-aminoamide, a chiral dimethylcyclopropylproline analog (P2), (*S*)-*tert*-Leucine, and a *tert*-butylcarbamoyl group. The 3-azabicyclo[3.1.0]hexane structure of P2 adopts a constrained conformation, so that the *gem*-dimethyl group has a fixed angle with respect to the bicyclic ring structure. The incorporation of the 3-azabicyclo[3.1.0]hexane moiety results in a 1000-fold increase in NS3 protease binding over proline in a pentapeptide scaffold [3].

P2 was traditionally prepared via a variety of methods, including cyclopropanation of a Δ^3-pyrroline derivative of pyroglutamic acid, desymmetrization of caronic anhydride combined with diastereospecific cyanation [4–8]. These routes are both long and suffer from low yield on expensive starting materials. Given these challenges and projected large demand for this intermediate, there were strong motivations to develop a more efficient approach.

In the reaction sequence shown in Figure 9.3 a superior process was envisioned where 6,6-dimethyl-3-azabicyclo[3.1.0]hexane **1** is desymmetrized enzymatically to provide the desired imine enantiomer **2**. We optimized both the enzyme and the chemical process to enable a practical chiral synthesis of P2 at large scale.

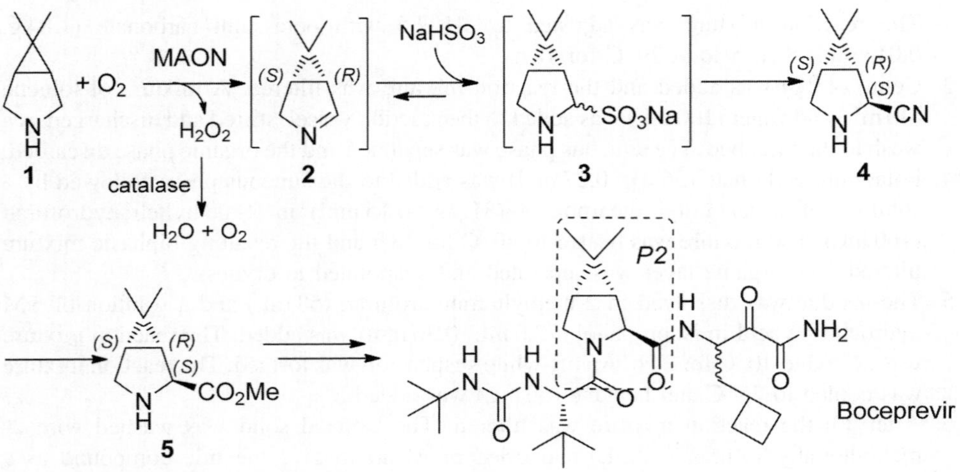

Figure 9.3 *Chemoenzymatic synthesis of (1R,2S,5S)-6,6-dimethyl-3-azabicyclo[3.1.0]hexane-2-carboxylic acid methyl ester **5** for Boceprevir, employing an enzymatic asymmetric oxidation of a prochiral amine.*

9.2.1   Procedure: Preparative-scale Preparation of 6,6-Dimethyl-3-azabicyclo[3.1.0] hexane-2-carboxylic Acid Methyl Ester

9.2.1.1   *Materials and Equipment*

- 6,6-Dimethyl-3-azabicyclo[3.1.0]hexane **1** (1.2 mL)
- Sodium metabisulfite ($Na_2S_2O_5$) (~23.6 g)
- Antifoam-204 (Sigma Aldrich, catalog number A-6426) (300 μL)
- *Aspergillus niger* catalase (Sigma Aldrich, catalog number C-3515) (600 μL)
- Monoamine oxidase evolved for this reaction (300 mg) [9]
- Substrate **1** (19.6 g)
- 2 N Sodium hydroxide
- Milli-Q water (120 mL)
- Distilled water (120 mL)
- Sodium cyanide (NaCN) (10.0 g)
- Methyl *tert*-butyl ether (MTBE) (600 mL)
- Celite 545® (6 g)
- 500 mL Four-neck flask
- Overhead stirrer
- Air sparging probe

9.2.1.2   *Procedure*

9.2.1.2.1   *Step 1: Enzymatic Oxidation*

1. To a 500-mL four-neck flask at room temperature (about 21 °C) fitted with an overhead stirrer (300 rpm) were added Milli-Q water (120 mL) and 6,6-dimethyl-3-azabicyclo

[3.1.0]hexane **1** (1.2 mL; 1.0 g; 9 mmol) to give a homogeneous solution with a pH of 11.8.

2. $Na_2S_2O_5$ (1.1 g; 12 mmol) was added in portions until the pH reached 7.5.
3. To the colorless solution were added Antifoam-204 (Sigma Aldrich, catalog number A-6426) (300 µL) and *Aspergillus niger* catalase (Sigma Aldrich, catalog number C-3515) (600 µL) to give a colorless solution at pH 7.5.
4. To this solution was added monoamine oxidase (300 mg) to provide a yellow solution and an air sparging probe (airflow rate of 60 mL · min^{-1}) was inserted. The pH of the reaction mixture began to drop immediately. The pH was maintained via feedback controlled addition of 2 N NaOH.
5. After approximately 40 min the pH remained stable and no further base addition occurred. After an additional 10 min (50 min total), 150 mL of substrate/NaHSO$_3$ solution (19.6 g of **1** with 120 mL of distilled water and 22.5 g of $Na_2S_2O_5$; pH ~7.2) was added to the reaction mixture at a rate of 0.12 mL · min^{-1}. The pH of the reaction mixture began to drop immediately and addition of 2 N NaOH resumed.
6. The substrate/NaHSO$_3$ addition was completed after approximately 21 h (approximately 22 h of total reaction time). Base addition continued at an increased rate after substrate addition was completed.
7. After 24 h, the reaction was judged to be complete by ^{1}H-NMR analysis.

9.2.1.2.2 Step 2: Cyanation

1. At the end of the enzymatic oxidation reaction, NaCN (10.0 g, 1.1 equiv.) was added to give a milky mixture (pH 9.9).
2. After 30 min the reaction was extracted with MTBE (300 mL). The lower aqueous phase was drained off (~250 mL) and the upper organic layer filtered through Celite 545 (6 g; 2 inch diameter × ¼ inch height).
3. The Celite 545 pad was rinsed with MTBE (300 mL) and the MTBE used in the rinse was then used to extract the aqueous phase. The organic phases were combined and concentrated under reduced pressure using a rotatory evaporator at 40 °C for 1 h to give a white solid.
4. The white solid was further dried under reduced pressure for 30 mins to give 23.9 g (95% yield; >99% ee at ring fusion; >20:1 *trans:cis* at amino-nitrile) of (*1R,2S,5S*)-6,6-dimethyl-3-azabicyclo[3.1.0]hexane-2-carbonitrile **4**.

9.2.1.2.3 Step 3

1. (*1R,2S,5S*)-6,6-Dimethyl-3-azabicyclo[3.1.0]hexane-2-carbonitrile **4** was converted to the corresponding substantially enantiomerically pure (*1R,2S,5S*)-6,6-dimethyl-3-aza-bicyclo[3.1.0]hexane-2-carboxylic acid methyl ester **5** by the procedure described in patent WO 2007075790 [6].

9.2.1.3 Analytical Methods for the Oxidation of 6,6-Dimethyl-3-azabicyclo[3.1.0] hexane 1 to (1R,5S)-6,6-Dimethyl-3-azabicyclo[3.1.0]hex-2-ene 2

Enzymatic oxidation of 6,6-dimethyl-3-azabicyclo[3.1.0]hexane **1** and the stereomeric purity of the imine product were determined by chiral GC (column: Supelco Betadex 225 [part# 24348], 0.25 mm × 30 m × 0.25 µm; oven temperature: 85 °C isothermal; carrier flow: $1.1 \, mL \cdot min^{-1}$; injection volume: $4 \, \mu L$; injection split: 100:1; injector temperature: $200 \,^{\circ}C$; FID detection). The order of elution was: substrate **1** with a retention time of approximately 7 min, the desired (*1R,5S*)-imine **2** at approximately 10.6 min, and the undesired (*1S,5R*) imine at approximately 11.0 min. Alternatively, the enantiomeric purity of compound **2** was determined by HPLC using a Chiralpak AD-H column ($4.6 \times 250 \, mm$, Diacel Chemical Industries Ltd) at $25 \,^{\circ}C$. The mobile phase was a mixture of heptane (90%), ethanol (10%), and 0.1% trifluoroacetic acid at flowrate of $1 \, mL \cdot min^{-1}$. The retention time for undesired (*1S, 5R*)-imine was 7.1 min, and the retention time for the desired (*1R, 5S*)-imine **2** was 8.4 min. The *trans:cis* ratio of the amino nitrile was determined by ^{1}H-NMR in $CDCl_3$. *Trans* (C-*H*)-CN 3.27 ppm (m, 1H); *cis* (C-*H*)-CN 3.49 ppm (m, 1H).

9.2.2 Summary

The one-pot synthesis of a substituted proline analog containing three chiral centers starting from an achiral molecule has been described. An enzymatic oxidation was employed to set the two chiral centers in the imine intermediate **2**, which subsequently induces the desired chirality at the third center. Molecular oxygen was used as the oxidant and the hydrogen peroxide byproduct was decomposed by a commercial catalase. Enzyme inactivation by the imine **2** was alleviated by *in situ* trapping as the bisulfite adduct **3**. Oxidation of bisulfite and substrate inhibition were circumvented by running the reaction in a fed-batch mode. The bisulfite adduct was readily telescoped to the aminonitrile **4**, the direct precursor to the desired product **5**. The overall reaction is the net conversion of a prochiral C−H bond to an optically pure C−CN bond.

This process has been scaled-up to batches of hundreds of kilograms and provides significant environmental benefits [9,10].

9.3 Focused Carbonyl Reductase Screening for Rapid Gram Supply of Highly Enantioenriched Secondary Alcohol Libraries

Andrew S. Rowan[1], Thomas S. Moody[1], Roger M. Howard[2], and Toby J. Underwood[3]

[1] *Almac, UK*
[2] *Pfizer Ltd, Chemical Research & Development, UK*
[3] *Royal Society of Chemistry, UK*

Enzymatic reduction of ketones using selectAZyme™ carbonyl reductase (CRED) technology was used to synthesize libraries of highly enantioenriched secondary alcohols in both enantiomeric forms. First, microscale screening matched ketone substrate classes to subsets of highly enantioselective CREDs (Figure 9.4). Only six selectAZyme™ CREDs were required to gain coverage of all seven substrate classes tested.

Libraries of ketones belonging to these substrate classes were then screened against their CRED subsets. Excellent results were observed, identifying both Prelog and anti-Prelog hits (defined as $\geq 95\%$ ee, $\geq 5\%$ conversion) for 33 of the 41 substrate ketones tested. It was also observed that the majority of target alcohols (68%) were generated in $\geq 99\%$ ee.

Finally, the best Prelog and anti-Prelog hits for each substrate were employed in preparative bioreductions, allowing the rapid generation of 100–1500 mg amounts of

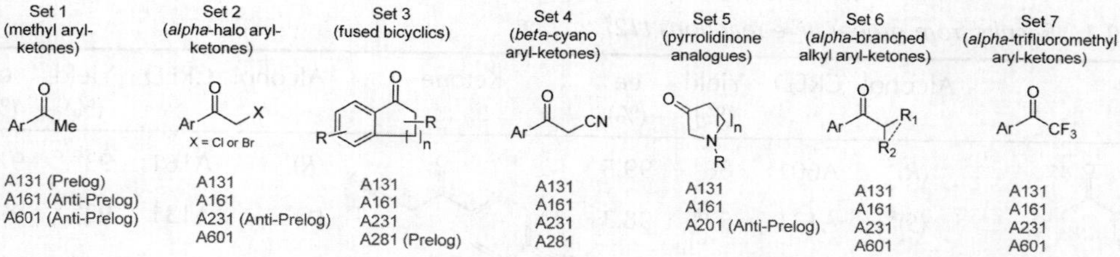

Set 1 (methyl aryl-ketones)	Set 2 (*alpha*-halo aryl-ketones)	Set 3 (fused bicyclics)	Set 4 (*beta*-cyano aryl-ketones)	Set 5 (pyrrolidinone analogues)	Set 6 (*alpha*-branched alkyl aryl-ketones)	Set 7 (*alpha*-trifluoromethyl aryl-ketones)
A131 (Prelog) A161 (Anti-Prelog) A601 (Anti-Prelog)	A131 A161 A231 (Anti-Prelog) A601	A131 A161 A231 A281 (Prelog)	A131 A161 A231 A281	A131 A161 A201 (Anti-Prelog)	A131 A161 A231 A601	A131 A161 A231 A601

Figure 9.4 *Structural motifs of the substrate classes involved in the study. High-performing carbonyl reductases associated with each class are given underneath with their typical selectivity in brackets.* [11]

each enantiomer of each chiral alcohol in typically >95% ee and the majority (66%) in ≥99.0% ee (Table 9.1) [12]. The isolated yields were respectable for unoptimised one-time reactions with an average of 68%. In the cases where low yields were observed, improvement of the work-up procedure would likely have increased the yield considerably.

9.3.1 Screening Procedure

The method outlined here describes the initial screen used to match ketone substrate classes to subsets of highly enantioselective CREDs. Once the composition of the CRED subset had been determined for a certain substrate class, further substrates from the same class could be screened by the same method against only the subset.

9.3.1.1 Materials and Equipment

- Almac selectAZyme™ CESK-6000 CRED panel (lyophilized cell free extracts)
- 0.1 M Potassium dihydrogen phosphate (KH_2PO_4) buffer (pH 7) (96 mL)
- Nicotinamide adenine dinucleotide phosphate (NADP) (20 mg.mL^{-1} stock in buffer)
- Nicotinamide adenine dinucleotide (NAD) (20 mg.mL^{-1} stock in buffer)
- Glucose dehydrogenase (GDH) (40 mg.mL^{-1} stock in buffer)
- Glucose (250 mg.mL^{-1} stock in buffer)
- Ketone substrate of choice (1.2 g)
- Dimethyl sulfoxide (DMSO) (3 mL)
- MTBE (180 mL)
- 4-mL Glass vials

9.3.1.2 Procedure

1. A set of 60 glass vials was labelled with codes corresponding to the different CREDs in the panel.
2. CRED (5–10 mg) was added to each vial.
3. 0.1 M KH_2PO_4 buffer (pH 7.0, 1.6 mL) was added to each vial.
4. NADP or NAD (depending on the enzyme's co-factor preference) was added to each vial (50 μL from 20 mg · mL^{-1} stock solution, corresponding to 1 mg).
5. GDH was added to each vial (50 μL of a 40 mg.mL^{-1} stock solution, corresponding to 2 mg).

Table 9.1 Results from preparative reactions [12].

Left half:

Ketone	Alcohol	CRED	Yield (%)	ee (%)
(4-fluoro-3-methylacetophenone)	(R)[#]	A601	88[a]	99.5
	(S)[#]	A131	52[a]	98.3
(4-(1,2,4-triazol-1-yl)acetophenone)	(R)[†]	A601	55[b]	100
	(S)[†]	A131	79[b]	99.9
(2-acetylpyrazine)	(R)[#]	A601	45[b]	99.8
	(S)[#]	A131	59[b]	99.8
(3-chloro-4-acetylpyridine)	(R)[#]	A601	71[a]	95.4
	(S)[#]	A131	54[a]	99.9
(3-sulfamoylacetophenone)	(R)[‡]	A161	89[a]	98.3
	(S)[‡]	A131	90[a]	99.8
(methyl-acetylpyrazine)	(R)[#]	A161	68[a]	99.5
	(S)[#]	A131	25[a]	99.7
(methyl-acetylthiazole)	(R)[#]	A161	47[b,c]	99.6
	(S)[#]	A131	72[b]	99.8
(F₃CO-phenacyl bromide)	(R)[‡]	A131	54[b,c]	98.8
	(S)[‡]	A601	49[b,c]	96.9
(MeO,F-phenacyl bromide)	(R)[‡]	A131	26[a,b]	99.6
	(S)[‡]	A161	24[a,b]	99.2

Right half:

Ketone	Alcohol	CRED	Yield (%)	ee (%)
(4-bromophenyl cyanomethyl ketone)	(R)*	A161	93[b,c]	97.7
	(S)*	A131	93[b,c]	99.4
(phenyl cyanomethyl ketone)	(R)*	A161	92[b]	99.6
	(S)*	A131	80[a]	100
(pyridinyl cyanomethyl ketone)	(R)[#]	A161	54[a]	99.8
	(S)[#]	A131	44[c]	99.6
(1-(pyrimidin-2-yl)piperidin-3-one)	(R)[#]	A201	39[a]	94.8
	(S)[#]	A131	31[b]	99.2
(1-(pyridin-2-yl)pyrrolidin-3-one)	(R)[#]	A161	56[a]	95.6
	(S)[#]	A131	56[b]	99.9
(1-(pyrimidin-2-yl)pyrrolidin-3-one)	(R)	No hit	-	-
	(S)[#]	A131	40[b]	99.9
(4-bromophenyl cyclopropyl ketone)	(R)[#]	A231	60[a]	99.4
	(S)[#]	A131	58[a]	99.0
(pyridin-4-yl cyclopropyl ketone)	(R)[#]	A601	52[a]	99.7
	(S)[#]	A131	54[b]	99.9
(pyridin-3-yl isopropyl ketone)	(R)[#]	A161	93[b]	99.1
	(S)[#]	A131	94[b]	98.1

Table 9.1 (*Continued*)

Ketone	Alcohol	CRED	Yield (%)	ee (%)	Ketone	Alcohol	CRED	Yield (%)	ee (%)
	(R)[#]	A131	25[a]	99.9		(R)[#]	A161	58[a]	99.5
	(S)[#]	A601	42[b]	97.6		(S)[#]	A131	76[b]	97.7
	(R)[*]	A131	78[b]	100		(R)[#]	A131	71[c]	99.9
	(S)[*]	A601	81[b,c]	99.5		(S)[#]	A231	89[a]	99.9
	(R)[‡]	A131	93[b]	99.6		(R)[*,#]	A131	68[b]	99.8
	(S)[‡]	A161	91[b]	99.9		(S)[*,#]	A161	65[b]	98.7
	(R)[*]	A131	96[b]	99.1		(R)[*]	A131	89[b]	95.3
	(S)[*]	A161	96[b]	98.4		(S)[*]	A161	81[c]	99.6
	(R)[*]	A131	91[b]	99.9		(R)[*]	A131	90[c]	99.7
	(S)[*]	A231	76[b]	96.6		(S)[*]	A161	78[c]	93.4
	(R)[*]	A161	78[b]	99.8					
	(S)[*]	A131	76[a]	98.3					
	(R)[*]	A161	57[b]	99.1					
	(S)[*]	A131	44[a]	93.5					
	(R)[#]	A161	32[b]	99.3					
	(S)[#]	A131	80[a]	90.9					

(*continued*)

Table 9.1 (*Continued*)

Ketone	Alcohol	CRED	Yield (%)	ee (%)	Ketone	Alcohol	CRED	Yield (%)	ee (%)
	(R)‡	A231	91[a]	99.0					
	(S)‡	A281	62[a]	99.3					
	(R)#	A131	78[a]	95.9					
	(S)#	A231	93[a]	96.0					
	(R)*	A231	70[a]	96.3					
	(S)*	A131	79[b]	99.5					
	(R)‡	A131	99[b]	99.8					
	(S)‡	A231	92[a]	98.4					

* Stereopreference assigned by comparison to literature $[\alpha]_D^{20}$.
† Stereopreference assigned by comparison to the $[\alpha]_D^{20}$ of the commercially available compound.
# Stereopreference assigned or corroborated by vibrational circular dichroism (VCD) analysis.
‡ Stereopreference assigned by comparison to selectivities achieved for other members of this compound set.
[a] Experimental method A: CRED (200 mg), NADP or NAD (10 mg), GDH (20 mg), glucose (1.5 equiv.), ketone (900–1700 mg), an DMSO (2.5–5 mL) stirred in 0.1 M KH$_2$PO$_4$ buffer (pH 7, 50 mL) at 30 °C under pH stat control (pH 7, adjusted with 1 M NaOH solution
[b] Experimental method B: CRED (200 mg), NADP or NAD (10 mg), ketone (900–1700 mg) and DMSO (2.5–5 mL) stirred in a mixture 0.1 M KH$_2$PO$_4$ buffer (pH 7, 50 mL) and 2-propanol (7 mL). No pH control required.
[c] Experimental method C: Same as method B, but stirred under reduced pressure (500 mbar) to aid removal of acetone formed by IP oxidation.

6. Glucose was added to each vial (300 μL of a 250 mg.mL^{-1} stock solution, corresponding to 75 mg).
7. Ketone substrate (20 mg) in DMSO (50 μL) was added to each vial.
8. The reaction vials were sealed and shaken overnight at 30 °C.
9. The next day, the screening reactions were extracted with MTBE (2 mL).
10. A 200 μL sample of the extract was charged to a glass pipette containing a cotton wool plug plus anhydrous MgSO$_4$ then washed through to an HPLC vial with MTBE (1 mL).
11. The MTBE was evaporated from the sample in an oven and the residue redissolved in mobile phase and analyzed by chiral HPLC.

9.3.1.3 Analytical

Determination of the percentage conversion for screening reactions and enantiomeric excess of the alcohol products was achieved by chiral HPLC analysis. Method development was carried out for each substrate prior to screening.

9.3.2 Preparative-scale Production of (R)-1-(4-Bromophenyl)-2-chloroethanol

As an example, the bioreduction of *para*-bromophenacyl chloride **1** to (R)-**2** is described (Scheme 9.1) [12]. In this case the best CRED identified for the reaction was A131, which

Scheme 9.1 *Bioreduction of* para-*bromophenacyl chloride 1 to give (R)-1-(4-bromophenyl)-2-chloroethanol (R)-2.*

was also able to utilize 2-propanol (IPA) for co-factor regeneration, precluding the requirement for GDH and, therefore, the need for pH control.

9.3.2.1 Materials and Equipment

- Ketone **1** (1.2 g)
- CRED A131 Lyophilized cell free extract (200 mg)
- NAD co-factor (10 mg)
- 2-Propanol (IPA) (7 mL)
- DMSO (2.5 mL)
- 0.1 M KH_2PO_4 Buffer (pH 7) (50 mL)
- MTBE (400 mL)
- Brine (100 mL)
- Anhydrous magnesium sulfate ($MgSO_4$) (10 g)
- 100-mL Round bottomed flask
- Stirrer hot plate connected to a temperature controller
- Oil bath
- Glass funnel
- Filter paper
- Rotary evaporator

9.3.2.2 Procedure

1. CRED A131 (200 mg) and NAD (10 mg) were measured into a 100-mL round bottomed flask.
2. Buffer (0.1 M KH_2PO_4, pH 7, 50 mL) was added.
3. IPA (7 mL) was added.
4. The solution was magnetically stirred at 30 °C in an oil bath.
5. A solution of ketone **1** (1.2 g) in DMSO (2.5 mL) was added.
6. The reaction mixture was stirred at 30 °C for 18 h.
7. Reaction progress was checked by ^{1}H NMR. The reaction was found to be complete.
8. The reaction mixture was extracted with MTBE (3 × 100 mL).
9. The combined MTBE extracts were washed with brine (100 mL) then dried over anhydrous $MgSO_4$ (10 g).
10. The dried MTBE extracts were filtered through filter paper in a glass funnel.
11. Additional MTBE (100 mL) was rinsed through the filter paper, and the combined MTBE fractions were concentrated on a rotary evaporator.

12. Total product isolated was 1.13 g of alcohol (R)-**2** as a white solid with >95% w/w purity determined by w/w NMR analysis. Yield was 93.4%.

9.3.2.3 Analytical

$[\alpha]_D^{20} = -39.4$ (c 1.0, CHCl$_3$); ^{1}H NMR (CDCl$_3$, 400 MHz): δ 7.51 (2H, m, $2 \times$ Ar–H), 7.27 (2H, m, $2 \times$ Ar–H), 4.87 (1H, m, CH–OH), 3.72 (1H, dd, $J = 11.2$, 3.4Hz, CH$_A$H$_B$), 3.60 (1H, dd, $J = 11.2$, 8.6 Hz, CH$_A$H$_B$). The enantiomeric excess was measured by chiral HPLC using a Chiralcel OJ column with eluent 9:1 hexane:IPA, flow rate 1 mL · min^{-1}, detection 254 nm and run time 16 min. Retention times for (S)-**2** and (R)-**2** were 13.6 min and 12.3 min, respectively. Measured ee of isolated (R)-**2** was 99.6%.

9.3.3 Conclusion

CRED technology has proven to be a fast and efficient tool for the generation of libraries of enantioenriched chiral alcohols with minimal screening and optimization effort required to transform gram amounts of their ketone precursors.

Acknowledgements

The authors wish to thank the UK Technology Strategy Board for its generous financial support of this project. Yanan He and Bo Wang (Biotools, Inc, Jupiter, FL, USA) are also thanked for their work on assignment of enantiomers by VCD. This study was part financed by the European Regional Development Fund under the European Sustainable Competitiveness Program for Northern Ireland.

9.4 A Rapid, Inexpensive and Colorimetric High-throughput Assay Format for Screening Commercial Ketoreductase Panels, Providing Indication of Substrate Scope, Co-factor Specificity and Enantioselectivity

Yuyin Qi[1,2], Mariacristina Bawn[1], Richard A. M. Duncan[1], Ruth M. Lloyd[1], James D. Finnigan[1], Gary W. Black[2], and Simon J. Charnock[1]

[1] *Prozomix Limited, UK*
[2] *Northumbria University, Department of Applied Science, UK*

Oxidoreductases have been employed in the synthesis of chiral alcohols from achiral ketones, primary alcohols from aldehydes, and in the oxidative direction towards carbonyl formation increasingly over approximately the last 30 years. [13] The class of enzymes involved, frequently referred to as ketoreductases (KREDs), comprises multiple protein families (PFAMs), including the medium-chain alcohol dehydrogenases (MDRs), short-chain alcohol dehydrogenases (SDRs) and aldo-keto reductases (AKRs). However, despite their biological ubiquity, long history of use, rising demand, and availability from numerous commercial suppliers, there has been no rapid and cost-effective assay format described to date for screening these enzymes against a desired biotransformation. Instead, both labor and equipment intensive methods are currently employed, such as (chiral) HPLC/GC. At the same time, assay methods based on diaphorase-mediated detection of NAD(P)H in the presence of iodonitrotetrazolium (INT) or other similar electron acceptors have enjoyed many applications over the last 35 years, from initial use in clinical diagnosis [14], through to current food/beverage testing [15]. However, with respect to biocatalysis, although this convenient method has been independently described with respect to the high-throughput screening of lipase/esterase mutants towards enantiomeric excess optimization [16], there has been no suggestion/attempt to utilize it towards either the characterization of novel KREDs against diverse sets of alcohols, or in the screening of commercial KREDs against desired biotransformations. Thus, towards this end, and in preparation for the future availability of more/larger KRED panels, as afforded by current genomic and metagenomics efforts, or via future protein evolution methodologies, a novel colorimetric assay format, "kREDy-to-go™", was developed. This rapid and very cost-effective method is suitable for manual or high-throughput applications, requiring only approximately 25 µg of crude enzyme powder and 100 µg of desired alcohol per assay, and, where both enantiomers of a desired alcohol are available, also preliminary assessment of enantioselectivity. The reaction pathway involved is:

$$\text{Alcohol} + \text{NAD(P)}^+ \xrightarrow{\text{(KRED)}} \text{NAD(P)H} + \text{carbonyl} + \text{H}^+ \tag{1}$$

$$\text{NAD(P)H} + \text{INT} + \text{H}^+ \xrightarrow{\text{(Diaphorase)}} \text{NAD(P)}^+ + \text{INT-formazan} \tag{2}$$

9.4.1 kREDy-to-go™ Assay Procedure

9.4.1.1 *Materials and Equipment*

- Commercial KRED panel (e.g., Prozomix Ltd, catalog number PRO-AKRP)
- Diaphorase suspension (e.g., Prozomix Ltd, catalog number PRO-DIA(001))
- Iodonitrotetrazolium chloride (INT) (e.g., Prozomix Ltd, catalog number PRO-INT)
- NAD$^+$ (e.g., Prozomix Ltd, catalog number PRO-NAD)
- NADP$^+$ (e.g., Prozomix Ltd, catalog number PRO-NADP)
- Sodium phosphate (e.g., Sigma, catalog number S1001 and S2002)
- Deionized (or better) water
- Standard 96-well microtiter plates (e.g., Sarstedt, catalog number 82.1581)
- Deep 96-well microtiter plates (e.g., Sarstedt, catalog number 95.1991.002)
- Single channel pipettes (e.g., Gilson; 20 µL, 200 µL, and 1000 µL)
- 8-Channel multipipette (*Optional*; e.g., 200 µL Finnpipette^R)
- Visible wavelength (340 nm) spectrophotometer (e.g., Jenway 6715)
- Freeze-drier (*Optional*; e.g., EDWARDS SuperModulyo)
- Chiral-HPLC system (*Optional*; e.g., U3000 + Regis cell 784104 column)
- Plate reader (*Optional*; e.g., Synergy H Multi-mode Microplate Reader)
- Microtiter plate sealing film (*Optional*; e.g., 4titude, catalog number 4ti-0566)
- Microtiter plate lids (*Optional*; e.g., Sarstedt, catalog number 82.1584)

- Resealable plastic bags (*Optional*; any supplier)
- 1 mL Plastic cuvettes (e.g., VWR, catalog number 634-2500)
- Standard agar plates (e.g., Sarstedt, catalog number 82.1473)
- Digital camera (any supplier)

9.4.1.2   General Procedure for Screening Target Alcohols Against a Commercial KRED Panel

1. Prepare assay reagent (50 mM sodium phosphate buffer, pH 7.5, containing 1.50 mM NAD$^+$, 1.35 mM NADP$^+$, and 15 µg · mL^{-1} diaphorase) sufficient for the number of enzymes to be screened (for example, 1 mL of assay reagent per enzyme in the panel is sufficient for the preparation of 10 kREDy-to-go^{TM} screening plates).
2. Using a deep format 96-well microtiter plate, aliquot 1 mL of assay reagent into each well (one well per individual KRED to be screened, working from A1 to A12, then B1 to B12, etc.).
3. Carefully prepare a 10 mg · mL^{-1} (in H$_2$O) solution of each KRED to be screened. *Note*: the assay is very sensitive, for instance 0.25 mL at 10 mg · mL^{-1} (i.e., 2.5 mg of KRED powder) is sufficient for 100 screens.
4. Add 25 µL of each 10 mg · mL^{-1} KRED solution *in turn* to individual wells of the deep format 96-well microtiter plate and mix thoroughly by repeated aspiration (the first KRED should be placed in well A1, the second in A2, the thirteenth in B1, and twenty-fourth in B12, etc.).
5. Using an 8-channel 200 µL multipipette (if available) transfer 100 µL aliquots from the deep format plate into the corresponding wells of standard format 96-well microtiter plates. If a large number of kREDy-to-go^{TM} screening plates are required, liquid handling apparatus, such as a MATRIX HYDRA or PlateMate Plus, should be used to speed up the process and reduce pipetting errors. These screening plates can either be used immediately or freeze-dried for long-term (>5 years) storage at −20 °C.
6. a. **Using kREDy-to-go^{TM} plates that were not freeze-dried:** prepare a solution comprising 0.75 mg · mL^{-1} INT (in H$_2$O) and 0.3 % (w/v or v/v) target alcohol to be screened. Using an 8-channel 200 µL multipipette (if available), as quickly and carefully as possible transfer 50 µL of the INT/alcohol solution into each well of the screening plate, mixing quickly and thoroughly (changing tips between additions). A single blank plate should also be set up at the same time, by adding 50 µL of 0.75 mg · mL^{-1} INT (containing no alcohol) to each well.
   b. **Using kREDy-to-go^{TM} plates that were freeze-dried:** prepare a solution comprising 0.25 mg · mL^{-1} INT (in H$_2$O) and 0.1 % (w/v or v/v) target alcohol to be screened. Using an 8-channel 200 µL multipipette (if available), as quickly and carefully as possible transfer 150 µL of the INT/alcohol solution into each well of the screening plate, mixing quickly and thoroughly (changing tips between additions). A single blank plate should also be setup at the same time, by adding 150 µL of 0.25 mg · mL^{-1} INT (containing no alcohol) to each well.
7. Where non-volatile alcohols are being screened, the plates can be covered by a standard microtiter plate lid and stacked together. However, should volatile alcohols be employed, or the environment of the laboratory contain significant volatile primary

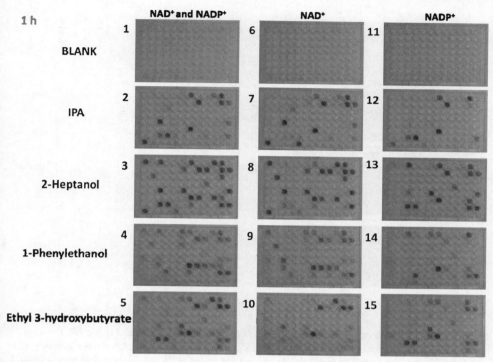

Figure 9.5 *Determination of substrate scope and co-factor specificity of 96 KREDs from a commercial enzyme panel from Prozomix Ltd. The alcohols employed were IPA, 2-heptanol, 1-phenylethanol and ethyl 3-hydroxybutyrate as indicated. Images were recorded digitally after 1 h. Plates 1–5 were formulated with both NAD$^+$ and NADP$^+$, plates 6–10 with only NAD$^+$, and plates 11–15 with only NADP$^+$. All plates were incubated at ambient room temperature (approximately 23 °C).*

and/or secondary alcohols (such as ethanol, IPA, etc.), then each plate should be placed in an individual resealable plastic bag.

8. Incubate the kREDy-to-go^{TM} screening plates (including one blank plate [no alcohol added to the INT solution] per experiment) in the dark at an ambient temperature (15–30 °C).

9. Take digital images and/or scan the plates (using a plate reader if available) at each time point, for example after 1, 3, 5 h, and, for completeness, overnight.

10. Typical results after 1 h for a series of standard alcohols are shown in Figure 9.5 (plates 1–5).

9.4.2 Procedure for the Determination of Co-factor Specificity of Enzymes from a Commercial KRED Panel

1. The entire assay procedure is performed as described previously for the general procedure, except that *three assay reagents* need to be prepared, one containing both NAD$^+$ and

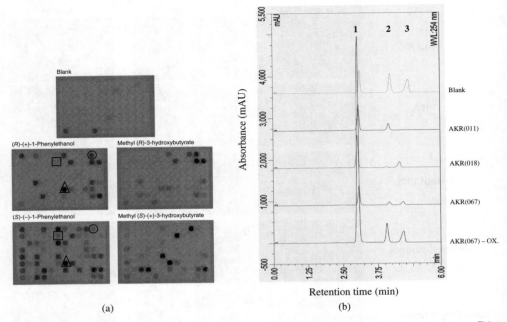

(a)   (b)

Figure 9.6 *Simultaneous substrate scope and chiral analyses performed using kREDy-to-go^{TM} and confirmed by chiral HPLC. (a) One blank plate and four assay plates were employed to screen a commercial KRED panel from Prozomix Ltd against (R)-(+)-1-phenylethanol, (S)-(-)-1-phenylethanol, methyl (R)-3-hydroxybutyrate and methyl (S)-(+)-3-hydroxybutyrate, as indicated. The 5h time point images are shown. (b) Observations of enantioselectivity made from the kREDy-to-go^{TM} analysis in (a) for 1-phenylethanol were confirmed by chiral HPLC (U3000 with Regis cell 784104 column running isocratically at 1.5 mL/min (96% hexane/4% ethanol) at 25°C and with detection at 254 nm). Reactions (1.092 mL) comprising 50 mM sodium phosphate buffer, pH 7.0, containing 0.28 mM NADH, 15.66 mM acetophenone, 598.8 mM IPA, and 0.458 mg·mL^{-1} KRED, were incubated at 25°C for 16 h, after which they were extracted by shaking with 0.5 mL 100% hexane and clarified by centrifugation at 18 000×g for 3 min immediately prior to analysis (20 µL injection). The open circles, open squares, and open triangles in (a) denote specific kREDy-to-go^{TM} enantioselectivity observations that were subsequently confirmed by chiral HPLC, as indicated by the chromatograms in (b), indicated by AKR(011), AKR (018), and AKR(067), respectively. The non-stereoselective oxidation of 1-phenylethanol to acetophenone by AKR(067) is illustrated by the chromatogram in (b), AKR(067)-OX. Standards are as follows: 1, acetophenone; 2, (R)-1-phenylethanol; 3, (S)-1-phenylethanol.*

NADP$^+$ (i.e., as previously described), one containing just NAD$^+$, and one containing just NADP$^+$, and three different sets of screening plates prepared accordingly.

2. Typical results after 1 h for a series standard alcohols are exhibited in Figure 9.5 (plates 1–15).

9.4.3 Procedure for Determination of Enantioselectivity of Enzymes from a Commercial KRED Panel

1. The assay procedure is performed as described previously for the general procedure, except that *three assay plates are required for each substrate*, one for the *R-*

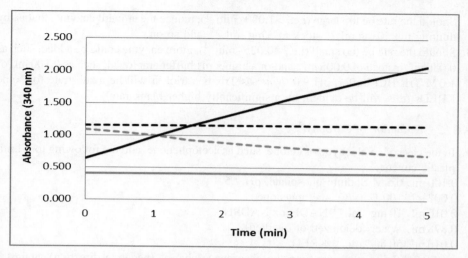

Figure 9.7 *Confirmation of kREDy-to-go*^{TM} *screening results at 340 nm using a spectrophotometer. The dashed grey line represents confirmation of the reduction of acetophenone to 1-phenylethanol by AKR(020) as described in the text (where the dashed black line is the corresponding blank reaction). The solid black line represents confirmation of the oxidation of 1-phenylethanol to acetophenone by AKR(020) as descirbed in the text (where the solid grey line is the corresponding blank reaction).*

enantiomer, and one for the *S*-enantiomer (a single blank plate is required as normal).

2. Typical results after 5 h for 1-phenylethanol and methyl-3-hydroxybutyrate, are shown in Figure 9.6(a).

9.4.4 Confirmation of kREDy-to-go^{TM} Screening Results Using a Spectrophotometer at 340 nm

Regardless of the direction of the desired biotransformation, the kREDy-to-go^{TM} screening results can be confirmed by either oxidation of alcohol or reduction of carbonyl at 340 nm, using a common spectrophotometer (Figure 9.7).

9.4.4.1 Oxidation of Alcohol

1. In the case of an alcohol such as 1-phenylethanol, add the following to a 1 mL plastic cuvette:

0.100 mL 0.5 M Sodium phosphate, pH 7.5
0.002 mL 100 % (v/v) 1-Phenylethanol
0.100 mL 50 mg.mL^{-1} NAD$^+$ *or* NADP$^+$
0.793 mL Water (deionized or distilled)
0.005 mL 50 mg · mL^{-1} KRED.

After mixing well, measure the absorbance change (positive direction) against a blank cuvette set up as above, but replacing the alcohol with water.

2. Should the rate be too large (i.e., $>1.000 \cdot \text{min}^{-1}$), reduce the amount enzyme, that is, by diluting the 50 mg.mL^{-1} stock to 5 mg $\cdot$ mL^{-1}, and so on.
3. Should the rate be too small (i.e., $<0.025 \cdot \text{min}^{-1}$), more enzyme could be added, such as 0.050 mL instead of 0.005 mL, and/or a higher pH buffer employed, such as 0.100 mL of 1.0 M Tris/HCl buffer, pH 9.0. *Note*: as Tris is a triol, it will be a substrate for some KREDs (this will be evident by a significantly higher blank rate).

9.4.4.2   Reduction of Carbonyl

1. In the case of an aldehyde or ketone, such as acetophenone, add the following to a 1 mL plastic cuvette:
   0.100 mL 0.5 M Sodium phosphate, pH 7.5
   0.002 mL 100 % (v/v) Acetophenone
   0.010 mL 10 mg $\cdot$ mL^{-1} NADH *or* NADPH
   0.878 mL Water (deionized or distilled)
   0.010 mL 50 mg $\cdot$ mL^{-1} KRED.
   After mixing well, measure the absorbance change (negative direction) against a blank cuvette set up as above, but replacing the aldehyde or ketone with water.
2. Should the rate be too large (i.e., $>1.000 \cdot \text{min}^{-1}$), reduce the amount of enzyme, that is, by diluting the 50 mg $\cdot$ mL^{-1} stock to 5 mg $\cdot$ mL^{-1}, and so on.
3. Should the rate be too small (i.e., $<0.025 \cdot \text{min}^{-1}$), more enzyme could be added, such as 0.050 mL, and/or a lower pH buffer employed, such as 0.100 mL of 0.5 M sodium phosphate buffer, pH 6.0.
4. It should be noted that crude enzyme preparations frequently contain high NADH oxidase activity and thus high blank rates with NADH as co-factor should be expected and accounted for accordingly.

9.4.5   Confirmation of kREDy-to-go™ Screening Results by Chiral HPLC

Where the target compound is a chiral alcohol, confirmation by chiral HPLC/GC analysis will be desired, and can be conducted as illustrated in Figure 9.6.

9.4.6   Conclusion

With respect to screening of existing commercial KRED panels against particular compounds of interest, the kREDy-to-go™ assay format offers for the first time the ability to rapidly and cost-effectively screen these key biocatalysts without the need for expensive and/or time consuming techniques such as chiral HPLC/GC. Moreover, the method is very sensitive, meaning 40 assays can be performed for each 1 mg of crude cell-free extract KRED powder, and thus the cost per preliminary screening reaction is reduced by up to 40-fold (as typically 1 mg is used per screening reaction currently). Additionally, where both chiral enantiomers of a target alcohol are available, the method can also identify which KREDs possess desired enantioselectivity during preliminary screening. The method is also scalable, meaning that as the numbers of KREDs in commercial panels inevitably expands significantly over the coming years, due to mining of the expanding genomic/metagenomic databases and ever efficient/rapid enzyme evolution methodologies, it will be possible to screen the resultant novel biocatalysts quickly and economically.

Acknowledgements

The authors wish to thank Innovate UK for its generous funding. The research leading to these results has also received funding from the European Union's Seventh Framework Programme for research, technological development and demonstration under grant agreement number 613849 ("BIOOX").

9.5 Stereoselective Production of (*R*)-3-quinuclidinol Using Recombinant *Escherichia coli* Whole Cells Overexpressing 3-Quinuclidinone Reductase and a Co-factor Regeneration System

Teresa Pellicer[1], Francisco Marquillas[1], Xavier Pérez Javierre[2], and Antoni Planas[2]
[1] *Interquim SA, R&D Department, Spain*
[2] *Universitat Ramon Llull, Institut Químic de Sarrià, Laboratory of Biochemistry, Spain*

(*R*)-(−)-3-Quinuclidinol **2** is an important building block of a number of antimuscarinic drugs, such as Aclidinium bromide [17] and Solifenacin [18], for the treatment of chronic obstructive pulmonary disease (COPD) or urinary incontinence caused by an overactive bladder. For the stereoselective production of **2**, chemical and biochemical methods have been developed, most of them involving the resolution of a racemic mixture of 3-quinuclidinol with a maximal theoretical yield of 50%.

A number of enzymes have been reported as catalysts for the bioreduction of 3-quinuclidinone **1** to **2**. Uzura and colleagues described a stereoselective biocatalytic method to produce **2** with >99.9% enantiomeric excess within 21 h [19], using *Escherichia coli* cells co-expressing a protein with 3-quinuclidinone reductase activity and a co-factor regeneration system in two independent plasmids. Isotani and co-workers described a bioreduction system for synthesizing **2** using two recombinant *Escherichia coli* cell biocatalysts possessing a 3-quinuclidinone reductase from *Microbacterium luteolum* and the *Leifsonia sp.* alcohol dehydrogenase (LSADH), respectively [20]. The combination of crude extracts of the resulting *E. coli* biocatalysts, exhibited a level of production of **2** of 150 mg.mL^{-1} and high enantioselectivity. More recently, a new system where the bacterial reductase from *Agrobacterium radiobacter* and the *Bacillus megaterium* glucose dehydrogenase are

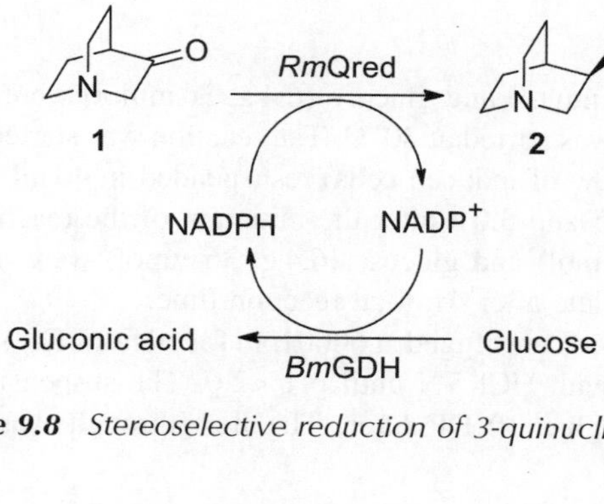

Figure 9.8 *Stereoselective reduction of 3-quinuclidinone **1**.*

co-expressed from the same plasmid has been described [21]. The systems described hitherto require the expression of two independent polypeptides in one or two independent *E. coli* cultures and/or the addition of expensive nicotinamide-derived co-factors (i.e., NADH or NAPDH). We decided to attempt the production of a biocatalyst composed of *E. coli* whole cells expressing a single polypeptide containing 3-quinuclidinone reductase activity and a co-factor regeneration system, ruling out the need for more than one culture and the addition of co-factors.

Quinuclidinone reductase of *Rhodotorula mucilaginosa* (*Rm*Qred) is a nicotinamide adenine dinucleotide phosphate (NADPH)-dependent carbonyl reductase that catalyzes the stereospecific reduction of **1** to **2** [19]. Thus, Qred requires NADPH as co-factor for the reaction. The *Bacillus megaterium* glucose 1-dehydrogenase (*Bm*GDH), catalyst for the chemical reaction of *beta*-D-glucose + NADP to yield gluconic acid + NADPH, was used as a co-factor regeneration enzyme (Figure 9.8). *Bm*GDH and *Rm*Qred were in-frame cloned and separated by an 18-mer amino acid sequence in an inducible plasmidic vector. The resulting chimeric protein contained both glucose dehydrogenase and 3-quinuclidinone reductase enzymatic activities. *E. coli* whole cells expressing the inducible chimeric construction were used as biocatalyst and were able to efficiently and selectively produce **2** (conversion >99%; ee >99%) from **1** and consuming glucose.

9.5.1 Materials and Equipment

- 3-Quinuclidinone **1** (64 g, 100 mmol)
- Glucose (25.6 g)
- Phosphate buffer 50 mM, pH 7.0 (200 mL)
- *Escherichia coli* cells expressing *Bm*GDH and *Rm*Qred (15 g wet cell weight [wcw])

9.5.2 Biocatalyst Production

E. coli cells transformed with the inducible chimeric construct coding the *Bm*GDH-*Rm*Qred polypeptide were grown in a bioreactor in glycerol-based defined medium and protein induction was triggered by the addition of 0.5 mM IPTG. After 18 h of induction, cells were harvested and the cell pellet was resuspended in 50 mM phosphate buffer, pH 7.0 [375 mg (wcw) $\cdot$ mL^{-1} buffer].

9.5.3 Procedure

1. Substrate **1** (16 g, 100 mmol) and glucose (6.4 g, 36 mmol) were dissolved in 160 mL phosphate buffer that was stirred at 30 °C. The reaction was started with the addition of 15 g of biocatalyst (wcw of induced cells) resuspended in 40 mL of phosphate buffer.
2. At time points 0.5, 1.5 and 3.0 h after the initiation of the reaction, more 3-quinuclidinone **1** (16 g, 100 mmol) and glucose (6.4 g, 36 mmol) were added to the mixture. Conversion was complete after 4 h total reaction time.
3. The reaction mixture was centrifuged at 6000 rpm for 30 min. The supernatant liquid was filtered and acidified with HCl 5 N until pH <2.0. The suspension was subsequently vacuum filtered through a PVDF filter. The flow-through was treated with NaOH

solution until pH >12. This was rotavaporated to dryness with the help of acetonitrile and the dry pellet was resuspended in dichloromethane and vacuum filtered. Flow-through was recovered and rotavaporated to dryness.

4. Toluene was added and the mixture was warmed until complete solubility. 3-Quinu-clidinol **2** crystals appeared after slow cooling of the solution. Crystals were vacuum dried and the powdered product analyzed by infrared spectrophotometry, polarimetry, nuclear magnetic resonance, and mass spectroscopy.

9.5.4 Sample analyses

For the screening of experimental conditions, a HPLC/MS method was set up in order to determine conversion rates. Supernatant liquids from samples were diluted 1000-fold in mQ H_2O and analyzed by HPLC using a HILIC column (XBridge™ HILIC [3.5 µm, 2.1 × 100 mm]; mobile phase [acetonitrile/ammonium acetate 10 mM, pH 4 = 80/20]; column temperature: 30 °C; flow rate: 0.3 mL · min^{-1}; detection: single quadrupole equipped with an ESI interface operating in positive mode, SIM [m/z] 3-quinuclidinone **1** [126], 3-quinuclidinol **2** [128]). Approximate retention time: 3-quinuclidinone **1** 2.3 min, (R,S)-3-quinuclidinol **2** 3 min.

Enantiomeric purity determination for the bioconversion of 3-quinuclidinone **1** into (R)-3-quinuclidinol **2** was carried out by chiral gas chromatography. Supernatant (1 mL) from samples was evaporated at 80 °C under a nitrogen stream for 30 min in 1.5 ml HPLC-type glass vials. The dried pellet was resuspended in isobutanol. The soluble fraction was analyzed by GC using a chiral column [Cyclosil B (30 m × 0.25 mm × 0.25 µm); flow 1.4 mL · min^{-1}; oven program: 140 °C for 18 min, 20 °C · min^{-1} to 240 °C for 4 min), injector 280 °C, detection: FID 250 °C). Approximate retention time: 3-quinuclidinone **1** 8.8 min, (S)-3-quinuclidinol 14.2 min, (R)-3-quinuclidinol **2** 14.8 min.

9.5.5 Conclusion

E. coli whole cells expressing a single polypeptide containing 3-quinuclidinone reductase and glucose dehydrogenase activities produce (R)-3-quinuclidinol **2** in a selective and efficient manner. The design of this new biocatalyst rules out the need for more than one culture and avoids microbial cell disruption. Moreover, the addition of expensive co-factors is not required because intracellular concentration of NADPH is sufficient for stereo-selective reduction of 3-quinuclidinone **1**.

9.6 Preparation of *N*-Boc-D-Serine Using a Coupled D-Acylase/Racemase Enzyme System

Anna Fryszkowska[1], Pieter de Koning[2], and Karen Holt-Tiffin[2]
[1]*Merck Research Laboratories, USA.*
[2]*Dr Reddy's Laboratories Ltd, Chirotech Technology Centre, UK*

Optically active amino acids are useful building blocks for the enantioselective synthesis of many natural products and active pharmaceutical ingredients. *N*-Boc-D-serine is a potential intermediate in the synthesis of Lacosamide, a medication developed for the adjunctive treatment of partial-onset seizures and diabetic neuropathic pain, marketed as Vimpat (Scheme 9.2).

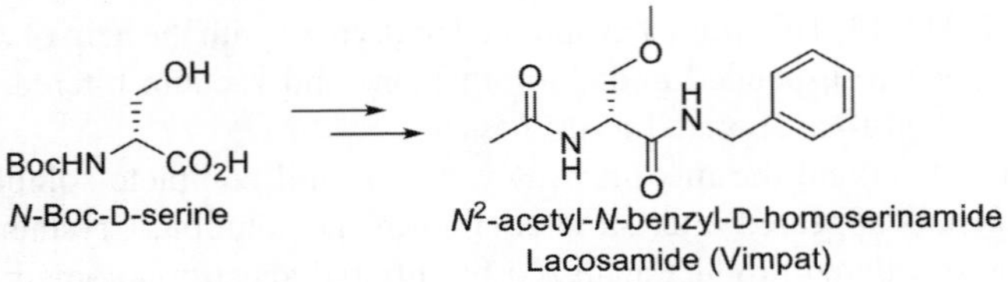

BocHN CO₂H

N-Boc-D-serine

N²-acetyl-N-benzyl-D-homoserinamide
Lacosamide (Vimpat)

Scheme 9.2 *Vimpat synthesis via N-Boc-D-serine intermediate.*

Kinetic resolution of easily made *N*-acetyl amino acids using amino acylase enzymes is an established method for the synthesis of single isomer amino acids; however, the maximum possible yield is 50%. Dynamic kinetic resolutions allow synthesis of the desired enantiomer in yields approaching 100%, hence they are an attractive alternative to classic kinetic resolution processes. One such approach is the use of an *N*-acetyl amino acid racemase (NAAAR)/acylase-coupled system on the racemic *N*-acetyl amino acid derivative. A neat extension is the use of this coupled enzyme system to convert the natural isomer *N*-acetyl-L-amino acid to the unnatural D-amino acid.

A key factor for the successful development of biocatalytic processes is fast access to inexpensive biocatalysts at large scale with suitable properties, such as high activity, high selectivity, and high stability. Chirotech has previously developed L- and D-amino acylases with industrially-desirable properties, namely L-aminoacylase from *Thermococcus litoralis* and D-aminoacylase from *Alcaligenes sp.* [22], which were cloned and overexpressed into both *E. coli* and *Pseudomonas* expression platforms. Recently, as a result of a collaboration with the University of Edinburgh, a novel, improved NAAAR from *Amycolatopsis sp. Ts-1-60* was developed *via* directed evolution [23]. Two single point mutations gave a significantly improved variant, NAAAR G291D F323Y – the new protein has up to sevenfold higher activity with a broad spectrum of substrates and shows no substrate inhibition up to a concentration of 300–500 mM.

Since the cost of L-serine is less than the racemate at large scale, it was decided to employ the racemase to bring about the inversion of the L-isomer to the desired D-isomer. Thus, a three-step procedure was successfully developed to convert L-serine to *N*-Boc-D-serine as shown in Scheme 9.3, utilizing the NAAAR/D-acylase coupled enzyme system as the key step.

9.6.1 Procedure

9.6.1.1 Materials and Equipment

Step 1

- Sodium hydroxide (14.4 g, 360 mmol)
- Deionized water (25 mL)
- L-Serine (15.0 g, 142.7 mmol)

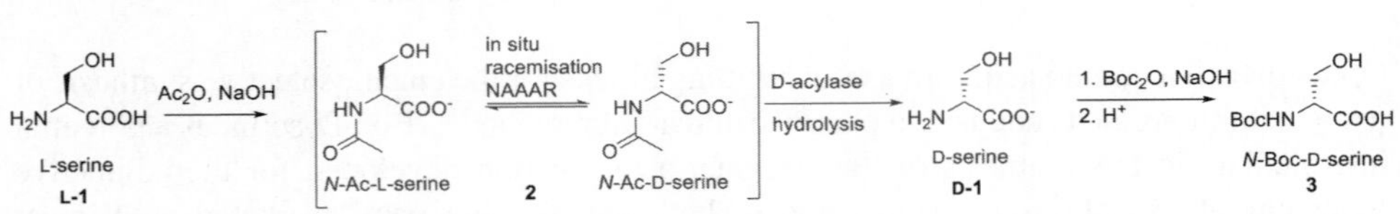

Scheme 9.3 *Three-step synthesis of N-Boc-D-serine from L-serine.*

- Acetic anhydride (18 mL)
- Ethanol (180 mL)
- Methanol (120 mL)
- HCl (conc.)
- Toluene (100 mL)
- 250-mL jacketed reactor equipped with overhead stirrer

Step 2

- Deionized water (500 mL)
- Cobalt chloride (609 mg)
- Magnesium sulfate (300 mg)
- Cell free extract of *Amycolatopsis sp. Ts-1-60* NAAAR G291D F323Y mutant (~1500 U)
- Concentrated solution of D-acylase (275 μL, 286 U)
- 2 M NaOH
- 3 M HCl
- 1-L Jacketed reactor equipped with overhead stirrer and pH-stat

Step 3

- Ethyl acetate (AcOEt) (2×100 mL)
- 46% NaOH
- Boc_2O (38.6 g, 177.1 mmol)
- Acetone (100 mL)
- 1 M $KHSO_4$
- Brine
- AcOEt (3×500 mL)
- MTBE
- Heptane
- 1-L Jacketed reactor equipped with overhead stirrer and pH-stat

9.6.1.2   Procedure

Step 1

1. In a 250-mL jacketed reactor with overhead stirring, NaOH (14.4 g, 360 mmol) was dissolved in water (25 mL) and L-serine (15.0 g, 142.7 mmol) was added.
2. The solution was cooled to 0 °C and acetic anhydride (15 mL, 16.2 g, 158.8 mmol) added dropwise, maintaining the temperature <25 °C. After 1 h, additional Ac_2O (3 mL) was added to allow full conversion and the mixture was stirred for a further 15 min.
3. EtOH (100 mL) was added and the slurry was stirred for 1 h at room temperature to decompose unreacted Ac_2O. White precipitate (AcONa) was filtered off and the filtrate was evaporated *in vacuo* to give crude *N*-acetyl-L-serine sodium salt containing AcONa (1.7 equiv. by NMR).
4. The residue was redissolved in a mixture of MeOH (120 mL) and EtOH (80 mL) and acidified to pH 4.5 with concentrated HCl upon cooling in an ice bath.
5. White precipitate (NaCl, 25 g) was filtered off. Toluene (100 mL) was added to the filtrate and the residue was evaporated to azeotrope part of the acetic acid (down to 1.2 equiv.).

Step 2

1. The crude *N*-acetyl-L-serine thus obtained was dissolved in deionized water (500 mL) in a 1-L jacketed vessel equipped with overhead stirring and pH meter.
2. Cobalt chloride (609 mg) and magnesium sulfate (300 mg) were added, the mixture was warmed up to 40 °C and the pH was adjusted to 8.0.
3. NAAAR G291D F323Y CFE (30 mL, 1500 U) was added and the reaction was monitored by chiral HPLC using a Chirobiotic T column.
4. After 3 h the racemization reached 50% conversion and D-acylase was added (260 U, 250 μL).
5. The mixture was stirred at 40 °C overnight, with pH maintained at 7.8–8.0 using 2 M NaOH. After 21 h the conversion reached approximately 80% (by NMR). Accumulation of *N*-acetyl-D-serine was observed by chiral HPLC.
6. Another aliquot of D-acylase (26 U, 25 μL) was added and the reaction was stirred for further 3 h to reach 90% conversion.
7. The mixture was cooled to 5 °C and the pH was brought to 2 using 3 M HCl.
8. The enzyme debris was removed by centrifugation (30 min, 8000 rpm).

Step 3

1. The aqueous phase was washed with AcOEt (2 × 100 mL) to remove unreacted *N*-acetyl-serine, then transferred to a 1-L jacketed reactor with overhead stirring and cooled to 5 °C.
2. The pH was adjusted to 12 using 46% NaOH and Boc anhydride (38.6 g, 177.1 mmol, 1.2 eq) dissolved in acetone (100 mL) was added dropwise, keeping the temperature <10 °C.
3. The mixture was stirred overnight at room temperature.
4. The aqueous layer was cooled to 5 °C and acidified to pH 3 using 1 M KHSO$_4$ (gas evolution). The product was extracted with AcOEt (3 × 500 mL) and the organic layer was washed with brine, dried over magnesium sulfate and evaporated to dryness to give a transparent oil of crude *N*-Boc-D-serine **3** (94% ee).
5. Crystallization from MTBE/heptane upon seeding gave *N*-Boc-D-serine **3** (16.42 g) as white crystals in 96% ee and 56% overall yield for three steps from L-serine.

9.6.1.3 *Analytical – Chiral HPLC Methods*

Enantiomeric excess of N-acetyl-serine and serine

Column:	Chirobiotic T (50 mm × 4.6 mm, SN#15684)
Flow:	1 mL · min^{-1}
Detector:	ELSD or UV (λ 210 nm)
Mobile phase:	80:20 MeOH:0.1% TEAA (triethylamine acetate pH 4)
Run time:	4 min
Retention times:	L-(**1**) – 1.2 min, D-(**1**) – 1.7 min, L-(**2**) – 2.1 min, D-(**2**) – 2.8 min

Enantiomeric excess of N-Boc-serine

Column:	Chirobiotic R (25 cm × 4.6 mm, SN#17284)
Flow:	1 mL · min⁻¹
Detector:	UV (λ 200 nm)
Mobile phase:	40:60 MeOH:0.1% TEAA (triethylamine acetate pH 4)
Run time:	15 min
Retention times:	L-(**3**) – 8 min, D-(**3**) – 10 min

9.6.2 Conclusion

N-Boc-D-serine has been prepared from L-serine using a three-step procedure enabled by a coupled D-acylase/NAAAR enzyme system. Quantities of D-acylase and NAAAR have to be carefully tailored, so that the activity of racemase exceeds the activity of D-acylase, which displays imperfect enantioselectivity. The *N*-Boc-D-serine product is obtained in good yield and 96% ee purity. If required, recrystallization can be employed to give single enantiomer product.

9.7 Scale-up of a Biocatalytic Oxidase in a Dynamically Mixed Tubular Flow Reactor

Gilda Gasparini[1], Ian Archer[2], Ed Jones[3], and Robert Ashe[1]

[1] *AM Technology, UK*
[2] *Ingenza Ltd, Roslin BioCentre, UK*
[3] *C-Tech Innovation Ltd, UK*

Biocatalytic oxidase processes can benefit from improved mixing and mass transfer as can all multiphase processes. Flow reactors are established for their mixing capabilities but there are practical challenges in terms of slurry and gas/liquid handling. The oxidase of a DL-amino acid (Figure 9.9) was studied and scaled up in flow in a Coflore system and

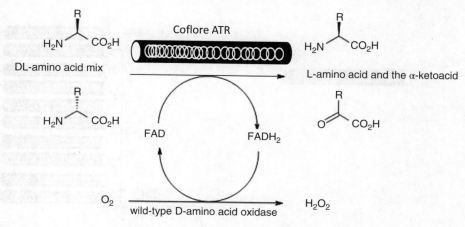

Figure 9.9 *Biocatalytic resolution of DL-alanine.*

compared to batch processing conditions. The improved mass transfer under flow conditions resulted in a reduction in reaction time, enzyme consumption and pressure drop.

9.7.1 Procedure 1: 1-L and 4-L Batch Operation

9.7.1.1 Materials and Equipment

- Oxygen, $0.25\,L \cdot min^{-1}$, added via a sparged gas inlet
- Alanine (89.09 g, 1 mol) per liter of water
- Biocatalyst is produced by fermentation of *Pichia pastoris* expressing the DAAO enzyme, $21\,g \cdot L^{-1}$
- 1-L Glass reactor with a 4 cm impeller agitator rotating at 400 rpm
- 4-L Glass reactor with same impeller as above

9.7.1.2 Procedure

1. Alanine solution and catalyst were added in a 1-L glass reactor at 25 °C. Oxygen was then added and the reaction was started.
2. At time points 0.5, 1, 2, 3, 4, 5, and 24 h after the initiation of the reaction, samples were taken to be analyzed.
3. The same procedure was followed for the 4-L batch and samples were taken at 0.5, 1, 2, 3, 4, 5, 24, and 29 h.

Product conversion was analyzed by HPLC

9.7.2 Procedure 2: 1-L and 10-L Continuous Operation

9.7.2.1 Materials and Equipment

- Oxygen, added directly to the Coflore tube
- Alanine (89.09 g, 1 mol) per liter of water
- Biocatalyst is produced by fermentation of *Pichia pastoris* expressing the DAAO enzyme, $21\,g \cdot L^{-1}$

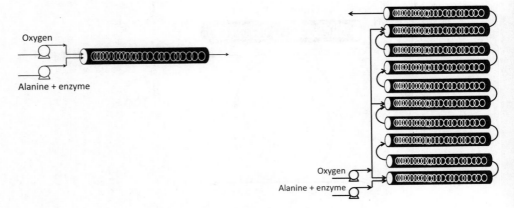

Figure 9.10 *Experimental set-up in flow.*

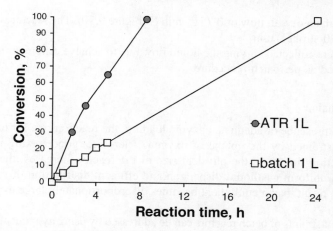

Figure 9.11 *Effect of continuous operation.*

- 1-L Single tube Coflore reactor
- Watson Marlow peristaltic pump

9.7.2.2 Procedure

1. A slurry of catalyst and alanine was prepared and kept under an inert atmosphere.
2. The slurry was pumped to one end of the single 1-L Coflore tube. The oxygen was introduced via a second inlet at the same end of the tube at a flow rate of $0.25\,L \cdot min^{-1}$. The reactor was agitated at $2\,Hz$, or $120\,strokes \cdot min^{-1}$.
3. For the 10-L test, all of the 10 tubes of the Coflore reactor were connected in series. The slurry was injected at the first tube while the oxygen was added in three stages at tubes 1,

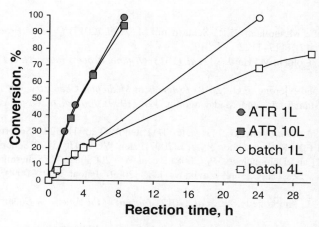

Figure 9.12 *Effect of scale-up in continuous and batch.*

5, and 9 with an overall flow of $0.75\,L \cdot min^{-1}$ (Figure 9.10). The reactor was agitated at $2\,Hz$, or $120\,strokes \cdot min^{-1}$.

4. Samples were collected varying the liquid flow rate to achieve different residence times and analysed as per batch procedure.

9.7.3 Conclusion

Subject to the presence of adequate enzyme loading, the reaction rate for this reaction is primarily constrained by the uptake of oxygen. Therefore, good mixing is required to speed up reaction times. As the physical size of the reactor increases, the problems of maintaining a uniform gas/liquid dispersion and efficient distribution of mixing energy increase. As a result, batch equipment becomes disproportionately large as throughput is increased.

The scale constraints of batch reactors can be addressed by using dynamically mixed flow reactors. The Coflore reactor is a dynamically mixed flow reactor that uses transverse mixing by lateral movement of the agitators. Even at the 1-L scale (Figure 9.11), the reaction rate in a 1-L Coflore tubular reactor is three times faster than a 1-L batch reactor. When the Coflore reactor is scaled up to 10-L (Figure 9.12), the reaction rate remains substantially unchanged. It is also worth noting that the 10-L Coflore reactor used 70% less oxygen than the 1-L system due to experimental constraints. This highlights the fact that in batch operation oxygen is used in large excess and wasted.

By employing a dynamically mixed Coflore flow reactor, it was demonstrated that transverse mixing was not only substantially more effective than traditional rotational mixing but that throughput can be increased by an order of magnitude with negligible impact on reaction rate. Despite the presence of live cells and organic debris, no problems of blockage were encountered with the Coflore ATR unit. The commercial implications of this system are reduced cost of capital equipment, lower operating costs, and reduced catalyst consumption, due to faster throughput, for manufacturing processes.

References

1. Meadows, R.E., Mulholland, K.R., Schurmann, M., *et al.* (2013) *Organic Process Research & Development*, **17**, 1117–1122.
2. Frodsham, L., Golden, M., Hard, S., *et al.* (2013) *Organic Process Research & Development*, **17**, 1123–1130.
3. Chen, K.X., and Njoroge, F.G. (2010) *Progress in Medicinal Chemistry*, **49**, 1–36.
4. Zhang, R., Mamai, A., and Madalengoitia, J.S. (1999) *Journal of Organic Chemistry*, **64**, 547–555.
5. Park, J., Sudhakar, A., Wong, G. S., *et al.* (2004) Patent WO 2004113295 (Schering Corporation).
6. Wu, G., Chen, F.X., Rashatasakhon, P., *et al.* (2007) Patent WO 2007075790 (Schering Corporation).
7. Berranger, T., and Demonchaux, P. (2008) Patent WO 2008082508 (Schering Corporation).
8. Kwok, D.-L., Lee, H.-C., and Zavialov, I.A. (2009) Patent WO 2009073380 (Schering Corporation).
9. Li, T., Liang, J., Ambrogelly, A., *et al.* (2012) *Journal of the American Chemical Society*, **134**, 6467–6472.
10. Mijts, B., Muley, S., Liang, J., *et al.* (2012) US Patent 8,178,333 (Codexis Inc).

11. For an explanation of Prelog's rule for ketone bioreduction enantioselectivity see: (a) Pscheidt, B., and Glieder, A. (2008) *Microbial Cell Factories*, **7** (25) [online]. doi: 10.1186/1475-2859-7-25. (b) Prelog, V. (1962) *Pure and Applied Chemistry*, **9**, 119–130.

12. For full analytical and experimental details see: Rowan, A.S., Moody, T.S., Howard, R.M., *et al.* (2013) *Tetrahedron: Asymmetry*, **24**, 1369–1381.

13. Gorisch, H., Boland, W., and Jaenicke, L. (1984) *Journal of Applied Biochemistry*, **6**, 103.

14. Gella, F.J., Olivella, M.T., Pegueroles, F., and Gener, J. (1981) *Journal of Clinical Chemistry*, **27**, 1686.

15. McCleary, B.V., and Charnock, S.J. (2006) Patent Application WO/2006/064488.

16. Bustos-Jaimes, I., Hummel, W., Eggert, T., *et al.* (2009) *ChemCatChem*, **1**, 445.

17. Prat, M., Fernandez, D., Buil, M.A., *et al.* (2009) *Journal of Medicinal Chemistry*, **52**, 5076–5092.

18. Baumann, M., and Baxendale, I.R. (2013) *Beilstein. Journal of Organic Chemistry*, **9**, 2265–2319.

19. Uzura, A., Nomoto, F., Sakoda, A., *et al.* (2009) *Applied Microbiology and Biotechnology*, **83**, 617–626.

20. Isotani, K., Kurokawa, J., and Itoh, N. (2012) *International Journal of Molecular Sciences*, **13**, 13542–13553.

21. Zhang, W.-X., Xu, G.-C., Huang, L., *et al.* (2013) *Organic Letters*, **15**, 4917–4919.

22. Taylor, I.N., Brown, R.C., Bycroft, M., *et al.* (2004) *Biochemical Society Transactions*, **32**, 290–292.

23. Baxter, S., Royer, S., Grogan, G., *et al.* (2012) *Journal of the American Chemical Society*, **134**, 19310–19313. doi: 10.1021/ja305438y.

Index

Practical Methods for Biocatalysis and Biotransformations 3, First Edition.
Edited by John Whittall, Peter W. Sutton, and Wolfgang Kroutil.
© 2016 John Wiley & Sons, Ltd. Published 2016 by John Wiley & Sons, Ltd.